Industrial Systems Engineering

Industrial Systems Engineering

Editor: Malinda Lynn

NY RESEARCH PRESS

New York

Published by NY Research Press
118-35 Queens Blvd., Suite 400,
Forest Hills, NY 11375, USA
www.nyresearchpress.com

Industrial Systems Engineering
Edited by Malinda Lynn

International Standard Book Number: 978-1-63238-655-7 (Hardback)

Cataloging-in-Publication Data

Industrial systems engineering / edited by Malinda Lynn.
p. cm.
Includes bibliographical references and index.
ISBN 978-1-63238-655-7
1. Industrial engineering. 2. Systems engineering. I. Lynn, Malinda.
T56.24 .I53 2019
658--dc23

Contents

Permissions

List of Contributors

Index

Preface

Industrial systems engineering is a specialized branch of engineering, which is concerned with the optimization of complex processes, systems and organizations to achieve an effective utilization of time, human resources and material resources. The various fields of study inherent to this discipline are manufacture engineering, engineering management, process engineering, systems engineering, software engineering, etc. Modern industrial systems engineering uses computer simulation and mathematical optimization for the in-depth study of organizational and industrial systems as well as for systems analysis, evaluation and optimization. Besides these, the tools of data science and machine learning are also of importance in the study of systems. The applications of this field are diverse. It is used in flow process charting, strategizing for operational logistics, developing a new algorithm in a financial process, supply chain management, etc. This book elucidates the key aspects of the field of industrial and systems engineering and presents the technological advances in a comprehensive way. It unfolds the innovative aspects of industrial engineering which will be crucial for the progress of this field in the future. It strives to be a complete reference guide for students and experts working in this domain.

The information shared in this book is based on empirical researches made by veterans in this field of study. The elaborative information provided in this book will help the readers further their scope of knowledge leading to advancements in this field.

Finally, I would like to thank my fellow researchers who gave constructive feedback and my family members who supported me at every step of my research.

Editor

1

Optimization of multi-response dynamic systems using multiple regression-based weighted signal-to-noise ratio

Susanta Kumar Gauri[a*] and Surajit Pal[b]

[a]SQC & OR Unit, Indian Statistical Institute, 203, B. T. Road, Kolkata-700108, India
[b]SQC & OR Unit, Indian Statistical Institute, 110, Nelson Manickam Road, Chennai- 600029, India

CHRONICLE	ABSTRACT
Keywords: *Dynamic system* *Multiple responses* *Optimization* *Composite desirability function* *Multiple regression* *Weighted signal-to-noise ratio*	A dynamic system differs from a static system in that it contains signal factor and the target value depends on the level of the signal factor set by the system operator. The aim of optimizing a multi-response dynamic system is to find a setting combination of input controllable factors that would result in optimum values of all response variables at all signal levels. The most commonly used performance metric for optimizing a multi-response dynamic system is the composite desirability function (CDF). The advantage of using CDF is that it is a simple unit less measure and it has a good foundation in statistical practice. However, the problem with the CDF is that it does not consider the variability of the individual response variables. Moreover, if the specification limits for the response variables are not provided the CDF cannot be computed. In this paper, a new performance metric for multi-response dynamic system, called multiple regression-based weighted signal-to-noise ratio (MRWSN) is proposed, which overcome the limitations of CDF. Two sets of experimental data on multi-response dynamic systems, taken from literature, are analysed using both CDF-based and the proposed MRWSN-based approaches for optimization. The results show that the MRWSN-based approach also results in substantially better optimization performance than the CDF-based approach.

1. Introduction

The usefulness of Taguchi method (Taguchi, 1986) in optimizing the parameter design in static as well as dynamic system has been well established. In a static system, a response variable representing the output characteristic of the system has a fixed target value. But in a dynamic system, the target value of a response variable depends on the level of the signal factor set by the system operator. For example, signal factor may be the steering angle in the steering mechanism of an automobile or the speed control setting of a fan. In other words, a dynamic system has multiple target values of a response variable depending on the setting of signal variable of the system. Most of the modern manufacturing processes have several response variables and the process needs to be optimized for all response variables. Extensive research works have been carried out aiming to resolve the multi-response optimization

* Corresponding author.
E-mail: susantagauri@hotmail.com (S. K. Gauri)

problem in a static system (Derringer & Suich, 1980; Khuri & Conlon, 1981; Pignatiello, 1993; Su & Tong, 1997; Tong & Hsieh, 2000; Wu, 2004; Liao, 2006; Tong et al., 2007; Jeong & Kim, 2009; Pal & Gauri, 2010; Wang et al., 2016). Product/process design with a dynamic system offers the flexibility needed to satisfy customer requirements and can enhance a manufacturer's competitiveness. In recent time, therefore, many researchers (Miller & Wu, 1996; Wasserman, 1996; McCaskey & Tsui, 1997; Tsui, 2001; Joseph & Wu, 2002; Chen, 2003; Lesperance & Park, 2003; Su et al., 2005; Bae & Tsui, 2006) have been motivated to study the robust design problem concerning the dynamic systems. However, all these research articles are focused on optimization of a single-response dynamic system.

Industry has increasingly emphasized developing procedures capable of simultaneously optimizing the multi-response dynamic systems in light of the increasing complexity of modern product design. To cope with the need of the modern industries, several studies (Tong et al., 2001; Tong et al., 2004; Hsieh et al., 2005; Wang & Tong, 2004; Wu & Yeh, 2005; Chang, 2006; Wang, 2007; Chang, 2008; Tong et al., 2008; Wu, 2009; Chang & Chen, 2011; Gauri, 2014) have proposed different procedures for optimizing a multi-response dynamic system. The goal of optimizing a multi-response dynamic system is to find a setting combination of control factors (controllable variables) that would result in the optimum values of all response variables at all signal levels. Generally, it is very difficult to obtain such a combination, because optimum values of one response variable may lead to non-optimum values for the remaining response variables. Hence, it is desirable to find a best setting combination of control factor levels that would result in an optimal compromise of response variables. Here optimal compromise means each response variable is as close as possible to its target value at each signal level and with minimum variability around that target value.

Most of the researchers have attempted to optimize multi-response dynamic system using Derringer and Suich's (1980) composite desirability function (CDF) as a performance metric. Tong et al (2001), Hsieh et al (2005), and Wu (2009) have modelled the response variables using response surface methodology (RSM) and then determined the optimal settings of the control factors by maximizing an overall performance measure (OPI), which is essentially the CDF. On the other hand, Chang (2006), Chang (2008) and Chang and Chen (2011) used artificial neural networks (ANN) for modelling the response functions and then obtained the optimal settings of the control factors by considering OPI, which is essentially CDF, as the performance metric.

The basic advantage of using CDF as performance metric is that it is a simple unit less measure and it has a good foundation in statistical practice. However, if the specification limits for the response variables are not provided, the CDF cannot be computed. Another disadvantage with this metric is that it does not take into consideration the variability of individual response variables. Hence, a CDF-based approach may produce an optimal solution where the expected means of the response variables at a signal level is very close to their target values, but variability of one or more response variables around the target is very high, which may not be acceptable by the engineers.

Pal and Gauri (2010) have shown that many limitations of the CDF-based approach, encountered during optimization of a multi-response static system, can be overcome by using multiple regression-based weighted signal-to-noise ratio (MRWSN) as the performance metric. The advantages of MRWSN as a performance measure are that (a) signal-to-noise (SN) ratio for a response variable can be computed even when the specification limits and target values for the response variable are unknown, (b) SN ratio takes care of both location (mean) and dispersion (variability) of a response variable and (c) since SN ratios are always expressed in decibels (dB) whatever be the units of measurements of the individual responses, there is no problem in summing the SN ratios of the individual responses.

The aim of the current research is to develop an appropriate procedure for optimizing the multi-response dynamic systems using MRWSN as the performance metric and evaluate the effectiveness of the MRWSN-based optimization approach. The article is organized as follows: the second section outlines briefly about the dynamic systems and reported various approaches for its optimization. The third section describes formulation of commonly used performance metric, called CDF, for the multi-response dynamic system. The formulation of the proposed performance metric, called MRWSN, for the multi-response dynamic system is presented in the fourth section. The procedure for implementation of the MRWSN-based optimization approach is described in section five. In the sixth section, analysis of two experimental datasets taken from literature and related results are presented. We conclude in the final section.

2. Dynamic Systems and its Optimization

Dynamic systems are those where the response variable does not have a fixed target value and the target value of the response variable depends on the level of a control factor (called signal factor) set by the system operator. For example, the steering mechanism of an automobile or speed controller of a fan is a dynamic system. In the case of an automobile steering system, the signal may be the angle of the steering wheel and the response may be the direction of motion or turning radius of the car. In a dynamic system, a response is expected to assume different target values for different levels of the signal factor and so it is often called multi-target system (Joseph & Wu, 2002). In case of a dynamic system, the signal-response relationship is of prime importance and therefore, it is also known as signal-response system (Miller & Wu, 1996). A single-response dynamic system contains only one response variable. On the other hand, a multi-response dynamic system contains more than one response variables and responses are expected to assume different target values as a result of changes in the levels of the signal factor.

2.1 Taguchi method and related works for optimizing a single-response dynamic system

Taguchi (1986) first took interest in designing robust dynamic systems and he considered only the single-response dynamic systems. For a dynamic system, according to Taguchi (1986), ideal quality is based on the ideal relationship between the signal factor and the response variable, and quality loss is caused by deviations from the ideal relationship. So, significant quality improvement can be achieved by first defining a system's ideal function and then using designed experiments to search for an optimal design which minimizes deviations from this ideal function. Taguchi (1986) assumed that a linear relationship exists between the response variable (Y) and the signal factor (M) of the system, and thus the ideal function can be expressed as follows:

$$Y = \beta M + \varepsilon,\tag{1}$$

where β is the slope or system sensitivity, and ε denotes the random error. Here ε is assumed to follow a normal distribution with a mean of zero and variance of σ_β^2. The deviation from the ideal function is represented by the variability of the dynamic system, i.e. σ_β^2. The objective is to determine the best combination of input controllable variables so that the system achieves the respective target value at each level of the signal factor and with minimum variability around the target value. For the purpose of optimization of a single-response dynamic system which has one response variable (Y) whose value are determined by p controllable variables $\mathbf{X} = (X_1, X_2, ..., X_p)$, a signal factor (M) and a noise factor (Z), Taguchi (1986) proposes the following guidelines for designing the experimental plan. Depending on the number of controllable factors (variables) and their levels, select first the most appropriate inner orthogonal array and accordingly determine various trial conditions or experimental runs. On the other hand, determine the noise factor and signal factor levels under which samples are to be tested. Then, conduct the experiments in such a way that different samples under each trial condition are exposed to different combinations of noise factor and signal factor levels. Let y_{jkl} ($j = 1, 2, ..., s$; $l = 1, 2, ..., n$) are the observed values of the response variable at the combination of jth level of signal factor (M_j) and l^{th}

level of noise factor (Z_l) under k^{th} trial condition i.e. vector of control factors levels $\mathbf{x}_k = (x_1, x_2, ..., x_p)$. Then, according to Taguchi (1986), the slope (β_k) and the variability around the slope ($\sigma^2_{\beta_k}$) under the kth trial condition can be obtained using the following equations respectively:

$$\beta_k = \frac{\sum_{j=1}^{s}\sum_{l=1}^{n}\left(y_{jkl}M_j\right)}{\sum_{j=1}^{s}M_j^2} \tag{2}$$

$$\sigma^2_{\beta_k} = \frac{1}{sn-1}\sum_{j=1}^{s}\sum_{l=1}^{n}\left(y_{jkl} - \beta_k M_j\right)^2 \tag{3}$$

The aim of robust design is to find the combination of controllable factors so that the effect of noise factors on the target response of the dynamic system is as small as possible. Taguchi, therefore, uses SN ratio to judge the performance under an experimental run or trial condition. For a dynamic system, the SN ratio under k^{th} experimental run is estimated as follows:

$$\mathrm{SNR}_k = 10\log_{10}\left(\frac{\beta_k^2}{\sigma^2_{\beta_k}}\right) \tag{4}$$

The larger SN ratio means the response has less deviation from its target. It may be noted that the target value of the slope (β) is different for different type of response variables. For dynamic nominal-the-best (DNTB), dynamic larger-the-better (DLTB), and dynamic smaller-the-better (DSTB) type response variables, the desired value of slopes are $0 < \beta < \infty$, $\beta = \infty$, and $\beta = 0$ respectively. In analysing the experimental data, Taguchi (1986) proposed a two-step procedure. In the first step, the settings of the controllable variables $X_1, X_2, ..., X_p$ are determined in such a way that SN ratio is maximized and in the second step, the slope (β) is adjusted by a suitable scaling factor to the desired slope. Any control factor that has a large effect on β but no effect on the variability (σ^2_{β}) is considered as a scaling factor. Miller and Wu (1996) observed that the conventional Taguchi approach for modelling the ideal relationship of a dynamic system lacks a solid basis. They proposed two strategies for modelling a dynamic system and analysing data. These are (i) performance measure modelling (PMM), which is termed as loss model (LM) by Tsui (1999) and (ii) response function modelling (RFM). In PMM approach, each performance measure (β and σ^2_{β}) is modelled as a function of control factors ($\mathbf{X}$) and in RFM approach, performance measures is modelled as a function of the control factors ($\mathbf{X}$) and signal factor (M). The advantage of the RFM approach is that it can reveal how specific control factor interact with specific noise factor. Tsui (1999) proposes a response modelling (RM) approach which directly models the response as a function of the control, noise and signal factors. Tsui (2001) investigated the performances of RM, PMM/LM and RFM approaches for optimizing a single-response dynamic system and concluded that the RM approach has more potential to reach to an optimal solution. Joseph and Wu (2002) have formulated the robust parameter design of dynamic system as a mathematical programming problem. Lesperance and Park (2003) have suggested to use a joint generalized linear model (GLM) so that model assumptions can be investigated using residual analysis. Chen (2003) has proposed a stochastic optimization modelling procedure for optimizing a single response dynamic system. Bae and Tsui (2006) have observed that the GLM-RM approach can provide more reliable results.

2.2 The approaches for optimizing a multi-response dynamic system

All the above research articles are focussed on optimization of a single-response dynamic system. Realizing the need of the modern industries, some authors took research interest in developing appropriate procedure for optimizing multi-response dynamic systems. Different researchers have advocated different approaches for modelling the multiple responses but most of them (Tong et al., 2001; Hsieh et al., 2004; Chang, 2006; Chang, 2008; Wu, 2009; Chang and Chen, 2011) have used CDF as the performance metric for optimization of the multi-response dynamic systems. Tong et al. (2004), Wang and Tong (2004), Wu and Yeh (2005), Wang (2007) and Gauri (2014) have used different performance metrics. While overall relative closeness to the ideal solution is considered as the performance metric by Tong et al. (2004), overall grey relational grade is considered as the performance metric by Wang and Tong (2004) and Wang (2007). Wu and Yeh (2005) have derived total quality loss and minimized it to determine the optimal settings for a multi-response dynamic system, and Gauri (2014) considered overall utility value as the objective function for optimization of multi-response dynamic systems. Among the various types of performance metrics for the multi-response dynamic system, the CDF is most popular among the researchers because it is a simple unit less measure and it has a good foundation in statistical practice.

3. Formulation of CDF for Multi-response Dynamic System

Suppose that in a multi-response dynamic system, there are r output responses $(Y_1, Y_2,..., Y_r)$, s signal factor settings $(M_1, M_2,..., M_s)$, p control factors $(X_1, X_2,..., X_p)$, and n noise factor settings $(Z_1, Z_2,..., Z_n)$. For analysing a dynamic system, the first requirement is to model the response variables appropriately. As reported by Tsui (2001), among the three approaches (RM, PMM, and RFM) for modelling the responses in a dynamic system, the RM approach has more potential to reach to an optimal solution. Most of the authors (Tong et al., 2001; Hsieh et al., 2005; Chang, 2008; Wu, 2009) using CDF as the performance metric, have also used RM approach for modelling the observed responses. Therefore, formulation of CDF is discussed here considering that RM approach is used for modelling the observed responses. Using RM approach, the observed responses in the multi-response dynamic system can be modelled as follows:

$$y_{ijkl} = f_i(M_j, \mathbf{X}_k, Z_l) + \varepsilon_{ijkl},$$ (5)

where $f_i(M_j, \mathbf{X}_k, Z_l)$ denotes the response function between the ith response and the corresponding setting at the jth level of signal factor (M_j) and the lth level of noise factor (Z_l) under k^{th} vector of control factors levels $\mathbf{x}_k = (x_1, x_2,..., x_p)$. The terms y_{ijkl} and ε_{ijkl} represent the values of ith response and error respectively at the jth level of signal factor and l^{th} level of noise factor under k^{th} vector of control factors levels. The error is assumed to follow a normal distribution with mean value as zero and a constant variance σ_ε^2. For each response, it is assumed that a linear form exists between the response and the signal factor as shown in Eq. (1). The exponential desirability function approach to evaluate the quality of a response was introduced by Harrington (1965). The exponential desirability functions normalize an estimated response $\hat{y}$ according to the system's desire and then use exponential functions to transform the normalized value to a scale-free value d, called desirability. It is a value between 0 and 1, and increases as the desirability of the response increases. Derringer and Suich (1980) presented an alternative form of desirability functions which are more flexible in the sense that these can assume a variety of shapes and they defined the CDF of m responses as the geometric mean of the individual desirability. Tong et al. (2001) and Hsieh et al. (2005) have used the same CDF as the performance measure for optimizing the multi-response dynamic systems. The lack of considering the correlation between quality characteristics is a disadvantage of Derringer and Suich's (1980) desirability function. Wu and Hamada (2000), therefore, defined different desirability functions, called double-exponential desirability functions. Chang (2008) and Wu (2009) applied double-exponential desirability functions for optimization of multi-response dynamic systems.

The double-exponential desirability functions for the responses in dynamic system can be better understood by defining first the double-exponential desirability functions for the NTB, LTB and STB type responses of a static system. According to Chang (2008), the double-exponential desirability function for the NTB, LTB and STB type responses under k^{th} trial condition in a static system can be formulated using the following equations, respectively.

$$d_k^{NTB} = \exp\left(-\left|\frac{2\hat{y}_k - (USL + LSL)}{USL - LSL}\right|\right), \tag{6}$$

$$d_k^{LTB} = \exp\left(-\exp\left(-\frac{\hat{y}_k - LSL}{LSL}\right)\right), \tag{7}$$

$$d_k^{STB} = \exp\left(-\left(1 + \frac{\hat{y}_k - USL}{USL}\right)\right), \tag{8}$$

where, $\hat{y}_k$ and d_k represent estimate and desirability of the response variable under k^{th} trial condition, and USL and LSL stand for upper and lower specification limits respectively. A dynamic system can be regarded as a system having multiple static targets which vary depending on their signal values (Joseph & Wu, 2002). Thus, for a dynamic response we can average the normalized values of all estimated responses in a specific experimental run to evaluate that run's desirability. Accordingly, Chang (2008) estimated the double-exponential desirability functions for DNTB, DLTB and DSTB type responses under k^{th} trial condition in a multi-response dynamic system using the following equations respectively:

$$d_{ik}^{DNTB} = \exp\left(-\frac{1}{sn}\sum_{j=1}^{s}\sum_{l=1}^{n}\left|\frac{2\hat{y}_{ijkl} - (USL_{ij} + LSL_{ij})}{USL_{ij} - LSL_{ij}}\right|\right), \tag{9}$$

$$d_{ik}^{DLTB} = \exp\left(-\exp\left(-\frac{1}{sn}\sum_{j=1}^{s}\sum_{l=1}^{n}\frac{\hat{y}_{ijkl} - LSL_{ij}}{LSL_{ij}}\right)\right), \tag{10}$$

$$d_{ik}^{DSTB} = \exp\left(-\left(1 + \frac{1}{sn}\sum_{j=1}^{s}\sum_{l=1}^{n}\frac{\hat{y}_{ijkl} - USL_{ij}}{USL_{ij}}\right)\right), \tag{11}$$

where $\hat{y}_{ijkl}$ denotes the estimate of ith response at j^{th} level of signal factor and l^{th} level of noise factor under k^{th} vector of control factors levels, and USL_{ij} and LSL_{ij} denotes the upper and lower specification limits for i^{th} response at j^{th} signal level respectively. Then, Chang (2008) evaluated the overall performance index (OPI) of the multi-response dynamic system having r response variables under the kth vector of control factors levels as follows:

$$OPI_k = \left(\prod_{i=1}^{r} d_{ik}\right)^{1/r}. \tag{12}$$

This OPI is essentially the CDF. In the CDF-based optimization approach for multi-response dynamic system, settings of the control factors that maximizes the OPI value is considered as the optimal settings.

4. Formulation of the Proposed Performance Metric for Multi-response Dynamic System

The proposed performance metric, MRWSN, is derived by integrating multiple regression technique and Taguchi's SN ratio concept. According to the conventional Taguchi method (1986), the SN ratios of the response variables for a specific experimental run in a dynamic system are obtained based on the observed values of the response variables in the experimental run under all possible signal × noise combinations. In the proposed method, it is suggested to predict first the values of the response variables for the experimental run under all possible signal×noise combinations based on the appropriately fitted multiple regression equations and then to compute the SN ratios for different response variables based on their predicted values instead of their observed values. It may be important to note that error (ε) in the response model represents a random error term which is assumed to follow a normal distribution with mean value as zero and a constant variance σ_ε^2 (Tsui, 1999). However, in general, it is expected that random error will have increased variability at higher signal levels instead of having constant variability at all signal levels. Lesperance and Park (2003), therefore, recommend that fitting such a model requires the use of GLM fitting techniques with a log-link or appropriate link function. Fitting a GLM with log-link function to the experimental data would be very difficult for many quality practitioners. For simplification, we suggest to use a logarithmic transformation of the response variables and then applying multiple regression techniques for fitting the response models. The fitted models, then, will partly ensure that the variability due to random error is more at higher signal levels. Suppose, each of the r response variables is related to p controllable variables $\mathbf{X} = (X_1, X_2,...,X_p)$, signal factor (M) and noise factor (N) by

$$\log_{10}(Y_i) = b_0 + \sum_{u=1}^{p} b_u X_u + \sum_{u,v} b_{uv} X_u X_v + M(\gamma_0 + \sum_{u=1}^{p} \gamma_u X_u) + N(\delta_0 + \sum_{u=1}^{p} \delta_u X_u) + \varepsilon, \quad (13)$$

where b_u, b_{uv}, γ_u, δ_u $(u = 1, 2,..., p; v = 1, 2, 3,..., p)$ are regression coefficients, X_u $(u = 0, 1, 2,..., p)$ are controllable variables and Y_i $(i = 1, 2,..., r)$ are i^{th} response variable. It may be noted that apart from the main factors, the quadratic terms of main factors may be included in the model when there is three or more number of levels in any control and signal factors. The interaction terms of controllable variables may also be included in the model. But, it is very important to include a few control × signal and control × noise interaction terms into the response model. If a few control × signal and control × noise interaction terms are not included in the response model, $\log_{10}(Y_i)$ will be additive at different signal levels independent of combinations of controllable factors $\mathbf{X}$, which implies that intercept only will change but slopes of the controllable variables $\mathbf{X}$ will remain constant. As a result, it will not be possible to obtain desired predicted values of the response variables Y_i $(i = 1, 2,..., r)$ at different signal levels. Let the combination of the control factors levels is $\mathbf{x}_k = (x_1, x_2,..., x_p)$ under k^{th} experimental run or trial condition. For the kth vector of control factors levels, the values of all the response variables at all possible combinations of signal × noise levels can be predicted using the fitted response models. Suppose, $\hat{y}_{ijkl}$ is the predicted value of ith response variable at j^{th} level of signal factor and l^{th} level of noise factor under k^{th} vector of control factors levels. Then, the slope (β_{ik}) and variance around slope $(\sigma_{\beta_{ik}}^2)$ of the ith response variable under the kth experimental run can be estimated as follows:

$$\hat{\beta}_{ik} = \frac{\sum_{j=1}^{s}\sum_{l=1}^{n} \hat{y}_{ijkl} M_j}{\sum_{j=1}^{s} M_j^2} \quad (14)$$

$$\hat{\sigma}_{\beta_{ik}}^2 = \frac{1}{sn-1} \sum_{j=1}^{s}\sum_{l=1}^{n} \left(\hat{y}_{ijkl} - \hat{\beta}_i M_j\right)^2 \quad (15)$$

Therefore, the SN ratio of the ith response variable in the experimental run can be obtained using Eq. (16) shown below. It is important to observe that the forms of Eq. (4) and Eq. (16) are the same. The two equations differ only with respect to usage of observed values or predicted values for making the required computations. In Eq. (4), β and σ_β^2 are calculated based on the observed values of a response variable under a trial condition. On the other hand, in Eq. (16), $\hat{\beta}$ and $\hat{\sigma}_\beta^2$ are calculated using the predicted values of a response variable based on the fitted multiple regression equation, under a trial condition. Therefore, the computed SN ratio that is obtained using Eq. (16) is called as multiple regression-based signal-to-noise ratio (MRSN). Eq. (16) gives estimate of the MRSN value of the ith response variable under k^{th} trial condition or vector of control factors levels.

$$MRSN_{ik} = 10\log_{10}\left(\frac{\hat{\beta}_{ik}^2}{\hat{\sigma}_{\beta_{ik}}^2}\right). \tag{16}$$

Likewise SN ratio, MRSN is always expressed in decibel (dB) unit and higher MRSN implies better quality. For a multi-response dynamic system, MRWSN can be taken as the overall performance measure. The MRWSN of the multi-response dynamic system under k^{th} trial condition can be obtained using the following equation:

$$MRWSN_k = \sum_{i=1}^{r} w_i \times MRSN_{ik} = \sum_{i=1}^{r} w_i \times \left[10\log_{10}\left(\frac{\hat{\beta}_{ik}^2}{\hat{\sigma}_{\beta_{ik}}^2}\right)\right], \tag{17}$$

where, $MRSN_{ik}$ is the multiple regression-based SN ratio of the ith response variable under k^{th} trial condition, w_i is the relative weight of the ith response variable, and $\sum_{i=1}^{r} w_i = 1$. It is suggested to consider $w_i = 1/r$, if the relative importance of the response variables are unknown. It may be noted that MRWSN is essentially a function of the input controllable variables or control factors. Since higher MRSN implies better quality, it is desired that the process conditions are set in such a way that it result in the maximum MRWSN value. It is worth to mention that MRWSN as a performance measure has the following advantages over the CDF: (1) MRWSN can be computed even when the specification limits and target values for one or more response variables are not known and (2) MRWSN take into consideration the variability of the response variables.

5. The Procedure for Implementation of the Proposed Optimization Approach

A multi-response dynamic system can be optimized using MRWSN as the performance metric in the following seven steps:

Step 1: Design the experimental plan, carry out the experimentation and record the experimental observations.

Step 2: Establish the most appropriate multiple regression equations (response models) for prediction of the response variables based on the values of the control factors, signal factor and noise factor levels.

For fitting the multiple regression equations for prediction of a response variable, it is suggested to consider the log transformed values of the response variable as the dependent variable to partly ensure that the variability due to random error is more at higher signal levels. The option of performing multiple linear regression analysis is available in Microsoft Excel as well as in many statistical software packages, e.g. MINITAB, STATISTICA, SPSS etc. The regression coefficients R^2 and adjusted R^2 are considered to drop unnecessary terms from the model and to include only those terms that have some contributions on the dependent variable. Diagnostic checks for validating the regression models must be performed. Using ANOVA and F-test for significance of regression, the adequacy of model can be checked. A residual analysis in terms of various plots, e.g. normality plot of residuals, plot of residual versus predicted values and plot of residual versus individual regression variable etc. should be examined to detect possible anomalies. If, after the diagnostic checks, no serious violations of model assumptions are detected, then the regression equation can be assumed to be adequate fit to predict the dependent variable. More details about fitting of multiple regression equations and the diagnostic checks are available in Montgomery et al. (2012).

Step 3: Choose an arbitrary setting combination of control factors levels (say, the existing combination) $\mathbf{x}_k$, and compute the predicted values of all response variables under the setting combination using respective regression equations for all possible signal$\times$noise levels.

Step 4: Compute slope ($\hat{\beta}_{ik}$), variance around the slope ($\hat{\sigma}^2_{\beta_{ik}}$) and MRSN$_{ik}$ ($i = 1,2,...,r$) under the vector of control factors levels $\mathbf{x}_k$ using Eqs. (14-16), respectively.

Step 5: Compute MRWSN$_k$ under the vector of control factors levels $\mathbf{x}_k$ using Eq. (17).

It may be noted that MRWSN$_k$ is essentially a function of the input controllable variables or control factors. Here the aim is to determine the level values of the input controllable variables that will maximize the MRWSN$_k$ value, which can be determined by changing level values of the input controllable variables and comparing the resulting MRWSN$_k$ values. It has been observed that if all the computations are carried out in Excel worksheet, this enumerative search for finding the optimal level values of the control factors can be performed very effectively using the 'Solver' tool of Microsoft Excel package. It is a kind of 'what if' analysis that finds the optimal value of a target cell by changing values in cells used to calculate the target cell. The 'Solver' tool employs the generalized reduced gradient (GRG) method for optimization, proposed by Del Castillo and Montgomery (1993). Examples on usage of 'Solver' tool is available in Pal and Gauri (2010).

Step 6: Determine the level values of the input controllable variables that maximize the MRWSN$_k$ value, using 'Solver' tool of Microsoft Excel package.

While running the 'Solver' tool it is necessary to specify the range of levels for the input variables. In certain cases, where one or more input variables take only discrete values, the integer restriction (IR) for those input variables need to be specified. Sometimes, one may have to add an additional constraint to keep the slope of the DNTB variable closer to its target slope value. Unless technically there is IR for a response variable, ideally, one should not restrict the optimization to only the actual experimental design settings because that may lead to a suboptimal solution. Because the MRWSN metric is defined based on regression models, the proposed method can provide an optimal solution over the entire experimental region of the input controllable variables while using the 'Solver' tool of Microsoft Excel package.

Step 7: Obtain the expected value of each response variable at each signal level at the derived optimal condition using the relevant multiple regression equation. Then carry out the confirmatory trial and verify that the actual results conform to the expected results.

6. Evaluation of Optimization Performance of the Proposed Method

Under the current research, there is no scope for collection of primary data from industry. Therefore, it is decided to analyse two sets of secondary data, i.e. published data in the literature as two separate case studies and compute the values of some appropriately defined utility measures for evaluation of the performance of the proposed MRWSN-based optimization method.

6.1 Utility measures for comparing optimization performance

From a process engineer's perspective, the best solution should result in the minimum total quality loss implying that maximum total SN ratio (TSN) according to Taguchi philosophy (1986). Therefore, it is decided that the expected TSN at the derived optimal process condition will be considered as an important utility metric for comparison of performance of CDF and MRWSN-based optimization methods. But many statisticians (Leon et al., 1987; Box, 1988) criticised Taguchi's SN ratio concept. Lin and Tu's (1995) mean square errors (MSE) function has a very good foundation in statistical practice. The MSE of ith response variable at j^{th} level of signal factor under k^{th} vector of control factors levels can be computed using the following equation:

$$\text{MSE}_{ijk} = \frac{1}{n}\sum_{l=1}^{n}\left(\hat{y}_{ijkl} - T_{ij}\right)^2 ,$$ (18)

where $\hat{y}_{ijkl}$ is the predicted value of ith response variable at j^{th} level of signal factor and l^{th} level of noise factor under k^{th} vector of control factors levels and T_{ij} is the target value of ith response variable at j^{th} level of signal factor. For a response variable in a dynamic system, the target values of the response variable change according to the signal factor levels. So, PMSE of ith response variable under k^{th} vector of control factors levels can be computed using the following equation:

$$\text{PMSE}_{ik} = \frac{1}{sn}\sum_{j=1}^{s}\sum_{l=1}^{n}\left(\hat{y}_{ijkl} - T_{ij}\right)^2 .$$ (19)

are very close to its target values with reasonably low variability at all the signal levels, the PMSE_{ik} will be quite small. It is important to note that for computation of PMSE_{ik} value, it is necessary to know the target values of ith response variable at all signal levels $(j = 1,2,...,s)$. However, often the target values are not specified for the DSTB and DLTB type responses variables. Therefore, PMSE_{ik} value cannot be computed straightway for these response variables. It is decided to make the following assumptions about the target values of DSTB and DLTB type variables at different signal levels to facilitate computations of PMSE_{ik} value for these variables. It can be found from Eq. (1) that the value of a response variable is equal to the slope β when $M = 1$, i.e. at the signal level M_1. Therefore, it is decided to compute the β values from the experimental observations at signal level M_1 in all the experimental runs, and then to consider the smallest β values among all the experimental runs as the target value of the DSTB variable at signal level M_1, and its multiples as the target values of the DSTB variable at signal levels M_2 and M_3 respectively. Similarly, the largest β value, computed from the experimental observations at signal level M_1 is considered as the target value of the DLTB variable at signal level M_1, and its multiples are considered as the target values of the DLTB variable at signal levels M_2 and M_3, respectively. The multivariate PMSE function (MPMSE) under an experimental run, which may be obtained by pooling the PMSE functions of the individual response variables under the experimental run, can be the most appropriate overall utility measure. But the problem with the MPMSE function is that it will be quite difficult to explain its unit because the PMSE for different response variables may have different units of measurements. Therefore, the PMSE values of the individual response variables are considered as another utility measure for comparison of optimization performance instead of MPMSE value.

6.2 The experimental data

Chang (2008) illustrated application of his proposed data mining approach for optimizing multi-response dynamic systems using a case study adopted from Chang (2006). This case involves simultaneous optimization of three dynamic response variables named Y_1, Y_2 and Y_3. Among these, Y_1 is DLTB type, Y_2 is DNTB type and Y_3 is DSTB type variable. In this case, six control factors A – F, each at three levels (1, 2, and 3), were considered and arranged in a standard L_{18} orthogonal array as shown in Table 1. The signal factor of the case had three levels named M_1, M_2 and M_3, the corresponding values of which were 10, 20 and 30, respectively. Two levels (N_1 and N_2) of noise factor were considered in this case. The specifications for the response variables and the experimental data of Chang (2006) are reproduced in Table 2 and Table 3 respectively. The experimental data of Chang (2006) are analysed as case study 1.

Table 1
Experimental design used by Chang (2006)

Trial no.	Factors and levels					
	A	B	C	D	E	F
1	1	1	1	1	1	1
2	2	2	2	2	2	2
3	3	3	3	3	3	3
4	1	1	2	2	3	3
5	2	2	3	3	1	1
6	3	3	1	1	2	2
7	1	2	1	3	2	3
8	2	3	2	1	3	1
9	3	1	3	2	1	2
10	1	3	3	2	2	1
11	2	1	1	3	3	2
12	3	2	2	1	1	3
13	1	2	3	1	3	2
14	2	3	1	2	1	3
15	3	1	2	3	2	1
16	1	3	2	3	1	2
17	2	1	3	1	2	3
18	3	2	1	2	3	1

Table 2
Specifications for the response variables specified by Chang (2006)

Signal levels	Responses				
	Y_1 (DLTB)	Y_2 (DNTB)			Y_3 (DSTB)
	LSL	LSL	Target	USL	USL
M_1	55	7	10	13	3
M_2	110	14	20	26	6
M_3	165	21	30	39	9

Table 3
Experimental Data of Chang (2006)

Trial no.	Noise factor	Responses																
		Y_1						Y_2						Y_3				
		M_1=10		M_2=20		M_3=30		M_1=10		M_2=20		M_3=30		M_1=10		M_2=20		M_3=30	
1	N_1	61.6	78.2	128.0	106.0	230.6	226.9	7.4	7.2	16.7	13.2	23.7	24.1	1.9	1.9	4.6	3.7	7.6	4.6
	N_2	70.8	57.1	137.3	160.3	282.2	252.5	9.1	10.2	22.8	17.8	26.2	26.7	2.0	2.1	3.9	4.8	4.7	4.3
2	N_1	88.3	93.6	175.2	181.5	259.7	304.5	10.1	8.8	23.4	22.6	29.6	30.3	1.8	2.0	4.0	2.8	6.0	3.3
	N_2	72.9	72.7	174.0	145.5	258.4	214.6	8.7	9.2	19.1	24.1	31.4	30.5	1.9	2.2	3.8	3.5	6.7	6.1
3	N_1	80.8	81.1	154.3	157.4	238.1	237.8	10.8	10.5	20.1	20.9	30.6	32.4	1.0	3.2	4.4	5.4	8.1	6.7
	N_2	77.2	83.3	167.1	159.0	251.8	257.9	10.6	10.9	21.8	23.7	30.5	32.7	1.4	2.6	3.8	3.9	8.0	2.7
4	N_1	65.9	71.3	179.2	151.5	196.1	221.6	7.6	7.2	15.3	14.8	22.5	22.2	1.7	2.2	4.1	4.9	5.9	6.0
	N_2	83.7	78.4	135.6	177.0	246.9	291.8	8.1	7.6	14.1	14.7	21.9	21.8	2.1	2.6	3.6	3.3	7.7	6.9
5	N_1	79.4	88.6	121.9	151.6	248.8	245.1	11.9	12.7	25.6	25.7	36.6	35.7	2.0	2.1	2.7	3.8	4.8	5.8
	N_2	67.8	87.3	113.6	141.3	171.5	244.7	10.5	11.8	25.8	26.2	39.1	33.1	2.6	1.7	4.1	3.6	5.5	5.2
6	N_1	90.5	87.0	161.8	169.4	286.9	236.5	10.2	10.2	23.7	21.8	32.0	32.8	1.8	2.2	2.5	4.1	4.9	5.4
	N_2	87.6	87.8	160.7	163.9	231.4	288.7	10.7	11.2	22.5	20.0	34.2	28.0	1.2	2.0	4.0	4.4	6.6	3.1
7	N_1	80.9	74.7	165.9	163.4	232.2	246.4	11.7	12.1	23.3	23.3	33.5	32.7	1.9	1.6	5.2	5.1	6.6	5.3
	N_2	69.9	78.7	141.7	159.1	260.4	239.7	11.6	11.6	22.0	22.5	33.8	34.0	1.9	1.7	3.6	5.1	5.8	5.1
8	N_1	92.3	71.7	185.7	154.3	233.1	240.5	8.3	8.4	16.9	18.5	28.1	27.3	2.8	1.5	3.4	3.4	6.2	3.7
	N_2	104.8	89.4	216.1	173.0	340.6	308.9	8.2	6.1	18.5	15.1	29.1	21.1	1.8	2.2	4.1	4.3	9.0	5.0
9	N_1	92.8	59.8	130.7	142.2	257.3	266.0	8.6	9.7	17.3	18.4	30.9	34.5	2.3	1.7	4.5	5.1	7.4	6.5
	N_2	82.1	87.0	175.1	138.2	161.4	274.4	8.2	8.4	18.6	18.0	31.3	30.9	2.2	1.2	4.0	4.0	7.2	6.7
10	N_1	86.0	100.0	179.1	175.4	246.9	244.9	6.9	6.6	15.8	16.4	22.2	24.9	1.7	1.7	3.9	5.8	5.2	8.4
	N_2	81.9	91.2	190.8	137.8	293.5	227.1	8.3	7.2	14.3	16.6	23.1	24.2	2.1	1.8	3.9	4.8	4.5	8.0
11	N_1	76.3	78.2	140.0	154.7	264.1	260.1	9.5	9.2	27.9	25.4	33.8	33.8	1.0	1.7	4.1	4.0	3.9	5.6
	N_2	67.1	76.0	169.5	175.4	239.2	251.8	10.8	12.2	23.6	20.4	26.8	29.2	2.1	2.3	5.0	4.6	6.7	4.5
12	N_1	91.4	81.8	160.1	167.2	238.5	197.2	11.8	10.5	22.9	25.9	36.9	36.8	2.1	1.9	4.0	3.7	7.4	6.2
	N_2	85.1	63.8	123.8	166.4	233.5	242.0	11.6	11.5	22.1	22.6	32.5	31.9	1.6	0.9	4.2	4.1	2.1	6.9
13	N_1	87.9	82.5	146.5	167.0	212.4	222.2	10.1	10.3	19.9	22.2	27.1	27.7	2.0	2.3	5.0	4.7	6.7	7.3
	N_2	57.4	78.5	91.6	182.3	250.3	207.3	10.6	10.1	20.1	19.2	27.8	24.6	2.2	2.1	4.5	4.7	7.4	6.2
14	N_1	88.1	78.1	156.7	170.1	239.1	215.2	12.1	10.0	24.1	20.2	28.0	35.6	1.7	2.1	3.8	3.3	7.2	4.9
	N_2	81.7	75.7	140.2	127.8	241.4	211.7	11.3	11.8	23.3	24.4	32.1	39.9	1.8	2.7	3.1	3.8	4.9	4.7
15	N_1	101.8	78.2	168.3	180.8	240.6	235.2	10.2	7.6	14.9	19.4	26.9	19.5	1.7	2.2	5.3	3.6	5.0	7.5
	N_2	80.4	76.5	206.7	222.6	325.1	285.2	8.7	7.2	16.8	14.1	26.6	25.1	2.2	1.9	4.8	2.6	3.8	5.5
16	N_1	77.4	75.4	171.7	159.0	201.3	219.7	10.4	10.6	20.7	22.2	34.4	30.0	1.9	2.3	3.1	4.2	5.8	3.3
	N_2	72.0	69.5	189.1	168.6	254.3	237.3	11.1	10.9	20.6	21.4	30.5	31.4	2.1	1.6	4.1	4.8	5.4	6.9
17	N_1	71.4	69.2	145.0	152.5	223.8	218.7	8.8	8.4	19.0	13.8	26.2	24.1	1.5	2.1	3.8	4.4	6.9	4.2
	N_2	77.0	70.5	158.4	154.0	218.4	224.1	9.2	9.0	16.7	17.4	27.1	26.2	1.6	1.8	3.7	4.4	4.6	4.2
18	N_1	82.8	67.8	183.7	175.5	276.1	254.4	10.7	9.2	19.8	20.1	27.3	31.0	2.5	2.2	3.1	3.4	7.7	7.4
	N_2	85.2	92.0	154.4	157.6	249.3	286.1	11.3	7.7	19.3	22.6	29.4	24.6	1.6	1.8	3.9	4.7	6.0	7.3

Chang and Chen (2011) proposed a neuro-genetic approach for optimizing a multi-response dynamic system and they illustrated the effectiveness of their proposed approach using a simulated example of a dynamic system containing multiple responses. This simulated example data involves three dynamic response variables named Y_1 (DLTB type), Y_2 (DNTB type) and Y_3 (DSTB type). In this study, six control factors A – F, each at three levels, were considered in an L_{18} orthogonal array design. Further, a signal factor with three levels M_1, M_2 and M_3, and a noise factor with two levels N_1 and N_2 were considered. The experimental design used by Chang and Chen (2011) is the same as given in Table 1. The specification limits for the response variables and the simulated experimental data are available in Chang and Chen (2011). The experimental data of Chang and Chen (2011) are analysed as case study 2.

6.3 Analysis and results

6.3.1 Case study 1

It may be noted from Table 2 that Chang (2006) specified the target value of Y_2 only. However, for computation of PMSE value of a response variable, it is necessary to know its target values at all signal levels $(j = 1,2,...,s)$. Therefore, the target values Y_1 and Y_3 at all the signal levels are assumed as per the procedure described in section 6.1. The β values for Y_1 and Y_3 are computed first from the experimental observations at signal level M_1. It was found that the largest β value for Y_1 and the smallest β value for Y_3 among all the experimental runs are 86 and 1.735 respectively. So the target values for Y_1 at the signal levels M_1, M_2 and M_3 are chosen as 86, 2×86 and 3×86 respectively, and the target values for Y_3 at the signal levels M_1, M_2 and M_3 are chosen as 1.735, 2×1.735 and 3×1.735 respectively. Table 4 shows the target values and specification limits for the three response variables.

Table 4
Target values and specification limits for the response variables (case study 1)

| Signal levels | Responses | | | | | | |
| | Y_1(DLTB) | | Y_2 (DNTB) | | | Y_3 (DSTB) | |
	LSL	Target	LSL	Target	USL	Target	USL
M_1	55	86	7	10	13	1.735	3
M_2	110	172	14	20	26	3.470	6
M_3	165	258	21	30	39	5.205	9

The first requirement for application of the proposed optimization method is to establish the most appropriate multiple regression equations for prediction of the response variables (see step 2 of section 5). Variability at the higher signal level is expected to be higher and therefore, it is decided to establish the multiple regression equations considering log transformed values of the responses as the dependent variables. Logarithm transformation of each response variable is made and then the response model is developed for each response variable as a function of controllable variables (A – F), signal variable (M) and noise variable (N) using MINITAB software. The developed response models for the three response variables are given below:

$$\log_{10}(Y_1) = 1.1489 + 0.01338A + 0.0450B + 0.1216C + 0.0042D + 0.0958E + 0.0021F + 0.5084M$$
$$- 0.0562M^2 + 0.0253N - 0.0243C^2 - 0.0249E^2 - 0.0098MB - 0.0097MC - 0.008NB$$
$$- 0.0068NC + 0.0126NE - 0.01NF \tag{20}$$
$$\left[R^2 = 0.943 \text{ and adjusted } R^2 = 0.938 \right]$$

$$\log_{10}(Y_2) = 0.2784 + 0.1258A + 0.2846B - 0.1459C - 0.1315D - 0.1074E + 0.1531F + 0.5486M$$
$$- 0.0771M^2 + 0.0273N - 0.0266A^2 - 0.0651B^2 + 0.0324C^2 + 0.0388D^2 + 0.0259E^2$$
$$- 0.0323F^2 - 0.0137NE \tag{21}$$
$$\left[R^2 = 0.967 \text{ and adjusted } R^2 = 0.965 \right]$$

$$\log_{10}(Y_3) = -0.2005 - 0.0729A - 0.0038B + 0.0191C + 0.0626D - 0.1014E + 0.0304F + 0.5667M$$

$$-0.0905M^2 + 0.0689N + 0.0274A^2 - 0.0207D^2 + 0.0176E^2 + 0.0077MC + 0.0085ME$$

$$-0.03334NA - 0.0154NC + 0.0152ND + 0.0194NE - 0.0242NF \tag{22}$$

$$\left[R^2 = 0.822 \text{ and adjusted } R^2 = 0.804 \right]$$

Using ANOVA and F-test for significance of regression, the adequacy of the models is checked. Residual analysis in terms of normality plot of residuals, plot of residual versus predicted values and plot of residual versus individual regression variable are carried out and found satisfactory. According to step 3 in section 5, an arbitrary setting combination of control factors levels $\mathbf{x}_k = A_2B_2C_2D_2E_2F_2$ is selected. For the setting combination $\mathbf{x}_k$, the predicted values of Y_1 at all signal $\times$ noise levels, i.e. $\hat{y}_{11k1}, \hat{y}_{11k2}, \hat{y}_{12k1}, \hat{y}_{12k2}, \hat{y}_{13k1}, \hat{y}_{13k2}$ are obtained using Eq. (20). According to step 4 in section 5, the slope and variance around the slope for the response variable Y_1 under the setting combination $\mathbf{x}_k$, i.e. $\hat{\beta}_{1k}$ and $\hat{\sigma}_{1k}^2$ are computed using Eqs. (15-16), respectively. Then the MRSN for the response variable Y_1 under the setting combination $\mathbf{x}_k$, i.e. MRSN_{1k} is obtained using Eq. (17). Similarly, MRSN_{2k} and MRSN_{3k} for the response variables Y_2 and Y_3 respectively are obtained. Then, according to step 5 in section 5, the MRWSN under the setting combination $\mathbf{x}_k$, i.e. MRWSN_k is computed using Eq. (18). As the relative importance of the three response variables is unknown, equal weights are given to all three response variables while MRWSN_k value is computed. All these computations are carried out in Excel worksheet.

Now according to step-6 in section 5, the 'Solver' tool of Microsoft Excel is applied to maximize the MRWSN_k value. If integer restrictions (IR) are specified for the level values of input controllable variables, the 'Solver' tool will determine the optimal combination among the actual experimental design settings and if IR are not specified for the level values of input variables, the 'Solver' tool will determine an optimal solution over the entire experimental region of the input variables. It is decided to determine the optimal solutions under both the conditions, i.e. under IR and under no IR for the level values of the input variables. While optimizing, in both the cases, an additional constraint (absolute difference between predicted slope and target slope is less than 0.10) is added to keep the slope of the DNTB variable (Y_2) closer to its target slope value 10. In the first case where IR are specified for the level values of input variables, the optimal solution is found to be $A_2B_1C_3D_3E_2F_2$ and in the second case where IR are not specified for the level values of input variables, the optimal solution is found to be as follows: A = 1.93, B = 2.55, C = 2.17, D = 1.0, E = 2.06 and F = 1.68. The expected values of each response variable at all signal $\times$ noise levels under these two optimal conditions are computed using the relevant fitted response models and then, the slope ($\hat{\beta}$), variance around slope ($\hat{\sigma}_\beta^2$), MRSN and PMSE values for each response variable are estimated using Eqs. (14-16) and Eq. (19), respectively.

With the aim to compare the optimization performance of the proposed MRWSN-based approach with the CDF-based approach, the same experimental data are analysed again considering the OPI (which is essentially a CDF) as the objective function for the optimization. Y_1, Y_2 and Y_3 are DLTB, DNTB and DSTB type response variables respectively. The predicted values of the three variables at the chosen arbitrary setting are converted into appropriate double-exponential desirability functions first using in Eqs. (9-11) respectively. Then, the OPI value at the arbitrary setting is obtained using Eq. (12). The 'Solver' tool of Microsoft Excel is applied again to maximize the OPI value under both the conditions, i.e. with IR and without IR for the level values of the input variables. The optimal solution is found to be $A_3B_2C_2D_3E_2F_1$ when IR for the level values of the controllable variables are specified, and the optimal solution is found to be A = 2.56, B = 3.0, C = 1.64, D = 3.0, E = 2.02 and F = 1.20 when no IR is specified. The expected values of each response variable at all signal $\times$ noise levels under these two optimal conditions are computed and then the expected SN ratio and PMSE for each response variable are obtained using Eq. (16) and Eq. (19), respectively. Table 5 displays the expected values of the two utility measures, i.e. TSN and PMSE of the individual response variables at different optimal conditions obtained by the proposed MRWSN-based approach and the CDF-based approach.

It can be noted from Table 5 that the derived optimal solutions using MRWSN-based approach result in higher TSN values in both the cases (i.e. under IR and under no IR for the level values of the control factors) than the optimal solutions derived by the CDF-based approach. This implies that the proposed MRWSN-based approach leads to better optimal solution than the CDF-based approach with respect to Taguchi's philosophy. Table 5 further reveals that the PMSE value for Y_1 at MRWSN-based optimal solutions is lesser than the PMSE value for Y_1 at the CDF-based optimal solution. The PMSE values for Y_2 and Y_3 at MRWSN-based optimal solutions is lesser or equal to the PMSE values for Y_2 and Y_3 at the CDF-based optimal solutions. Therefore, it may be concluded that MRWSN-based approach results in better optimal solution in proper statistical sense also.

Table 5
Expected TSN and PMSE values under different optimal conditions (case study 1)

Optimization Method	Optimal solution	SN ratio (in dB)			TSN	PMSE		
		Y_1	Y_2	Y_3		Y_1	Y_2	Y_3
MRWSN-based Approach (under IR for input variables)	$A_2 B_1 C_3 D_3 E_2 F_2$	29.71	29.55	23.95	83.21	207.92	0.13	0.08
CDF-based Approach (under IR for input variables)	$A_3 B_2 C_2 D_3 E_2 F_1$	27.22	29.55	22.72	79.49	250.58	0.18	0.09
MRWSN-based optimization (under no IR for input variables)	A = 1.93, B = 2.55, C = 2.17, D = 1.0, E = 2.06 and F = 1.68	37.19	29.53	22.00	88.72	6.93	0.093	0.024
CDF-based Optimization (under no IR for input variables)	A = 2.56, B = 3.0, C = 1.64, D = 3.0, E = 2.02 and F = 1.20	34.89	29.55	21.12	85.58	244.72	0.093	0.024

6.3.2 Case study 2

Chang and Chen (2011) also did not specify the target values of Y_1 and Y_3. However, to facilitate computation of MSE values for Y_1 and Y_3 at different signal levels, necessary computations were made from the experimental data and then target values of these response variables were assumed as per the procedure described in section 6.1. Table 6 shows the target values and specification limits for the three response variables.

Table 6
Target values and specification limits for the response variables (case study 2)

Signal levels	Y_1(DLTB)		Y_2 (DNTB)			Y_3 (DSTB)	
	LSL	Target	LSL	Target	USL	Target	USL
M_1	4.8	9.2	0.6	1.0	1.4	16	28
M_2	9.6	18.4	1.2	2.0	2.8	32	56
M_3	14.4	27.6	1.8	3.0	4.2	48	84

At first, the response model for each response variable is developed as a function of controllable variables $(A - F)$, signal variable (M) and noise variable (N) using MINITAB. The developed response models for the three response variables are given below:

$$\log_{10}(Y_1) = 0.3012 - 0.0329A - 0.0032B + 0.105C - 0.002D + 0.1077E + 0.0062F + 0.4571M$$
$$-0.0504M^2 + 0.0621N - 0.0186C^2 - 0.0191E^2 - 0.0147AC + 0.0235AF + 0.0067MA \quad (23)$$
$$-0.0124MF - 0.0123NC + 0.0134ND - 0.0121NE - 0.0235NF$$
$$\left[R^2 = 0.957 \text{ and adjusted } R^2 = 0.948 \right]$$

$$\log_{10}(Y_2) = -0.6249 + 0.1643A + 0.2717B - 0.2169C - 0.1038D - 0.1220E + 0.190F + 0.4930M$$
$$- 0.0636M^2 - 0.0016N - 0.0377A^2 - 0.064B^2 + 0.0488C^2 + 0.0323D^2$$
$$+ 0.0242E^2 - 0.0421BF$$
$$\left[R^2 = 0.967 \text{ and adjusted } R^2 = 0.962 \right]$$

(24)

$$\log_{10}(Y_3) = 0.6642 - 0.1384A + 0.0836B - 0.1411C - 0.0172D + 0.0725E + 0.109F + 0.6209M$$
$$- 0.068M^2 + 0.024N + 0.0383A^2 + 0.0359C^2 - 0.0549AB + 0.0229AC - 0.0235MC$$
$$- 0.0263MF + 0.0332NB - 0.0231NE - 0.0246NF$$
$$\left[R^2 = 0.842 \text{ and adjusted } R^2 = 0.810 \right]$$

(25)

The adequacy of the developed models is checked using ANOVA and F-test for significance of regression. Residual analysis in terms of normality plot of residuals, plot of residual versus predicted values and plot of residual versus individual regression variable also are carried out and found satisfactory. An arbitrary setting combination of control factor levels $\mathbf{x}_k = A_2B_2C_2D_2E_2F_2$ is chosen. At this setting combination, the values of the all the three response variables are predicted using the fitted regression models for all signal × noise levels. Using these predicted values $\hat{\beta}_{1k}$, $\hat{\beta}_{2k}$, $\hat{\beta}_{3k}$, $\hat{\sigma}^2_{\beta_{1k}}$, $\hat{\sigma}^2_{\beta_{2k}}$ and $\hat{\sigma}^2_{\beta_{3k}}$ are computed using the relevant equations. Then, MRSN_{ik} ($i = 1,2,3$) are computed using Eq. (16) and the performance metric MRWSN_k is computed using Eq. (17) considering equal weight for all the response variables (since relative importance of the response variable are unknown). Now the 'Solver' tool of Microsoft Excel is applied to maximize the MRWSN_k value considering two cases (IR are present and IR are not present for the level values of input variables). While optimizing, in both the cases, an additional constraint (absolute difference between predicted slope and target slope is less than 0.05) was added to keep the slope of the DNTB variable (Y_2) closer to its target slope value 1.0. In the first case, the optimal setting combination is found to be $A_3B_3C_1D_2E_2F_3$ and in the second case, the optimal setting combination is found to be as follows: A = 2.47, B = 3.0, C = 1.53, D = 3.0, E = 2.43 and F = 2.73. The expected values of each response variable at all signal × noise levels under the two optimal conditions are computed using the relevant fitted response models and then, the slope ($\hat{\beta}$), variance around slope ($\hat{\sigma}^2_\beta$), SN ratio and PMSE values for each response variable are estimated using the relevant equations.

For the purpose of comparison of the optimization performance of the proposed MRWSN-based approach and CDF-based approach, the same experimental data are analysed again considering the OPI (which is essentially a CFD) as the objective function for the optimization. Y_1, Y_2 and Y_3 are DLTB, DNTB and DSTB type response variables respectively. The predicted values of these response variables at the chosen arbitrary setting are converted into appropriate double-exponential desirability functions using in Eq. (9), Eq. (10) and Eq. (11), respectively, and then, the OPI value at the arbitrary setting is obtained using Eq. (12). The 'Solver' tool of Microsoft Excel is applied again to maximize the OPI value under both the conditions, i.e. under IR and under no IR for the level values of the input controllable variables. Under the condition of IR for the level values of the input variables, the optimal solution is found to be $A_3B_3C_1D_3E_2F_2$, and under the condition of no IR for the level values of the input variables, the optimal solution is found to be: A = 1.77, B = 1.0, C = 1.95, D = 3.0, E = 1.80, F = 1.46. The expected values of each response variable at all signal × noise levels under these two optimal conditions are computed and then the expected SN ratio and PMSE for each response variable are obtained using the relevant equations. Table 7 displays the expected values of the two utility measures, i.e. TSN and PMSE of the individual response variable at different optimal conditions obtained by the proposed MRWSN-based approach and the CDF-based approach.

Table 7

Expected TSN and PMSE values under different optimal conditions (case study 2)

Optimization Method	Optimal solution	SN ratio (in dB)			TSN	PMSE		
		Y_1	Y_2	Y_3		Y_1	Y_2	Y_3
MRWSN-based Approach (under IR for input variables)	$A_3B_3C_1D_2E_2F_3$	24.87	45.93	32.64	103.44	0.280	0.001	2.056
CDF-based Approach (under IR for input variables)	$A_3B_3C_1D_3E_2F_2$	21.79	45.93	17.74	85.46	0.504	0.004	3.627
MRWSN-based optimization (under no IR for input variables)	A = 2.47, B = 3.0, C = 1.53, D = 3.0, E = 2.43 and F = 2.73	28.09	45.93	38.92	112.94	0.136	0.000	3.700
CDF-based Optimization (under no IR for input variables)	A = 1.77, B = 1.0, C = 1.95, D = 3.0, E = 1.80 and F = 1.46	21.50	45.93	20.38	87.51	0.514	0.000	4.123

Table 7 reveals that the derived optimal solutions based on MRWSN-based approach result in higher expected TSN values in both the cases (i.e. under IR and under no IR for the level values of the control factors) than the optimal solutions derived based on CDF-based approach. This implies that the proposed MRWSN-based approach leads to better optimal solution than the CDF-based approach with respect to Taguchi's philosophy. Table 7 further reveals that the PMSE values for Y_1 and Y_3 at MRWSN-based optimal solutions are lesser than the PMSE values for Y_1 and Y_3 at the CDF-based optimal solutions. The PMSE value for Y_2 at MRWSN-based optimal solution is lesser or equal to the PMSE value for Y_2 at the CDF-based optimal solution. Therefore, it may be concluded that MRWSN-based approach results in better optimal solution in proper statistical sense also. In this research, there is no scope to carry out the confirmatory trials with the optimal factor-level combinations. However, the results of analysis of the two case studies are indicative that proposed MRWSN-based approach for optimizing a multi-response dynamic system is promising because it not only overcome the limitations of the CDF-based approach but also results in better optimal solution with respect to utility measures like TSN and PMSE for the individual response variables.

7. Conclusions

Industries are increasingly emphasizing optimization of dynamic multi-response problems in the light of increasing complexities of modern manufacturing systems. CDF-based approach has gained popularity in recent years for optimization of multi-response dynamic systems. A major disadvantage with the desirability index is that it does not consider the variability of a response variable. Moreover, if the specification limits of a response variable are not provided, then the CDF cannot be computed. This article presents a new method, called multiple regression-based weighted signal-to-noise ratio (MRWSN) method, for optimization of a multi-response dynamic system. The proposed method not only overcome the limitations of CDF-based approach but also results in better optimization performance. In this method, at first appropriate multiple regression equations are fitted based on the experimental observations for prediction of the response variables, and then the values of the slope, variance around the slope and MRSN for different response variables are computed based on their predicted values instead of their observed values. The optimal setting of the input controllable variables is then determined by maximizing the MRWSN. Two sets of experimental data taken from the literature are analysed using the proposed method and the CDF-based approach. The results show that the proposed method is superior to the CDF-based approach with respect to TSN as well as PMSE values of the individual responses.

References

Bae, S. J., & Tsui, K. L. (2006). Analysis of dynamic robust design experiment with explicit & hidden noise variables. *Quality Technology & Quantitative Management*, 3(1), 55-75.

Box, G. (1988). Signal-to-noise ratios, performance criteria, and transformations. *Technometrics*, *30*(1), 1-17.

Chang, H. H. (2006). Dynamic multi-response experiments by backpropagation networks and desirability functions. *Journal of the Chinese Institute of Industrial Engineers*, *23*(4), 280-288.

Chang, H. H. (2008). A data mining approach to dynamic multiple responses in Taguchi experimental design. *Expert Systems with Applications*, *35*(3), 1095-1103.

Chang, H. H., & Chen, Y. K. (2011). Neuro-genetic approach to optimize parameter design of dynamic multiresponse experiments. *Applied Soft Computing*, *11*(1), 436-442.

Chen, S. P. (2003). Robust design with dynamic characteristics using stochastic sequential quadratic programming. *Engineering Optimization*, *35*(1), 79-89.

Del Castillo, E., & Montgomery, D. C. (1993). A nonlinear programming solution to the dual response problem. *Journal of Quality Technology*, *25*(3), 199-204.

Derringer, G., & Suich, R. (1980). Simultaneous optimization of several response variables. *Journal of Quality Technology*, *12*(4), 214–219.

Gauri, S. K. (2014). Optimization of multi-response dynamic systems using principal component analysis (PCA)-based utility theory approach. *International Journal of Industrial Engineering Computations*, *5*(1), 101-114.

Harrington, E. C. (1965). The desirability function. *Industrial quality control*, 21(10), 494-498.

Hsieh, K. L., Tong, L. I., Chiu, H. P., & Yeh, H. Y. (2005). Optimization of a multiresponse problem in Taguchi's dynamic system. *Computers & Industrial Engineering*, *49*(4), 556-571.

Jeong, I. J., & Kim, K. J. (2009). An interactive desirability function method to multi-response optimization. *European Journal of Operational Research*, *195*(2), 412-426.

Joseph, V.R., & Wu, C.F.J. (2002). Robust parameter design of multiple-target systems. *Technometrics*, *44*(4), 338-346.

Khuri, A. I., & Conlon, M. (1981). Simultaneous optimization of multiple responses represented by polynomial regression functions. *Technometrics*, *23*(4), 363–375.

León, R. V., Shoemaker, A. C., & Kacker, R. N. (1987). Performance measures independent of adjustment: an explanation and extension of Taguchi's signal-to-noise ratios. *Technometrics*, *29*(3), 253-265.

Lesperance, M. L., & Sung-Min, P. (2003). GLMs for the analysis of robust designs with dynamic characteristics. *Journal of Quality Technology*, *35*(3), 253-263.

Liao, H. C. (2006). Multi-response optimization using weighted principal component. *The International Journal of Advanced Manufacturing Technology*, *27*(7-8), 720-725.

Lin, D. K., & Tu, W. (1995). Dual response surface optimization. *Journal of Quality Technology*, *27*(1), 34-39.

McCaskey, S. D., & Tsui, K. L. (1997). Analysis of dynamic robust design experiments. *International Journal of Production Research*, *35*(6), 1561-1574.

Miller, A., & Wu, C. F. J. (1996). Parameter design for signal-response systems: A different look at Taguchi's dynamic parameter design. *Statistical Science*, *11*(2), 122-136.

Montgomery, D. C., Peck, E. A., & Vining, G. G. (2015). *Introduction to linear regression analysis*. New York: John Wiley & Sons.

Pal, S., & Gauri, S. K. (2010). Multi-Response Optimization Using Multiple Regression–Based Weighted Signal-to-Noise Ratio (MRWSN). *Quality Engineering*, *22*(4), 336-350.

Pignatiello, Jr., & Joseph, J. (1993). Strategies for robust multi-response quality engineering. *IIE Transactions*, *25*(3), 5-15.

Su, C. T., Chen, M. C., & Chan, H. L. (2005). Applying neural network and scatter search to optimize parameter design with dynamic characteristics. *Journal of the Operational Research Society*, *56*(10), 1132-1140.

Su, C. T., & Tong, L. I. (1997). Multi-response robust design by principal component analysis. *Total Quality Management*, *8*(6), 409–416.

Taguchi, G. (1986). *Introduction to quality engineering: designing quality into products and processes*. Tokyo, Japan: Asian Productivity Organization.

Tong, L. I., Chen, C. C., & Wang, C. H. (2007). Optimization of multi-response processes using the VIKOR method. *The International Journal of Advanced Manufacturing Technology, 31*(11-12), 1049-1057.

Tong, L.I., & Hsieh, K.L. (2000). A novel means of applying artificial neural networks to optimize multi-response problem. *Quality Engineering, 13*(1), 11–18.

Tong, L. I., Wang, C. H., Houng, J. Y., & Chen, J. Y. (2001). Optimizing dynamic multi-response problems using the dual-response-surface method. *Quality Engineering, 14*(1), 115-125.

Tong, L. I., Wang, C. H., Chen, C. C., & Chen, C. T. (2004). Dynamic multiple responses by ideal solution analysis. *European Journal of Operational Research, 156*(2), 433-444.

Tong, L. I., Wang, C. H., & Tsai, C. W. (2008). Robust design for multiple dynamic quality characteristics using data envelopment analysis. *Quality and Reliability Engineering International, 24*(5), 557-571.

Tsui, K. L. (1999). Modeling and analysis of dynamic robust design experiments. *IIE Transactions, 31*(12), 1113-1122.

Tsui, K. (2001). *Response model analysis of dynamic robust design experiments*, in: Lenz, H.J., Wilrich, P.T. (Eds.) Frontiers in Statistical Quality Control 6. Physica-Verlag Heidelberg, New York, pp. 360-370.

Wang, C. H., & Tong, L. I. (2004). Optimization of dynamic multi-response problems using grey multiple attribute decision making. *Quality Engineering, 17*(1), 1-9.

Wang, C. H. (2007). Dynamic multi-response optimization using principal component analysis and multiple criteria evaluation of the grey relation model. *The International Journal of Advanced Manufacturing Technology, 32*(5-6), 617-624.

Wang, J., Ma, Y., Ouyang, L., & Tu, Y. (2016). A new Bayesian approach to multi-response surface optimization integrating loss function with posterior probability. *European Journal of Operational Research, 249*(1), 231-237.

Wasserman, G. S. (1996). Parameter design with dynamic characteristics: A regression perspective. *Quality and Reliability Engineering International, 12*(2), 113-117.

Wu, C. F. J. (2004). Optimization of correlated multiple quality characteristics using desirability function. *Quality Engineering, 17*(1), 119-126.

Wu, C.F.J., & Hamada, M. (2000). *Experiments: Planning, analysis, and parameter design optimization.* New York: Wiley-Interscience.

Wu, F. C., & Yeh, C. H. (2005). Robust design of multiple dynamic quality characteristics. *The International Journal of Advanced Manufacturing Technology, 25*(5-6), 579-588.

Wu, F. C. (2009). Robust design of nonlinear multiple dynamic quality characteristics. *Computers & Industrial Engineering, 56*(4), 1328-1332.

2

Green open location-routing problem considering economic and environmental costs

Eliana M. Toro[a], John F. Franco[b], Mauricio Granada Echeverri[c]*, Frederico G. Guimarães[d] and Ramón A. Gallego Rendón[e]

[a]Facultad de Ingeniería Industrial, Universidad Tecnológica de Pereira. Pereira, Colombia
[b]Universidade Estadual Paulista Júlio de Mesquita Filho, UNESP, Ilha Solteira, Brazil
[c]Programa de Ingeniería Eléctrica, Facultad de Ingenierías, Universidad Tecnológica de Pereira., Pereira, Colombia
[d]Department of Electrical Engineering, Universidade Federal de Minas Gerais, UFMG, Belo Horizonte, Brazil
[e]Programa de Ingeniería Eléctrica, Facultad de Ingenierías, Universidad Tecnológica de Pereira., Pereira, Colombia

CHRONICLE

ABSTRACT

This paper introduces a new bi-objective vehicle routing problem that integrates the Open Location Routing Problem (OLRP), recently presented in the literature, coupled with the growing need for fuel consumption minimization, named Green OLRP (G-OLRP). Open routing problems (ORP) are known to be NP-hard problems, in which vehicles start from the set of existing depots and are not required to return to the starting depot after completing their service. The OLRP is a strategic-level problem involving the selection of one or many depots from a set of candidate locations and the planning of delivery radial routes from the selected depots to a set of customers. The concept of radial paths allows us to use a set of constraints focused on maintaining the radiality condition of the paths, which significantly simplifies the set of constraints associated with the connectivity and capacity requirements and provides a suitable alternative when compared with the elimination problem of sub-tours traditionally addressed in the literature. The emphasis in the paper will be placed on modeling rather than solution methods. The model proposed is formulated as a bi-objective problem, considering the minimization of operational costs and the minimization of environmental effects, and it is solved by using the epsilon constraint technique. The results illustrate that the proposed model is able to generate a set of trade-off solutions leading to interesting conclusions about the relationship between operational costs and environmental impact.

Keywords:
Open Location-Routing Problem
Green Vehicle Routing Problem
Green logistics
Mixed-Integer Linear Programming
Vehicle Routing Problem

1. Introduction

The open vehicle routing problem (OVRP) was first proposed in the early 1980s when there were cases where a delivery company did not own a vehicle fleet or its private fleet was inadequate for fully satisfying customer demand (Schrage, 1981; Bodin et al., 1983). Therefore, contractors who were not employees of the delivery company used their own vehicles for deliveries. In these cases, vehicles were not required to return to the central depot after their deliveries because the company was only concerned

* Corresponding author
E-mail: magra@utp.edu.co (M. G. Echeverri)

with reaching the last customer. Compensation was not given for any driving outside of meeting this goal. Thus, the goal of the OVRP is to design a set of Hamiltonian paths to satisfy customer demand. During the last decade, consumers, businesses and governments have increased their attentions to the environment. Society in general is becoming increasingly aware and concerned of the environmental impacts of human activities and the indiscriminate use of natural resources. A growing interest is being perceived in companies to assess and reduce the environmental impacts of their products and services (Daniel et al., 1997; Frota Neto et al., 2009). In this context, the transportation industry has a significant effect on the planet, because of the large quantity of fuel used in its regular operations and the environmental consequences and greenhouse effects of fuel consumption and pollution. As a consequence, Green Logistics and Green Transportation have emerged in all levels of supply chain management, with growing value to researchers and organizations, motivated by the fact that current logistics centered on economic costs without accounting for the negative impacts on the environment is not sustainable in the long term (Lin et al., 2014).

During the last few years, many logistics and operations research problems have been extended to include environmental issues and costs related to the environmental impacts of industrial and transportation activities (Bektas &d Laporte, 2011; Erdogan & Miller-Hooks, 2012; Demir et al., 2014; Lin et al., 2014). In this paper, we present a bi-objective mathematical model that integrates Open Location Routing Problem (OLRP) and the minimization of fuel consumption, named Green OLRP (G-OLRP). Considering that the vehicles do not return to the depots, we assert that the solution to the OLRP should be formed by radial paths, which allows us to propose a set of new constraints focused on maintaining the radiality condition of the paths, hence simplifying the set of constraints associated with the connectivity and capacity requirements.

Briefly, the OLRP can be stated as the following graph theoretic problem. Let $G = (V, A)$ be a complete and directed graph, where $V = I \cup J$ is the vertex set and A is the arc set. Vertex set $I = \{1, 2, ..., m\}$ represents the set of candidate capacitated depots to be installed. Vertex set $J = \{m + 1, m + 2, ..., m + n\}$ represents the customers to be served. Each customer $j \in J$ is associated with a known non-negative demand of goods D_j to be delivered. Each candidate depot $i \in I$ has a fictitious demand $D_i = 0$ and has an unlimited fleet of identical vehicles with the same positive capacity, denoted as Q. Note that for feasibility, the vehicle must meet the criterion that $0 < D_j \leq Q$ for all customers j. In addition, each candidate depot i has limited storage capacity of goods and a set-up cost, respectively denoted as W_i and O_i. A non-negative traveling cost C_{ij} and non-negative traveling distance T_{ij} are associated with each arc $(i,j) \in A$, and a non-negative fixed cost F is associated with the use of each vehicle. Thus, the OLRP consists of finding a collection of routes of minimum travelling cost connected to a set of depots with minimum setup cost to satisfy the following:

- Each customer vertex is visited by exactly one route.
- Each route begins at a depot and ends at a client, thus forming radial paths only.
- Each customer has to be fully served when visited (no fractional deliveries).
- The sum of the demands of the vertices visited by a route cannot exceed the vehicle capacity.

In the G-OLRP, in addition to the operational costs, a second objective function is included, which considers emissions generated due to fuel consumption in the routes performed. The mathematical model for the computation of fuel consumption and total emissions is based on the forces acting on each vehicle during its operation. Given the inherent difficulty of aggregating these objectives into a global criterion, we formulate the problem as a bi-objective one. The Pareto optimal solutions for the problem can be found and then analyzed a posteriori by decision-makers. This model corresponds to a mixed integer linear formulation and was implemented in AMPL (Fourer et al., 2002) and solved with CPLEX (called with the optimality gap option equal to 0%), using a work station with an Intel i7 870 processor.

The main contributions of this paper are:

- The OLRP is extended in the proposed G-OLRP, considering the environmental impact in terms of fuel consumption minimization.
- Because it is an approach that focuses on radiality of the paths, the proposed model is well suited to ORP, and the results are of high quality and performance.
- This paper presents a contribution to the discussion of OLRP, which is a new problem insufficiently addressed in the literature.
- This paper presents a contribution to the discussion of green VRP, by considering the integrated location of multiple deposits and open routing of multiple vehicles.
- The radial characteristic of the solution-paths may be extended to other types of problems, such as the VRP with backhauls (VRPB) and the electric vehicle routing problem (EVRP), which are all in the context of Green Transportation.

2. Literature Review

In practice, the OVRP formulation represents situations, such as: home delivery of packages and newspapers, school bus routing, routing of coal mines material or shipment of hazardous materials (Braekers et al., 2015). When considering that the vehicles depart from a set of available depots and finish their deliveries at their last customers, the OVRP is generalized to the case known as Multi-Depot OVRP (MDOVRP). Tarantilis and Kiranoudis (2002) proposed the MDOVRP for the first time within the context of food distribution. Since this problem is a generalization of the OVRP, the MDOVRP belongs to the NP-hard class; therefore, heuristic methods are commonly presented in the literature to find good quality solutions quickly (Dueck & Scheuer, 1990; Ho et al., 2008; Mirabi et al., 2010; Liu & Jiang, 2012; Yao et al., 2014).

Liu and Jiang (2012) proposed a hybrid genetic algorithm for the MDOVRP; however, the authors additionally proposed a mixed integer programming (MIP) formulation, which is the first formal mathematical definition of the problem.The literature on exact approaches for the MDOVRP is sparse and has received very little attention from researchers. The most recent MIP formulation was reported by Lalla-Ruiz et al. (2016), in which a new mixed integer programming formulation was presented and compared with the unique MIP formulation given in the literature at that time (Liu & Jiang, 2012). The authors focused on the improvements and extensions of the Miller-Tucker-Zemlin subtour elimination constraints, always considering that each vehicle departs from one of the available depots and finishes at the last customer it serves. As already mentioned above, when in the OVRP a set of candidate depots to be installed is considered, the resulting problem is known as the OLRP. This problem was recently introduced by Yu and Lin (2015), who also proposed an exact mathematical model to solve scenarios up to 50 clients and five depots with gaps close to zero. Additionally, they proposed a heuristic algorithm based on simulated annealing, which consists of two stages. First, an initial solution was constructed with a greedy approach, and second the initial solution was improved with a simulated annealing procedure with three search mechanisms: exchange, insertion and 2-opt movements. The proposal was also validated with adapted cases from Barreto et al. (2007), Prins et al. (2006) and Tuzun and Burke (1999).

About green issues in vehicle routing, the reduction of indirect green-house gases emissions, addressed in the vehicle routing problem, represents one of the most common objectives to be optimized. The cost of a route depends on several factors that can be divided into two sets. In the first set are included: distance, weight, speed, path conditions, a percentage of fuel that is generally associated to the unit of distance, and fuel costs. The second set of factors does not have direct relationship on the travel programming and includes tire and vehicle depreciation, maintenance, driver wages, taxes, among others (Palmer, 2007; Boriboonsomsin et al., 2010). Comparing the two sets, the first set of factors are directly related to fuel consumption and therefore can be considered as a variable cost or cost of fuel. Also, if other factors remain constant, fuel consumption depends mainly on the distance and load. Entities for environmental impact analysis in the transport sector indicate that there is a strong correlation between the gross vehicle weight and distance traveled using given amount of fuel, see for instance Xiao et al.

(2012). Kara et al. (2007) is one of the first publications that considers minimizing the fuel consumption where they define the Energy Minimizing Vehicle Routing Problem (EMVRP) as the CVRP in which the objective function is a product of the total load (including the weight of the empty vehicle) and the length of the arc. Figliozzi (2010) compares different levels of traffic congestion and vehicle speeds, to formulate and solve the problem called Emissions Vehicle Routing Problem (EVRP). The problem is an extension of the VRPTW. Greedy heuristic is the solution technique used by the author.

In Bektas and Laporte (2011), the authors consider factors such as speed, vehicle load and travel costs. The load and travel speed are factors that can be controlled. They have submitted four mathematical formulations for the Pollution Routing Problem (PRP) by considering time windows, speed, load, velocity. Branch-and-Cut is the solution technique chosen, using CPLEX 12.1. Suzuki (2011) develops an approach to the time-constrained, multiple stop, truck routing problem that minimizes the fuel consumption and pollutants emission to solve the traveling salesman problem with time windows (TSPTW). Their results suggest the approach may produce up to 6.9% in fuel savings over existing methods. The solution technique used was the compressed annealing. Xiao et al. (2012) define the Fuel Consumption Vehicle Routing Problem (FCVRP) and propose Fuel Consumption Rate (FCR) as a load dependent function, adding it to the classical CVRP to extend traditional studies on CVRP with the objective of minimizing fuel consumption. The methodology for solving the problem was based on the simulated annealing algorithm with a hybrid exchange rule to solve it. Their results show that the FCVRP model can reduce fuel consumption by 5% on average compared to the CVRP model.

Erdogan and Miller-Hooks (2012) introduce the Green Vehicle Routing Problem (G-VRP). The G-VRP is formulated as a mixed integer linear program. The solution method is based on two construction heuristics and the Modified Clarke and Wright Savings formulation of Bektas and Laporte (2011). Two construction heuristics, the Modified Clarke and Wright Savings heuristic and the Density-Based Clustering Algorithm, and a customized improvement technique, are developed. Results of numerical experiments show good performance of the heuristics. Moreover, problem feasibility depends on customer and station location configurations. Pradenas et al. (2013) formulate a model with emissions of greenhouse gases for the VRPB problem (VRP with Backhauls). Ubeda et al. (2011) present a case study considering environmental criteria based on real estimations. Other approximations that use metaheuristics can be found in Demir et al. (2012) and Jemai et al. (2012). Demir et al. (2014) propose the bi-objective Pollution Routing Problem (PRP), as an extension of the PRP, which consists of routing a number of vehicles to serve a set of customers, and determining their speed on each route segment. Two objective functions related to minimization of fuel consumption and driving time are proposed. Several multi-objective optimization techniques are developed and tested for the problem, finding trade-offs between fuel consumption and driver times.

Kücükoglu et al. (2013) present the G-CVRP optimization model, in which fuel consumption is computed considering the vehicle technical specifications, vehicle load and the distance. Fuel consumption equation is integrated to the model through a regression equation proportional to the distance and vehicle load. The GCVRP optimization model is validated by various instances with different number of customers. The authors present a mixed integer programming model, solving it with Gurobi 5.10. Recently, Lin et al. (2014) and Toro et al. (2016) presented an extensive literature review on Green Vehicle Routing Problems.

3. Proposed Model for the Green OLRP

In this section we introduce the proposed bi-objective formulation for the G-OLRP. In the first subsection, the mathematical model used to compute the fuel consumption of a vehicle between two nodes is presented. The model is developed based on the forces acting on the vehicle and a detailed extension of this model can be found in Golden et al. (2008). In the second subsection, the model for the G-OLRP is presented.

3.1 Computation of fuel consumption and total emission

In Toth and Vigo (2014), the analytical models for fuel consumption are classified into three classes, namely (i) emission factor models, (ii) average speed models, and (iii) modal models. In this paper an emission factor model is used which is expressed per unit of distance assuming constant speed. Eq. (1) represents one of the objectives of G-OLRP, consisting in the minimization of the fuel consumption and the total emission associated with this fuel consumption.

$$\Psi_2 = \alpha E \sum_{i,j \in V} T_{ij} s_{ij} + \gamma E \sum_{i,j \in V} T_{ij} l_{ij}. \tag{1}$$

The expression for the fuel consumption is formed by two parts. The first part, multiplied by α, corresponds to the amount of energy required considering the unloaded vehicle, where s_{ij} are the active arcs. The second term, multiplied by γ, corresponds to the amount of extra energy required considering the load carried on that arc, where l_{ij} represents the flow of goods transported by a vehicle through the arc $(i,j) \in A$. α is a parameter representing how much energy an unloaded vehicle spends in crossing the arc and is given in J/km. γ is a parameter representing the additional energy (per unit of load) that a loaded vehicle spends in that arc and is given in J/km-ton. Finally, both terms are multiplied by E, which is the total emission per unit of energy (kg of CO2/J), giving the total emission associated with the solution, which is calculated as:

$$E = E1 \times E2, \tag{2}$$

where E_1 is a conversion factor representing the amount of fuel required (gallons/J) and E_2 is another conversion factor representing the amount of emission per unit of fuel (kg of CO2/gallons). The developed model for computing emission is linear and defined as an objective in our formulation.

3.2 Problem formulation

The model is a two-index vehicle flow formulation that uses two binary decision variables:

- $s_{ij} = 1$ if arc $(i,j) \in A$ is used and takes value 0 otherwise
- $y_i = 1$ if a depot is set-up at vertex $i \in I$

Additionally, a real variable l_{ij} that represents the flow of goods transported by a vehicle through the arc $(i,j) \in A$ is used. The standard CLRP is often criticized because its objective function combines facility locations determined at a strategic level, while vehicle routes are optimized at the operational level. For this reason, we propose to put all the values in the same time horizon according to that opening cost and route cost by considering the Net Present Value (NPV) to the lessee, as expressed in (3), see (Trigeorgis, 1996):

$$NPV = V_0 - \sum_{t=0}^{N} \frac{I_t}{(1+r)^t} \equiv V_0 - I \tag{3}$$

with

$$I_t = L_t(1 - T) + D_t T \tag{4}$$
$$r = r_B(1 - T) \tag{5}$$

In these equations, V_0 is the current value (cost) of the leased asset, L_t is the lease rental payment at time t, D_t is the depreciation expense at time t, T is the lessee's effective corporate tax rate, r_B is the before tax cost of borrowing, and N is the life, i.e. maturity of the lease. In this way, O_i^{NPV} represents the NPV of the leasing cost associated to the use of a facility $i \in I$ and F^{NPV} is the NPV of the leasing cost associated to the use of a vehicle, assuming $F_k = F$, $\forall k \in K$. The two-index vehicle flow formulation for the bi-objective OLRP is defined as follows:

$$\min \Psi_1 = \sum_{\substack{i \in V \\ j \in V}} C_{ij} \cdot s_{ij} + \sum_{i \in I} O_i^{NPV} \cdot y_i + \sum_{\substack{i \in I \\ j \in V}} F^{NPV} \cdot s_{ij} \qquad (6)$$

$$\min \Psi_2 = \alpha E \sum_{i,j \in V} T_{ij} s_{ij} + \gamma E \sum_{i,j \in V} T_{ij} l_{ij} \qquad (7)$$

subject to :

$$\sum_{\substack{i \in V \\ j \in V}} s_{ij} = |J| \qquad (8)$$

$$\sum_{\substack{i \in V \\ i \neq j}} l_{ij} = \sum_{\substack{k \in V \\ k \neq j}} l_{jk} + D_j \qquad \forall j \in J \qquad (9)$$

$$\sum_{i \in V} s_{ij} = 1 \qquad \forall j \in J \qquad (10)$$

$$\sum_{k \in J} s_{jk} \leq \sum_{i \in V} s_{ij} \qquad \forall j \in J \qquad (11)$$

$$s_{ij} + s_{ji} \leq 1 \qquad \forall i, j \in V \qquad (12)$$

$$\sum_{i \in I} y_i \geq \sum_{j \in J} D_j \Big/ \sum_{i \in I} W_i \qquad (13)$$

$$\sum_{\substack{i \in I \\ j \in J}} s_{ij} \geq \sum_{j \in J} D_j \Big/ Q \qquad (14)$$

$$l_{ij} \leq Q \cdot s_{ij} \qquad \forall i, j \in V \qquad (15)$$

$$\sum_{j \in J} l_{ij} \leq W_i \cdot y_i \qquad \forall i \in I \qquad (16)$$

$$y_i \in \{0,1\} \qquad \forall i \in I \qquad (17)$$

$$s_{ij} \in \{0,1\} \qquad \forall i, j \in V \qquad (18)$$

$$l_{ij} \in R \qquad \forall i, j \in V \qquad (19)$$

The objective function (6) minimizes operating costs, which correspond to the total travelling cost of the radial routes used to deliver the goods to the costumers, the sum of the setup cost of depots and the fixed cost associated with use of each vehicle. The second objective (7) models the fuel consumption and the total emission associated to this fuel consumption. Note that in the optimal solution the routes have a radial configuration in which they start at the depot and end up at a client. Thus, the set of constraints (8) and (9) impose the radial connectivity requirements of the OLRP as described below.

The topology of open routing problems (ORPs) can be considered a graph consisting of x arcs and y nodes ($y = |V|$). As it is shown in (Lavorato et al., 2012; Bazaraa et al., 2011), in case there is only one depot, it is possible to compare the radial topology of an open problem with a tree, which is a subgraph connected with $x-1$ arcs. For m depots, the radial characteristic requires a subgraph connected with $x - m = |J|$ arcs. Therefore, it can be seen that the radial connectivity requirements impose that the cardinality of J (number of customers) must be equal the number of arcs used in the optimal solution, as is guaranteed by the constraint (8). However, this single condition is not sufficient for radial solutions, since it must ensure proper system connectivity through the set of constraints (9), which guarantees the balance of demand flow in each customer so that it is fully served when visited.

In general, one can state that the topology of an ORP with y nodes is radial if it satisfies the two following conditions:

– Condition 1: the solution must have $x - m$ arcs;
– Condition 2: the solution must be connected.

Constraint (10) imposes that exactly one arc enters to each vertex associated with a customer. Consequently, the outdegree constraint (11) imposes that exactly one arc leaves each vertex associated with a customer, except for those customers who are at the end of the route. The uni-directional constraint (12) ensures that only one of the two variables s_{ij} or s_{ji} must be used.

Constraints (13) and (14) impose both the depot and vehicle capacity requirements, respectively. Constraints (15) and (16) impose capacity requirements of flow of goods transported. Finally, constraints (17) and (18) define all binary decision variables, and constraint (19) defines the real variable.

3.3 Multi-objective optimization

Multi-Objective Optimization (MOO) is the computational process of simultaneously optimizing two or more conflicting objectives subject to a set of constraint functions. In MOO, the main focus is on to produce trade-off solutions representing the best possible compromises among different (possibly conflicting) objectives.

In this paper, we approach the proposed multi-objective problem using an *a posteriori* methodology, in which the optimization returns some Pareto-optimal solutions, leaving the decision-making process to a post-optimization stage (Marler and Arora, 2009).

In order to solve the multi-objective problem, one can adopt parameterized scalar problems, whose solution leads to a Pareto-optimal solution. For further discussion on multi-objective methods, we refer to (Miettinen, 1999; Ehrgott and Gandibleux, 2002; Marler and Arora, 2009). In the ò-constraint method, one objective is selected to be optimized, while the others are converted into inequality constraints by imposing upper bounds ò.

$$\min_{x} \Psi_1(x) \tag{20}$$

$$\Psi_k(x) \leq \varepsilon_k \qquad k = 2,...,m \tag{21}$$
$$x \in \Omega$$

Cohon and Marks (1975) show that the ò-constraint method can be derived from the Kuhn-Tucker conditions for optimality for a MOO problem. A systematic variation of the parameters ∂_k can yield Pareto optimal solutions (Marler and Arora, 2009). If it exists, a solution to the ò-constraint formulation is weakly Pareto optimal, as shown in (Miettinen, 1999). Moreover, if the solution is unique, then it is Pareto-optimal.

For more than two objectives, the ò-constraint formulation can lead to infeasibility problems, for some combinations of values of ∂_k. Nevertheless, for two objectives, as is the case in our formulation, the method can yield Pareto optimal solutions with a systematic variation of ċ. In order to generate points on the Pareto front, first we optimize each objective individually with the original constraints of the model and neglecting the other objective. This yields the minimum and maximum values of each objective that contain the Pareto front. Intermediate points on the front are obtained with discrete steps, varying ċ within the minimum and maximum range. The generation of the Pareto front is independent of which objective is chosen to be minimized. If we select Ψ_1 as objective and convert Ψ_2 into a constraint or vice-versa, the result is practically the same, the Pareto front is obtained regardless of the selected objective. Nonetheless, we have observed in our experiments that when the emission (Ψ_2) is used as objective and the operational costs (Ψ_1) are posed as a constraint, the model is solved much faster than when the converse is done, allowing the solution of larger instances.

4. Computational results

In this study the test scenarios correspond to instances of the literature to the capacitated location-routing problem presented in (Prins et al., 2007), which include 20, 50, 100 and 200 customers and from 5 to 10 deposits. The calculation of consumption of the vehicle was taken from the report of University of Michigan: Transportation Research Institute (2014), in which it is established that the average fuel consumption of a vehicle with these characteristics is 1 gallon per 15.81 km traveled. This value was used as a reference for calculating vehicle consumption at full load, which was estimated at 12 km per gallon. As for the amount of emissions per gallon of gasoline, we consider $8.70645 kg$ of CO_2 per gallon, this information is related to the fuel consumption guide (Gouvermment of Canada, 2015). In this report CO_2 emissions vary according to the type of fuel used and engine characteristics such as size, type, vehicle brand and optimum cruising speed. The cost of emissions is calculated based on quoted prices presented in (SENDECO2, 2014). The value of 0.009 USD per kg of CO_2. To quantify the price of a gallon of gasoline was consulted online[1]. This information is set according to the territory where the case study is contemplated. In our case, this parameter was 3.92 USD per gallon. The mathematical model was solved under CPLEX 12.5 (ILOG, 2008) on a computer Intel Core i7-4770 3.4 GHz, 16 GB of RAM and written in AMPL: A Modeling Language for Mathematical Programming (Fourer et al., 2002). The method used to generate the Pareto front is the ò-constraint method, using Ψ_2 as objective and Ψ_1 as a constraint. Sixteen instances used by Prins et al. (2007) are used in order to verify the efficacy of the proposed model and to confirm the observation that by increasing the number of vehicles the fuel consumption and hence total emission can be reduced, in the context of OLRP considering fuel consumption minimization.

In all the instances analyzed, we have obtained a Pareto front representing the trade-off between green-house gases emissions and operational costs. In order to better analyze the characteristics of the solutions on the Pareto front and to get better insights about the nature of this trade-off, we compare in Fig. 1 three representative points from the Pareto front obtained for the instance 100 10 _2b. Point A corresponds to the minimization of Ψ_1 (operational costs), point C corresponds to the minimization of Ψ_2 (environmental impact) and point B corresponds to the solution selected by the min-max criterion (or minimization of maximum regret), which usually corresponds to a middle point on the Pareto front, a point that minimizes the maximum deviation to the minimum of each objective. Interesting characteristics can be observed from the analysis of Fig. 1. Point A is a solution characterized by minimum number of depots and vehicles, leading to longer routes connecting the clients to the depots. This solution involves employing less vehicles, operating fully loaded under longer routes, which in turn generates greater fuel consumption and consequently greater emission of green-house gases. With less vehicles, the amount of cargo per vehicle increases, which affects directly the fuel consumption, leading to a high value in the objective function Ψ_2.

Point C, on the other extreme of the Pareto front, is related to the minimization of objective function Ψ_2 (environmental impact). This solution presents a greater number of depots and routes than the solution of point A. In this way, the routes that connect the customers to the depots are shorter on average, implying shorter paths and fast deliveries of the load. As a consequence, the vehicles presents less fuel consumption overall and less emission of green-house gases. It is intuitive that more fuel will be consumed if more vehicles are used. However, in the context of CLRP considering fuel consumption minimization, generally using more vehicles does not imply lengthening the travel distance. It is interesting to notice that by increasing the number of vehicles the fuel consumption and hence total emission can be reduced. Using few vehicles at full capacity does not necessarily imply in fuel economy, whereas using more vehicles, not fully loaded, can translate into reduced fuel cost and less environmental impact.

In the long term, by having more depots and more routes, the fuel economy can balance out the initial investment costs. Another aspect influencing the minimization of emission is that the vehicles tend to be dispatched to attend those clients with higher demand. Heavier loads are delivered earlier in the route, decreasing the weight of the vehicle and contributing to the reduction of emissions.

---

[1] http://es.globalpetrolprices.com/gasoline_prices/

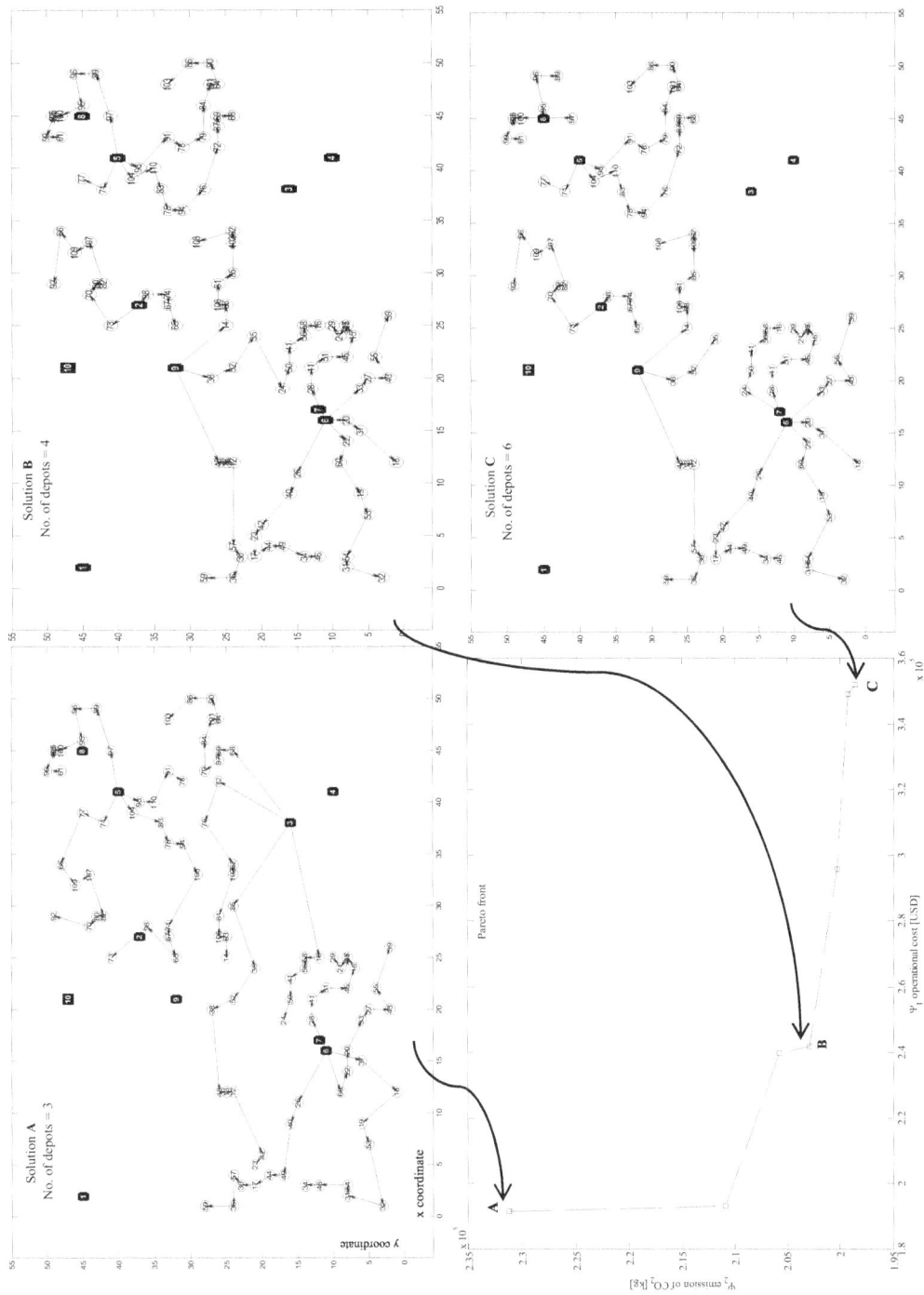

Fig. 1. Pareto front and three representative optimal solutions for the instance 100_10_2b used by Prins et al. (2007)

Point B was selected according to the min-max criterion. This solution is characterized by an medium number of depots and vehicles with medium length paths. Similarly to the previous cases, it is possible to note how the number of depots affects the length of the paths connecting customers and depots. This solution corresponds to a balance between operational costs and total emissions. The reduction of operational costs favors solutions with longer routes, few vehicles and depots. From the perspective of environmental impact, the reduction of emissions leads to solutions with more vehicles and shorter routes, usually attending clients with higher demand first. Other solutions along the Pareto front represent different trade-off configurations between these two objectives.

Table 1

Computational results for instances used by Prins et al. (2007)

Instance	No. of routes	No. of depots	Ψ1 (USD)	Ψ2 (kg CO2)	Cost of routes (USD)	Cost of depots (USD)	Cost of vehicles (USD)	Fuel cost (USD)	Fuell (Gallon)	Run time (seconds)	Optimal Pareto front
20_5_1a	5	3	43,849	121,683.43	17,691	21,158	5,000	96,787.28	13,986.60	9	
	6	3	44,608	120,023.93	17,450	21,158	6,000	95,467.31	13,795.85	30	
	6	3	45,351	94,930.16	13,802	25,549	6,000	75,507.67	10,911.51	1	
	6	3	46,455	92,351.68	13,427	27,028	6,000	73,456.74	10,615.14	2	
	7	3	47,292	91,230.23	13,264	27,028	7,000	72,564.74	10,486.23	12	
	7	4	52,317	83,897.99	12,198	33,119	7,000	66,732.66	9,643.45	3	
	7	5	63,110	83,567.84	12,150	43,960	7,000	66,470.05	9,605.50	2	
20_5_1b	3	2	33,564	103,630.63	15,067	15,497	3,000	82,428.04	11,911.57	4	
	4	2	33,729	97,886.13	14,232	15,497	4,000	77,858.86	11,251.28	2	
	4	2	37,980	97,633.45	14,195	19,785	4,000	77,657.87	11,222.24	2	
	4	3	45,909	97,157.09	14,126	27,783	4,000	77,278.97	11,167.48	1	
	4	3	46,267	96,152.93	13,980	28,287	4,000	76,480.26	11,052.06	1	
	5	3	47,018	94,440.23	13,731	28,287	5,000	75,117.98	10,855.21	4	
	5	4	59,231	93,938.15	13,658	40,573	5,000	74,718.62	10,797.49	4	
	5	3	41,125	91,864.45	13,356	22,769	5,000	73,069.20	10,559.13	2	
20_5_2a	6	3	41,798	89,615.08	13,029	22,769	6,000	71,280.04	10,300.58	10	
	6	3	42,528	84,820.59	12,332	24,196	6,000	67,466.49	9,749.49	7	
	8	3	44,198	82,549.56	12,002	24,196	8,000	65,660.11	9,488.45	8	
	10	4	53,727	78,284.89	11,382	32,345	10,000	62,267.98	8,998.26	2	
	9	5	66,594	77,026.29	11,199	46,395	9,000	61,266.89	8,853.60	1	
	10	5	67,527	76,565.04	11,132	46,395	10,000	60,900.01	8,800.58	2	
20_5_2b	3	2	32,520	107,359.49	15,609	13,911	3,000	85,393.99	12,340.17	0	
	4	2	33,041	104,061.39	15,130	13,911	4,000	82,770.67	11,961.08	1	
	4	2	39,361	93,689.21	13,622	21,739	4,000	74,520.62	10,768.88	1	
	5	2	40,228	92,774.04	13,489	21,739	5,000	73,792.69	10,663.68	0	
	6	3	53,439	83,557.15	12,149	35,290	6,000	66,461.55	9,604.27	0	
	7	3	54,330	82,807.37	12,040	35,290	7,000	65,865.17	9,518.09	0	
50_5_1a	12	2	64,217	253,340.32	36,832	15,385	12,000	201,507.48	29,119.58	30,226	
	14	3	69,763	208,553.33	30,321	25,442	14,000	165,883.80	23,971.65	2,616	
	13	4	78,100	205,754.10	29,914	35,186	13,000	163,657.29	23,649.90	703	
	14	4	79,061	205,485.40	29,875	35,186	14,000	163,443.56	23,619.01	491	
	14	5	92,652	205,100.22	29,819	48,833	14,000	163,137.19	23,574.74	538	
50_5_1b	6	2	49,114	190,721.62	27,729	15,385	6,000	151,700.42	21,922.03	138	
	8	2	51,074	190,446.22	27,689	15,385	8,000	151,481.36	21,890.37	305	
	9	3	59,418	171,782.80	24,976	25,442	9,000	136,636.44	19,745.15	11	
	9	4	73,009	171,397.62	24,920	39,089	9,000	136,330.06	19,700.88	9	
50_5_2a	12	3	68,121	184,350.98	26,802	29,319	12,000	146,633.19	21,189.77	20276	
	13	3	69,053	183,882.88	26,734	29,319	13,000	146,260.87	21,135.96	11,782	
	14	4	81,528	161,677.98	23,506	44,022	14,000	128,599.04	18,583.68	90	
	15	4	82,491	161,423.56	23,469	44,022	15,000	128,396.67	18,554.43	63	
	16	4	83,480	161,347.34	23,458	44,022	16,000	128,336.05	18,545.67	66	
50_5_2b	6	3	57,355	151,564.56	22,036	29,319	6,000	120,554.80	17,421.21	152	
	6	3	59,242	141,192.58	20,528	32,714	6,000	112,304.90	16,229.03		
	7	3	59,902	138,853.67	20,188	32,714	7,000	110,444.53	15,960.19	19	
	7	3	62,208	138,413.21	20,124	35,084	7,000	110,094.18	15,909.56	34	
	7	4	70,789	135,956.99	19,767	44,022	7,000	108,140.50	15,627.24		
	8	4	71,640	134,931.79	19,618	44,022	8,000	107,325.06	15,509.40	5	
	9	4	72,611	134,731.32	19,589	44,022	9,000	107,165.60	15,486.36		

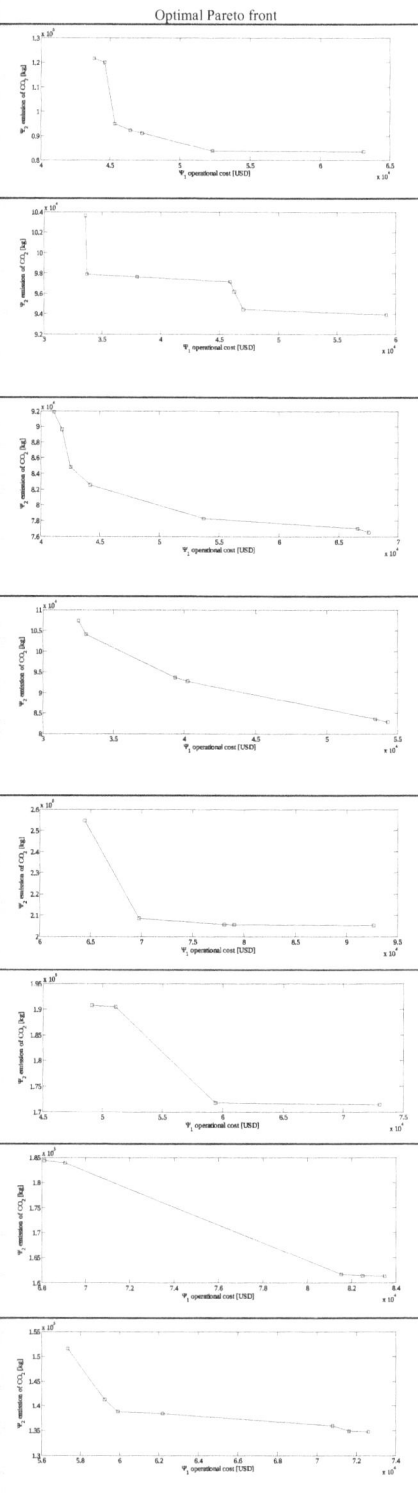

Table 1

Computational results for instances used by Prins et al. (2007) (Continued)

Instance	No. of routes	No. of depots	Ψ1 (USD)	Ψ2 (kg CO2)	Cost of routes (USD)	Cost of depots (USD)	Cost of vehicles (USD)	Fuel cost (USD)	Fuell (Gallon)	Run time (seconds)	Optimal Pareto front
50_5_2bis	12	3	60,052	194,432.67	28,267	19,785	12,000	154,652.19	22,348.58	1500	
	13	3	60,971	193,875.28	28,186	19,785	13,000	154,208.84	22,284.51	487	
50_5_2b bis	6	3	41,193	113,009.46	16,430	18,763	6,000	89,887.98	12,989.59	12	
	6	3	41,636	112,762.00	16,394	19,242	6,000	89,691.15	12,961.15	14	
	7	3	42,612	112,596.80	16,370	19,242	7,000	89,559.76	12,942.16	61	
	6	4	48,658	112,486.67	16,354	26,304	6,000	89,472.16	12,929.50	14	
50_5_3	12	2	62,581	217,489.78	31,620	18,961	12,000	172,991.87	24,998.83	4984	
	13	3	73,243	191,350.67	27,820	32,423	13,000	152,200.76	21,994.33	412	
	13	3	77,416	182,009.65	26,462	37,954	13,000	144,770.89	20,920.65	77	
	13	4	82,120	179,093.21	26,038	43,082	13,000	142,451.15	20,585.43	54	
	14	4	82,979	178,122.42	25,897	43,082	14,000	141,678.99	20,473.84	27	
50_5_3b	6	2	46,584	205,469.41	29,873	10,711	6,000	163,430.84	23,617.17	525	
	7	2	51,354	174,651.96	25,393	18,961	7,000	138,918.57	20,074.94	247	
	7	3	56,746	173,338.34	25,202	24,544	7,000	137,873.71	19,923.95	323	
	8	3	61,726	168,349.99	24,477	29,249	8,000	133,905.97	19,350.57	96	
	8	3	64,316	164,334.19	23,893	32,423	8,000	130,711.79	18,888.99	25	
	8	3	69,639	162,903.87	23,685	37,954	8,000	129,574.11	18,724.58	22	
	8	4	74,515	161,169.13	23,433	43,082	8,000	128,194.30	18,525.19	6	
100_5_1b	12	3	189,438	306,409.19	44,548	132,890	12,000	243,718.58	35,219.45	33,631	
	15	3	201,401	291,558.63	42,389	144,012	15,000	231,906.41	33,512.49	3,694	
	17	4	248,818	287,705.47	41,829	189,989	17,000	228,841.60	33,069.59	32,308	
	18	5	291,210	285,669.29	41,533	231,677	18,000	227,222.01	32,835.55	76,883	
100_10_2b	11	3	190,280	229,207.25	33,324	145,956	11,000	182,311.98	26,345.66	66,497	
	13	3	193,245	210,875.65	30,659	149,586	13,000	167,730.97	24,238.58	72,779	
	14	4	239,820	205,749.88	29,914	195,906	14,000	163,653.93	23,649.41	44,238	
	14	4	241,796	202,826.54	29,489	198,307	14,000	161,328.70	23,313.40	1,965	
	16	5	295,934	200,225.53	29,111	250,823	16,000	159,259.85	23,014.43	1,378	
	14	6	349,058	199,215.84	28,964	306,094	14,000	158,456.74	22,898.37	4,528	
	16	6	350,987	198,726.46	28,893	306,094	16,000	158,067.48	22,842.12	1,510	
	17	6	351,966	198,580.45	28,872	306,094	17,000	157,951.35	22,825.34	1,279	
200_10_1a	48	3	400,053	675,545.54	98,213	253,840	48,000	537,330.47	77,648.91	65,566	
	53	7	787,817	454,157.03	66,028	668,789	53,000	361,237.55	52,201.96	28,368	
	54	8	888,669	436,995.36	63,533	771,136	54,000	347,587.12	50,229.35	12,504	
	57	8	939,656	435,213.64	63,274	819,382	57,000	346,169.93	50,024.56	77,822	
	56	9	1,014,08	429,937.58	62,507	895,579	56,000	341,973.34	49,418.11	76,870	
	58	10	1,103,79	424,435.12	61,707	984,087	58,000	337,596.67	48,785.65	26,614	
200_10_2b	23	3	350,257	322,497.10	46,887	280,370	23,000	256,514.93	37,068.63	71,118	
	27	6	692,466	280,542.94	40,788	624,678	27,000	223,144.50	32,246.31	53,679	
	27	7	816,443	277,199.73	40,302	749,141	27,000	220,485.30	31,862.04	30,562	
	27	7	816,389	276,828.46	40,248	749,141	27,000	220,189.99	31,819.36	33,143	

Table 1 presents the results obtained with the proposed model in 16 instances. Columns 2-5 show information about the optimal solution in three different points on the Pareto front: number of routes, number of depots, value of objective function Ψ_1, value of objective function Ψ_2. Columns 6-8 present the discrimination of the three terms in the first objective of operational costs, which correspond to the total travelling cost of the radial routes used to deliver the goods to the costumers, the sum of the setup cost of depots and the fixed cost associated with the use of each vehicle, respectively. Columns 9-10 present the discrimination of the terms in the second objective of emission Ψ_2, which correspond to the fuel consumption and the total emission associated to this fuel consumption. Column 11 reports the required computational time and column 12 shows the points on the Pareto front obtained with the ϱ-constraint method. This last column also allows us to conclude that the proposed model is able to generate a set of trade-off solutions for the problem. In the results, we were able to obtain solutions with GAP equal to zero for instances up to 20 customers. In instances with 50 customers, most solutions obtained have GAP equal to zero and few solutions were obtained with GAP less than 3%. In instances with 100 customers, solutions were obtained with GAP inferior to 2% and in instances with 200 customers, with GAP less than 3%. We have observed in our experiments that when the emission (Ψ_2) is used as objective and the operational costs (Ψ_1) are posed as a constraint, the model is solved much faster than when the converse is done, allowing the solution of larger instances. Usually, when minimizing emissions, the number of selected depots is greater and therefore the complexity of the problem is lower, what reflects on the computation time required for the solution. Finally, it is interesting to highlight that the solutions obtained with the relaxation of (Ψ_1) correspond to the Pollution-Routing Problem (PRP). Therefore, the results reported here can be considered as a reference for comparison with the single objective PRP.

5. Conclusions and future work

This paper proposed a multi-objective model for the G-OLRP considering fuel consumption. The model proposed is a bi-objective problem, considering the minimization of operational costs and the minimization of environmental effects. The proposed model for the calculation of particle emission was developed under assumptions close to the real scenario, assuming constant speed. The model can be adjusted for more complicated scenarios, for instance, considering an average inclination between any pair of nodes, varying speed, conditions of the road, heterogeneous fleet, among others. These variations will affect the values of the parameters α and γ, possibly with different values for each arc.

The use of radiality constraints in the solution of open VRP has been very effective, which was motivated by the radial topology of open routes. The proposed mathematical model can serve as a reference to the solution of larger instances, in which other strategies can be employed, such as hybrid methods, heuristics, meta-heuristics, math-heuristics, set partitioning. The multi-objective approach to this problem is entirely appropriate, since the trade-off between economic aspects and environmental aspects can be selected from the Pareto front using a decision-making process. The fuel consumption is affected by the number of depots and the number of vehicles. The use of more vehicles implies a better distribution of the cargo per vehicle. However, when reducing emissions, one can find short routes, vehicles traveling with moderate load and delivering higher loads first. From the point of view of environmental impact, more vehicles performing shorter routes and serving as soon as possible those customers with higher demand seems to be the preferred strategy. In future work we should explore other methodologies for the calculation of fuel consumption and should consider aspects such as the slope of the road and vehicle speed in different edges.

Acknowledgements

 E.M. Toro, M. Granada and R.A. Gallego would like to thank the support given by the Technological University of Pereira, Colombia.

F. G. Guimarães would like to thank the support given by the National Council for Scientific and Technological Development (CNPq grant no. 312276/2013-3) and Fundação de Amparo à Pesquisa do Estado de Minas Gerais (FAPEMIG), Brazil.

References

Barreto, S., Ferreira, C., Paixao, J., & Santos, B. S. (2007). Using clustering analysis in a capacitated location-routing problem. *European Journal of Operational Research*, *179*(3), 968-977.

Bazaraa, M. S., Jarvis, J. J., & Sherali, H. D. (2011). *Linear programming and network flows*. John Wiley & Sons.

Bektaş, T., & Laporte, G. (2011). The pollution-routing problem. *Transportation Research Part B: Methodological*, *45*(8), 1232-1250.

Bodin, L., Golden, B., and Assad, A. (1983). Routing and scheduling of vehicles and crews: the state of the art. *Computers and Operations Research*, *10*(2), 63 – 211.

Boriboonsomsin, K., Vu, A., & Barth, M. (2010). Eco-driving: pilot evaluation of driving behavior changes among us drivers. *University of California Transportation Center.*

Braekers, K., Ramaekers, K., & Van Nieuwenhuyse, I. (2015). The vehicle routing problem: State of the art classification and review. *Computers & Industrial Engineering, 99*, 300-313.

Cohon, J. L., & Marks, D. H. (1975). A review and evaluation of multiobjective programing techniques. *Water Resources Research, 11*(2), 208-220.

Daniel, S. E., Diakoulaki, D. C., & Pappis, C. P. (1997). Operations research and environmental planning. *European Journal of Operational Research, 102*(2), 248-263.

Demir, E., Bektaş, T., & Laporte, G. (2012). An adaptive large neighborhood search heuristic for the pollution-routing problem. *European Journal of Operational Research, 223*(2), 346-359.

Demir, E., Bektaş, T., & Laporte, G. (2014). The bi-objective pollution-routing problem. *European Journal of Operational Research, 232*(3), 464-478.

Dueck, G., & Scheuer, T. (1990). Threshold accepting: a general purpose optimization algorithm appearing superior to simulated annealing. *Journal of Computational Physics, 90*(1), 161-175.

Ehrgott, M., & Gandibleux, X. (2003). Multiobjective combinatorial optimization—theory, methodology, and applications. In *Multiple criteria optimization: State of the art annotated bibliographic surveys* (pp. 369-444). Springer US.

Erdogan, S. and Miller-Hooks, E. (2012). A green vehicle routing problem. *Transportation Research Part E: Logistics and Transportation Review, 48*(1), 100–114.

Figliozzi, M. A. (2010). An iterative route construction and improvement algorithm for the vehicle routing problem with soft time windows. *Transportation Research Part C: Emerging Technologies, 18*(5), 668-679.

Fourer, R., Gay, D. M., & Kernighan, B. W. (2002). *AMPL: A Modeling Language for Mathematical Programming*. Brooks/Cole-Thomson, 2nd edition.

Neto, J. Q. F., Walther, G., Bloemhof, J., Van Nunen, J. A. E. E., & Spengler, T. (2009). A methodology for assessing eco-efficiency in logistics networks.*European Journal of Operational Research, 193*(3), 670-682.

Golden, B. L., Raghavan, S., & Wasil, E. A. (Eds.). (2008). *The vehicle routing problem: latest advances and new challenges* (Vol. 43). Springer Science & Business Media.

Gouvermment of Canada (2015). Fuel consumption guide.

Ho, W., Ho, G. T., Ji, P., & Lau, H. C. (2008). A hybrid genetic algorithm for the multi-depot vehicle routing problem. *Engineering Applications of Artificial Intelligence, 21*(4), 548-557.

ILOG, S. (2008). CPLEX optimization subroutine library guide and reference.*System v11. 0 User's Guide.*

Jemai, J., Zekri, M., & Mellouli, K. (2012, April). An NSGA-II algorithm for the green vehicle routing problem. In *European Conference on Evolutionary Computation in Combinatorial Optimization* (pp. 37-48). Springer Berlin Heidelberg.

Kara, I., Kara, B. Y., & Yetis, M. K. (2007, August). Energy minimizing vehicle routing problem. In *International Conference on Combinatorial Optimization and Applications* (pp. 62-71). Springer Berlin Heidelberg.

Kucukoglu, I., Ene, S., Aksoy, A., & Ozturk, N. (2013). Green capacitated vehicle routing problem fuel consumption optimization model. *Computational Engineering Research, 3*, 16-23.

Lalla-Ruiz, E., Expósito-Izquierdo, C., Taheripour, S., & Voß, S. (2016). An improved formulation for the multi-depot open vehicle routing problem. *OR Spectrum, 38*(1), 175-187.

Lavorato, M., Franco, J. F., Rider, M. J., & Romero, R. (2012). Imposing radiality constraints in distribution system optimization problems. *IEEE Transactions on Power Systems, 27*(1), 172-180.

Lin, C., Choy, K. L., Ho, G. T., Chung, S. H., & Lam, H. Y. (2014). Survey of green vehicle routing problem: past and future trends. *Expert Systems with Applications, 41*(4), 1118-1138.

Liu, R., & Jiang, Z. (2012). The close–open mixed vehicle routing problem.*European Journal of Operational Research, 220*(2), 349-360.

Marler, R. T., & Arora, J. S. (2009). *Multi-objective optimization: concepts and methods for engineering.* VDM Publishing.

Miettinen, K. (1999). Nonlinear Multiobjective Optimization, volume 12 of International Series in Operations Research and Management Science.

Mirabi, M., Ghomi, S. F., & Jolai, F. (2010). Efficient stochastic hybrid heuristics for the multi-depot vehicle routing problem. *Robotics and Computer-Integrated Manufacturing*, *26*(6), 564-569.

Palmer, A. (2007). *The Development of an Integrated Routing and Carbon Dioxide Emissions Model for Goods Vehicles*. PhD thesis, School of Management, Cranfield University.

Pradenas, L., Oportus, B., & Parada, V. (2013). Mitigation of greenhouse gas emissions in vehicle routing problems with backhauling. *Expert Systems with Applications*, *40*(8), 2985-2991.

Prins, C., Prodhon, C., & Calvo, R. W. (2006). Solving the capacitated location-routing problem by a GRASP complemented by a learning process and a path relinking. *4OR*, *4*(3), 221-238.

Prins, C., Prodhon, C., Ruiz, A., Soriano, P., & Wolfler Calvo, R. (2007). Solving the capacitated location-routing problem by a cooperative Lagrangean relaxation-granular tabu search heuristic. *Transportation Science*, *41*(4), 470-483.

Schrage, L. (1981). Formulation and structure of more complex/realistic routing and scheduling problems. *Networks*, *11*(2), 229-232.

SENDECO2 (2014). The European bourse for European Unit Allowances (EUA) and Carbon Credits (CER's).

Suzuki, Y. (2011). A new truck-routing approach for reducing fuel consumption and pollutants emission. *Transportation Research Part D: Transport and Environment*, 16(1), 73-77.

Tarantilis, C. D., & Kiranoudis, C. T. (2002). Distribution of fresh meat. *Journal of Food Engineering, 51*(1), 85-91.

Toro, E., Escobar, A., & Granada, M. (2016). Literature Review on the Vehicle Routing Problem in the Green Transportation Context. *Luna Azul*, 42(1), 362 – 387.

Toth, P., & Vigo, D. (Eds.). (2014). Vehicle routing: problems, methods, and applications (Vol. 18). Siam.

Transportation Research Institute (2014). Large drop in fuel economy in september.

Trigeorgis, L. (1996). *Real options: Managerial flexibility and strategy in resource allocation*. MIT press.

Tuzun, D., & Burke, L. I. (1999). A two-phase tabu search approach to the location routing problem. *European journal of operational research*, *116*(1), 87-99.

Ubeda, S., Arcelus, F. J., & Faulin, J. (2011). Green logistics at Eroski: A case study. *International Journal of Production Economics*, *131*(1), 44-51.

Xiao, Y., Zhao, Q., Kaku, I., & Xu, Y. (2012). Development of a fuel consumption optimization model for the capacitated vehicle routing problem.*Computers & Operations Research*, *39*(7), 1419-1431.

Yao, B., Hu, P., Zhang, M., & Tian, X. (2014). Improved ant colony optimization for seafood product delivery routing problem. *PROMET-Traffic&Transportation*,*26*(1), 1-10.

Vincent, F. Y., & Lin, S. Y. (2015). A simulated annealing heuristic for the open location-routing problem. *Computers & Operations Research*, *62*, 184-196.

3

A hybrid metaheuristic for the time-dependent vehicle routing problem with hard time windows

N. Rincon-Garcia[a,b*], B.J. Waterson[a] and T.J. Cherrett[a]

[a] *Transportation Research Group, University of Southampton, Southampton, UK*
[b]*Department of Industrial Engineering, School of Engineering , Pontificia Universidad Javeriana, Bogota, Colombia*

CHRONICLE	ABSTRACT
	This article paper presents a hybrid metaheuristic algorithm to solve the time-dependent vehicle routing problem with hard time windows. Time-dependent travel times are influenced by different congestion levels experienced throughout the day. Vehicle scheduling without consideration of congestion might lead to underestimation of travel times and consequently missed deliveries. The algorithm presented in this paper makes use of Large Neighbourhood Search approaches and Variable Neighbourhood Search techniques to guide the search. A first stage is specifically designed to reduce the number of vehicles required in a search space by the reduction of penalties generated by time-window violations with Large Neighbourhood Search procedures. A second stage minimises the travel distance and travel time in an 'always feasible' search space. Comparison of results with available test instances shows that the proposed algorithm is capable of obtaining a reduction in the number of vehicles (4.15%), travel distance (10.88%) and travel time (12.00%) compared to previous implementations in reasonable time.
Keywords: *Vehicle routing problem* *Time-dependent travel time* *Hybrid metaheuristic algorithm*	

1. Introduction

In logistics operations, the routing and scheduling of collections and deliveries is vital to control costs and meet customer service levels (Chopra & Meindl, 2007). Effective planning improves productivity (DFT, 2010) and an important component of commercial routing and scheduling software commonly used by the industry is the model that supports the decision making process (Drexl, 2012). The impact of congestion has increased over the last 30 years, with the 101 largest US cities reporting that travel delay had increased from 1.1 billion hours in 1982 to 4.8 billion hours in 2011 (Chang et al., 2015). Fig. 1 shows the variation in the level of congestion by time-of-day that affects the freight transport in the US. Rising levels of traffic congestion mean that logistics providers face the challenge of maintaining time critical service levels whilst at the same time minimising the extra costs that congestion and delays impose.

* Corresponding author.
E-mail: nicolas.rincon@javeriana.edu.co (N. Rincon-Garcia)

The key impact congestion has on vehicle planning is that travel times between locations vary as a function of the changing traffic patterns. Failure to account for this in routing decisions leads to drivers running out of hours, additional overtime payments and missed deliveries (Haghani & Jung, 2005; Kok et al., 2012) with underestimation of travel times being reported as the most common problem by transport managers (Eglese et al., 2006; Ehmke et al., 2012).

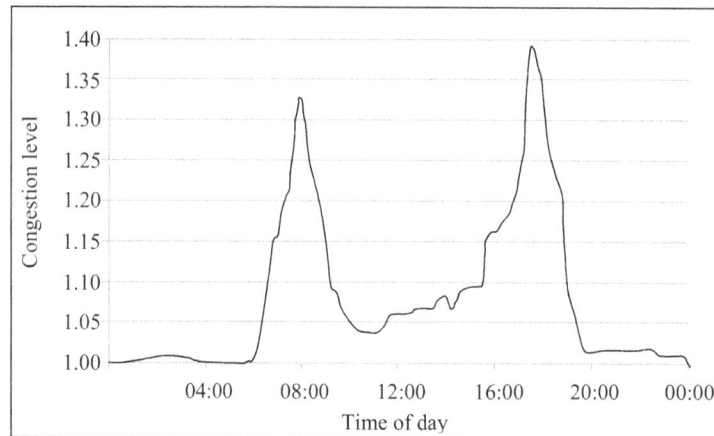

US-Department-Of-Transport (2003)
Fig. 1. Variation in Congestion by time-of-day

The time-dependent vehicle routing problem with hard time windows (TDVRPTW) is the variant where travel time between locations depends on the time of the day with a strict (non-negotiable) time window for the delivery being initially established by the customer (Figliozzi, 2012; Malandraki & Daskin, 1992). The primary objective is to reduce the number of vehicles required to complete the schedule whilst minimising travel distance and travel time (Figliozzi, 2012).

Variants of the vehicle routing problem (VRP) are NP-Hard and metaheuristic algorithms have been developed to solve the problem with trials suggesting significant improvement in performance over current schedules. Some implementations have achieved high accuracy (difference between best-known values and results of the particular algorithm for available test instances) with execution times that allow logistics planners to realistically use the approach as part of their everyday operations (Cordeau et al., 2002; Drexl, 2012).

In an industry where the profit margin is 3% (FTA, 2015), optimization procedures that account for congestion can mitigate its impact. Although managerial solutions are available to account for congestion such as planning vehicle schedules with average travel times, it might lead to poor solutions with missed deliveries and extra costs (e.g. more vehicles, duty time and distance) when compared to solutions based on the use of time-dependent VRP variants (Kok et al., 2012).

In a recent survey conducted with 19 companies in the UK that showed the identification of the most important improvements required by the freight industry for vehicle routing software, companies ranked at one of the two most important capabilities the optimisation of routes minimising the impact of congestion (Rincon-Garcia et al., 2015).

2. Literature review

The importance of providing effective and efficient solutions for city logistics is imperative bearing in mind the low profit margin in the sector (FTA, 2015) and the growing expectations of customers. Current retail trends show that on-line sales represent 14% of all UK brick-and-mortar stores and e-commerce and this is expected to rise up to 35% by 2020 (Javelin-Group, 2011; Visser et al., 2014). Currently most home delivery services do not provide a time window for the delivery, something that customers highly value. Moreover, for deliveries that require customer presence, there is a 12% chance of failure of delivery (IMRG, 2012). With current information and communication technologies is feasible to provide customers with more accurate delivery information (Visser et al., 2014), but the challenge of creating more accurate schedules is sizeable.

Congestion is an ever present problem affecting all road users and delays have significant negative impacts on the freight transport industry with models capable of mitigating the negative impacts being given greater credence (Kok et al., 2012). Analysis of schedules supported by VRP models with time-dependent travel times against models using constant speed illustrate the impact of not considering congestion on routing decisions when speed varies during deliveries (Kok et al., 2012). The solutions presented by constant-speed models have been shown to underestimate the actual travel time, provide unrealistic solutions and fail to honour customer delivery times (Donati et al., 2008; Fleischmann et al., 2004; Ichoua et al., 2003).

The first formal formulation of the time-dependent VRP (TDVRP) was presented by Malandraki (1989) and Malandraki and Daskin (1992). In the TDVRP, a fixed number of vehicles with limited capacity serve a number of customers with fixed demands with the overall objective of minimising total route time. Travel time between customers depends on distance and time of day and a single depot exists from where vehicles must depart and return to after completing the delivery tour. No split deliveries (multiple visits to a customer by the same of different vehicles) are allowed and service time windows can be present. Two different solution approaches were proposed (a set of 'greedy' heuristic algorithms and a cutting-plane algorithm) using up to 25 customers, randomly generated without time windows. Although the cutting-plane algorithm was more computationally expensive and was able to solve to optimality, it returned incumbent solutions with greater accuracy in 66% of the tests compared to the heuristic approach.

The TDVRP formulation by Malandraki and Daskin (1992) was based on a travel time step function where each period of time has a specific travel time between nodes associated with it. This has been criticised as being unrepresentative of reality as a vehicle with a later departure time might arrive earlier than a vehicle with an earlier departure time across the same link, as it is point out by Ichoua et al. (2003). Later work on the TDVRP has seen the implementation of travel time functions that respect the "first-in-first-out" principal (FIFO) based on continuous functions (Donati et al., 2008; Figliozzi, 2012; Fleischmann et al., 2004). Fig. 2 shows the difference between Malandraki and Daskin (1992) step travel time function and Ichoua et al. (2003) continuous travel time function for 5 periods of time in a day.

a. Step function   b. Continuous function

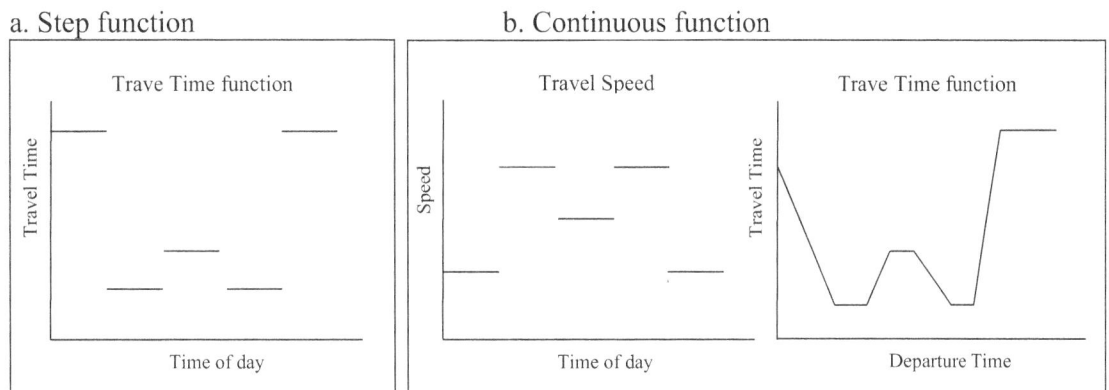

a. Malandraki and Daskin (1992)
b. Ichoua et al. (2003)

Fig. 2. Different Travel Time functions for constant distance with variable speed

Ichoua et al. (2003) studied the TDVRP with soft time windows, where violation of the required service time was permitted but with a resulting penalty in the objective function. The number of routes was fixed and vehicle capacity was not taken into consideration. The objective function was to minimise travel time and penalties incurred due to service time window violations. The solution approach was based on a Tabu Search metaheuristic algorithm with the resulting analysis against the constant-speed model suggesting that a time-dependent speed scenario lead to unrealistic solutions and greater overall travel times. Fleischmann et al. (2004) presented the TDVRP with and without time windows using scenarios based on real congestion patterns from a traffic information system in Berlin. The solution approach used different heuristic algorithms and local search techniques with the results suggesting that using constant-speed models might generate underestimates of travel time of 10%.

Donati et al. (2008) implemented an ant colony metaheuristic algorithm for TDVRP variants based on scenarios involving constant speed and where speeds were randomly assigned using Solomon (1987) instances in order to undertand the accuracy of the proposed algorithm by solving the vehicle routing problem with time windows and constant speed (VRPTW) with an algorithm capable to deal with time-dependent travel times. Solomon dataset provides the representation of deliveries for up to 100 customers with time windows and it is one of the most well-known set of instances to evaluate VRP algorithms, not only for the vehicle routing problem with time windows and constant speed (VRPTW) but for different VRP variants that consider additional restrictions to time windows (Figliozzi, 2012; Jiang et al., 2014; Liu & Shen, 1999; Toth & Vigo, 2001). A second scenario for the TDVRP without time windows was proposed based on the road network of Padua, Italy with congestion data taken from a traffic management system. Results from the constant-speed models when tested in a scenario with congested roads suggested that travel time was underestimated between 5% and 12%.

An exact algorithm for the TDVRPTW was presented by Dabia, Ropke, Van Woensel, and De Kok (2013) using a modified set of the well-known instances of Solomon (1987) for the VRPTW. Speed patterns in links between nodes were allocated randomly and the solution approach was based on a pricing algorithm utilising a column generation and a labelling algorithm. In trials, 63% of the 25 customer instances were solved, 38% of the 50 customer and 15% of the 100 customer. However, details about the categorization of links are not provided. Although Donati et al. (2008) and Fleischmann et al. (2004) used variations of Solomon set of instances to evaluate algorithms for the TDVRPTW, only Figliozzi (2012) provides the complete information to fully reproduce instances when speed variation is present. Figliozzi (2012) proposed an insertion heuristic (*IRCI*) to construct an initial feasible solution and a 'ruin-and-recreate' heuristic algorithm to improve the results (*IRCI-R&R*). Benchmarking for accuracy was performed with available best-known values for the VRPTW and executing the proposed algorithm for the TDVRPTW with constant-speed profiles. The VRPTW commonly has 2 objective functions, to minimise the number of vehicles and their travel distance. Results of Figliozzi (2012) approach suggested a 4.2% increase in vehicles required leading to an 8.6% increase in overall travel distance compared to the best-known values returned by the constant-speed case. Additional speed profiles are provided in order to analyse the impact of congestion patterns in results.

In some of the time-dependent variants the number of vehicles are fixed and optimisation is only based on travel time reduction across the fleet. However, finding the minimum number of vehicles required in the presence of hard time windows is in itself a complex problem, more computationally expensive with time-dependent travel times. Additionally, most current efficient methods for the VRP variants are intricate and difficult to reproduce (Vidal et al., 2013). Furthermore, some metaheuristic implementations have been tailored to work well in specific test instances by tuning parameters so specifically as considering the best random seed that provides high accuracy (Sörensen, 2015). There is clearly therefore a need for general and simple methods applicable to practical applications required by the industry (Vidal et al., 2013), such as effective algorithms that consider congestion and provide reliable schedules. However, the capability of current algorithms to provide high accurate solutions for time-dependent VRP variants is still not well understood due to the lack of adequate algorithm evaluation with previously proposed instances. Gendreau et al. (2015) conducted the most recent literature review in time-dependent routing problems. Some of the mentioned problems are the *"time-dependent point-to-point route planning"* (which is obtaining the optimal path between two locations in a road network) and the *"time-depend vehicle routing"* (time-dependent VRP variants). The challenge in the point-to-point route planning relies on providing efficient algorithms on-line for the next-generation web-based travel information systems that require results in milliseconds or microseconds. Although, it is required to use this problem to establish time-dependent travel times in the TDVRPTW, shortest paths might be determined in a pre-processing phase (Kok et al., 2012) prior to the execution of solving the actual schedule due to the fact that forecasted travel-times are used and there is no need of on-line applications when designing the routes for the following planning period. Additional VRP variants that consider fuel consumption are also mentioned. Gendreau et al. (2015) highlight the requirement of additional contributions of the Operational Research community in time-dependent routing problems, where techniques for constant-speed classic network optimization problems exist but it is required research for their time-dependent counterparts.

This research therefore presents the introduction of state-of-the-art techniques in VRP variants to provide efficient solutions to the industry by comparing quality of algorithm results. In the following subsections the set of restrictions that account for the TDVRPTW is presented along with the concepts of LNS and Variable Neighbourhood Search (VNS) used to guide the search in the algorithm that is proposed in this research. In order to solve VRP variants, exact methods (algorithms that can guarantee optimality) are only viable for small instances, and it is common to use metaheuristic algorithms capable to solve large instances in reasonable time (Drexl, 2012). However, it is important to understand the capability of algorithms to provide high accurate results. A test carried out by an academic group using different providers of software for vehicle routing showed significance difference in quality of solutions, up to 10% between the best vs. the worst schedule in instances of only 100 customers, where higher difference was found in larger problems (Bräysy & Hasle, 2014; Hallamaki et al., 2007). Therefore the importance to further investigate the capabilities of current time-dependent algorithms (Gendreau et al., 2015) to deliver improved performance across a variety of operational and traffic settings. Time-dependent models create new complexities for algorithm design related to tailoring existing search strategies designed specifically for constant-speed models. Common local search procedures require significant modification as alterations within a route as part of the search process could potentially affect the feasibility of the rest of the route. This might alter the departure times of subsequent visits to customers and consequently modify travel times. Route evaluation is considerably more computationally expensive with time-dependent travel times and accommodating hard time constrains for time windows also requires more computational resources than soft constraints where solutions with violations of time windows are allowed (Figliozzi, 2012). Large Neighborhood Search (LNS) is a search strategy that stands out in the range of concepts among state-of-the-art algorithms to solve vehicle routing models due to its simplicity and wider applicability and it has been extended to successfully address various variants (Vidal et al., 2013). In this research, a LNS algorithm is tailored to solve the complexities involved in the TDVRPTW with results compared against available test instances in order to understand its capabilities.

2.1. Problem definition

- Deliveries are requested by n customers
- let $G = (V, A)$ be a graph where vertex $V = (v_0, v_1, \ldots, v_n)$, v_0 is the depot and $v_1, \ldots, v_n$ the set of customers. Each element of V has an associate demand $q_i \geq 0$ (which must be fulfilled), a service time $g_i \geq 0$ and a service time window $[e_i, l_i]$. Note that $0 = q_0 = g_0$
- An undetermined number of identical vehicles each with maximum capacity q_{max} are available and stationed at v_0. Vehicles must depart from and return to the depot v_0 at the end of the delivery tour and their maximum capacities cannot be exceeded

- The departure time of any given vehicle from v_i is denoted b_i, its arrival time a_i
- Arrival time to customer v_i must be before l_i. If arrival time is before e_i, the vehicle have to wait until e_i. Each customer can only be visited once
- Let A be the set of arcs between elements of V, having constant distance d_{ij} between v_i and v_j
- For each arc $(i,j) \in A$ there exists a travel time $t_{ij}(b_i) \geq 0$ a function of departure time b_i, (e.g. of a form as proposed by Ichoua, Gendreau and Potvin (Ichoua et al., 2003) see appendix A).

The primary objective is to minimise the number of vehicles and the second objective is to minimise the sum of travel distance, travel time or duty time

2.2. Large Neighbourhood Search

LNS is an algorithm for neighbourhood exploration introduced by Shaw (1997) utilising a very similar concept to the 'ruin-and-recreate' algorithm introduced by Schrimpf, Schneider, Stamm-Wilbrandt, and Dueck (2000) (Ropke & Pisinger, 2006; Shaw, 1998). A number of partial-destruction procedures are used to remove customers from the solution and a different set of reconstruction procedures are used to create a new solution by inserting removed customers in a smart way, see Fig. 3.

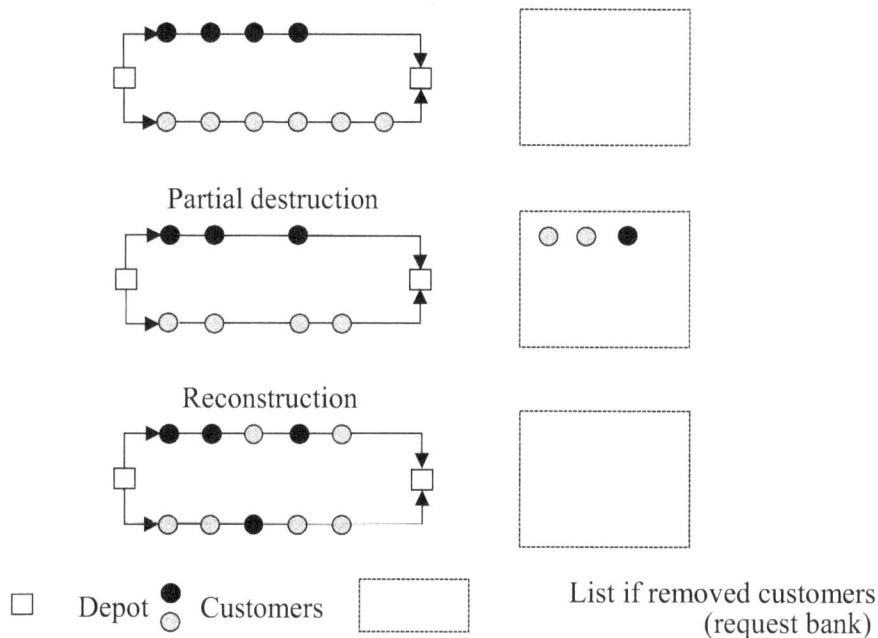

Partial destruction

Reconstruction

☐ Depot ● Customers [   ]   List if removed customers
          ○   (request bank)

Fig. 3. Large Neighbourhood Search movement

Schrimpf et al. (2000) proposed some basic procedures for the VRPTW that were extended by Ropke and Pisinger (2006), Pisinger and Ropke (2007), and Mattos Ribeiro and Laporte (2012) for VRP variants, some are presented below:
Destruction procedures

- *Random-Ruin*: Randomly select and remove w customers from all customers in the solution.
- *Radial-Ruin*: Randomly select a customer v_i. Remove v_i and $w - 1$ closest customers to v_i.
- *Sequential-Ruin*: Select a random route k (vehicle k) and select a random customer v_i in k. Remove a chain of consecutive customers of length w in k starting with v_i.
- *Worst-removal*: Remove the customers that have the most negative impact according to a function *removal-cost*(v_i).

Recreation procedure

- *Basic-greedy heuristic*: Given the list of removed customers U, calculate an *insertion-cost*(v_i, k, p) for all $v_i \in U$ in all possible routes and positions when v_i is inserted in route k in position p, and insert v_i with the lowest *insertion-cost*(v_i, k, p) in the solution. Repeat the procedure until all $v_i \in U$ are inserted or no feasible insertion exists.

- One characteristic of the LNS for VRP variants is that the request bank is an entity that allows the search process for the exploration of infeasible solutions (Ropke & Pisinger, 2006) without directly calculating the violation of restrictions. In the case of the TDVRPTW, any solution with unscheduled customers is infeasible. Additionally, insertion procedures are quite myopic, in order to avoid stagnating search processes, where destruction and recreation procedures keep performing the same modification to a solution, providing diversification in different levels of the search process might improve accuracy of solutions (Mattos Ribeiro & Laporte, 2012).

- Previous LNS implementations have made use of other metaheuristic algorithms at the master level to guide the search to new regions and accept improved solutions such as Simulated Annealing (Mattos Ribeiro & Laporte, 2012; Ropke & Pisinger, 2006). In the neighbourhood exploration, applying noise in recreation procedures also avoids stagnation e.g., by using randomisation in the insertion evaluation function in recreation procedures (Ropke & Pisinger, 2006), or tailoring recreation procedures to ensure diversification (Mattos Ribeiro & Laporte, 2012).

2.3. Variable Neighbourhood Search

VNS is a metaheuristic algorithm introduced by Mladenović and Hansen (1997) and Hansen and Mladenović (2001). VNS has been previously implemented in a range of combinatorial problems (Hansen et al., 2010) including VRP models (Bräysy, 2003; Kytöjoki, Nuortio et al., 2007) and the TDVRP with soft time windows (Kritzinger, Tricoire, Doerner, & Hartl, 2011). VNS uses local search neighbourhoods and avoids local optima with specially designed procedures called "*Shaking*" which usually have random elements (Hansen & Mladenović, 2001; Hansen et al., 2010). An additional element of VNS is the concept that a local optima in a neighbourhood is not necessarily a local optima for other neighbourhoods, therefore changing neighbourhoods might avoid local optima. The pseudo code that illustrates the basic concept of VNS is presented as follows:

Algorithm 1: Basic concepts of Variable Neighbourhood Search

Start
1. Initialization by selecting H neighbourhood structures $H = \{h_1, ..., h_{max}\}$
2. Initialize Incumbent solution
3. Current solution $\leftarrow$ Incumbent solution
4. $h \leftarrow 1$
5. **Repeat**
6.   Current solution $\leftarrow$ *Shaking* with h^{th} neighbourhood (Incumbent solution)
7.   Current solution $\leftarrow$ *Local search* (Current solution)
8.   **If** (Current solution < Incumbent solution) **then**
9.   Incumbent solution $\leftarrow$ Current solution
10. $h \leftarrow 1$
11.   **Else**
12. $h \leftarrow h+1$
13.   **EndIf**
14. **Until** $h = h_{max}$
End

A characteristic of the presented basic VNS concept is that it works on an 'only-descendent' approach. It changes the search space region when an improvement has been found, (lines 8 to 10). However, it may be easily transformed to a 'descendent–ascent' approach by introducing some selection criteria to allow exploration of regions with a deteriorated solution, e.g., randomness (Hansen & Mladenović, 2001) or a threshold acceptance value (Kritzinger et al., 2011). An additional characteristic of VNS is that it does not follow a trajectory, but it explores increasingly distant regions, the set of procedures for "*Shaking*" is at the core of VNS and provides a balance between obtaining a sufficiently large perturbation of the incumbent solution while still making sure desired attributes of "good" solutions are maintained. In order to guide the search, a metric in the "*Shaking*" procedure is introduced (Hansen & Mladenović, 2001), (lines 1 and 6). Local search, line 7, is a set of procedures that allow the exploration of the local search space. An example of VNS for the multi-depot VRPTW is presented by Polacek, Hartl, Doerner, and Reimann (2004), as initialization of an incumbent solution with a cheap heuristic and fast running times was proposed. The set of procedures for "*Shaking*" is based on the CROSS-exchange operator where orientation of routes is preserved and the iCROSS-exchange operator where orientation of routes is reversed.

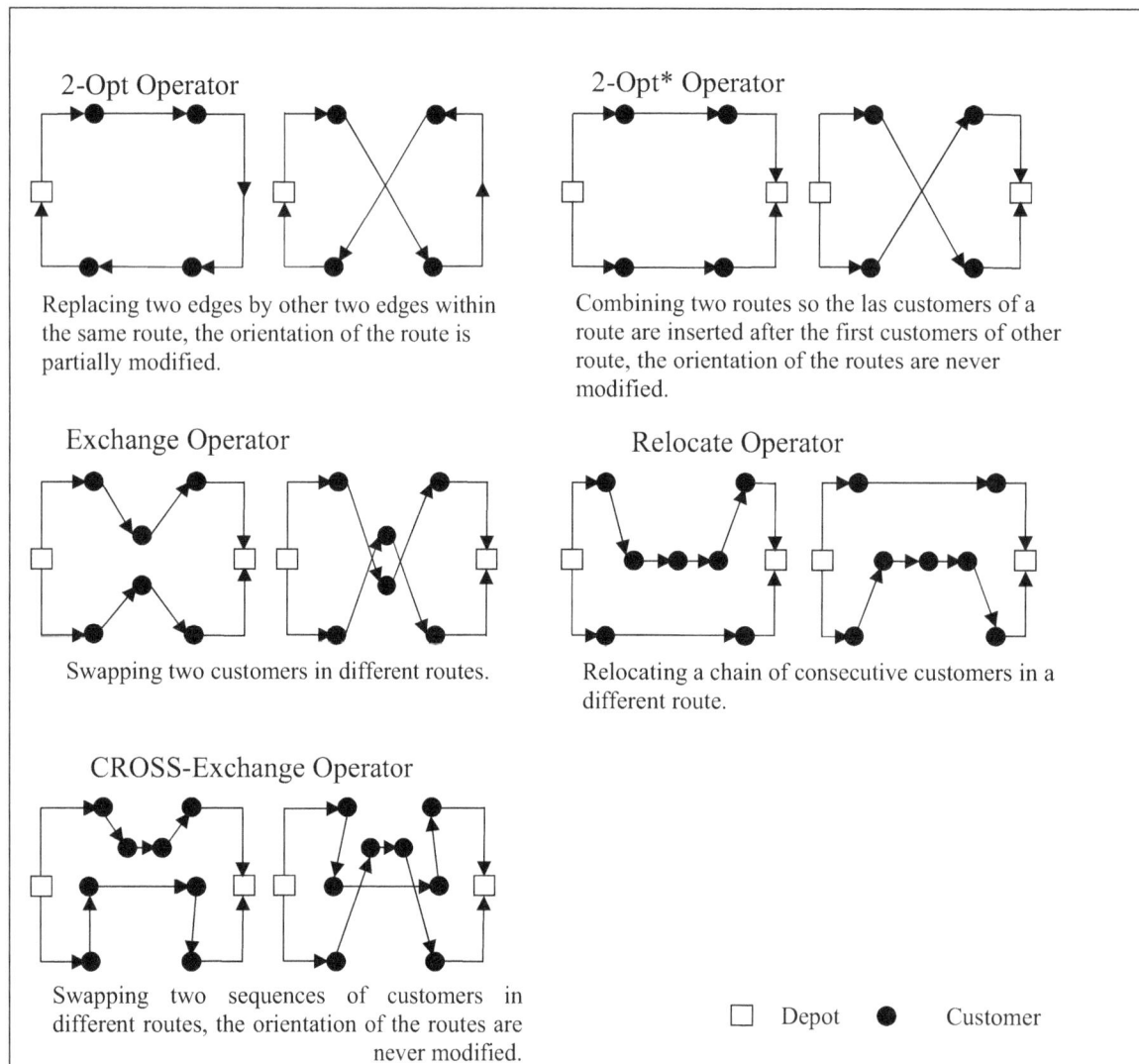

Adapted from: Bräysy and Gendreau (2005a)

Fig. 4. Well-known neighbourhood structures

Fig. 4 shows some well-known neighbourhood exploration procedures. The "*Shaking*" metric between solutions was established as the number of depots used to generate the new solution and the maximum length of the removed sequence in the neighbourhood construction. Local search was the 3-opt operator neighbourhood with reverse of route orientation not allowed and the length of the sequences to be exchanged bounded by an upper limit of three. Decision about moving the search to a new region follows the descendent–ascent approach with a threshold acceptance value. Analysis of results in terms of accuracy and speed showed that the proposed VNS was competitive to other metaheuristic algorithms and it was capable to improve some of the best-known results at that moment. A very similar VNS approach for the TDVRP with soft time windows was proposed by Kritzinger et al. (2011).

3. A hybrid metaheuristic for the TDVRPTW

Search procedures for the TDVRPTW are computationally expensive, with the proposed algorithm designed to guide the search to high accurate solutions in reasonable time. The search process is divided into two stages, the first is where an initial incumbent solution is created using a fast construction heuristic to undertake a reduction of vehicles. In the second stage the objective is either the reduction of travel distance or travel time.

3.1. Minimising the number of vehicles

In previous implementations of the LNS for constant-speed models, minimising the required number of vehicles relied on removing routes from an incumbent solution and placing customers in the request bank until a solution was found without unscheduled customers (some customers still in the request bank) (Pisinger & Ropke, 2007; Ropke & Pisinger, 2006). In the initial test of this approach for the TDVRPTW, lengthy computational time was required in order to get high accuracy due to the complexity in movement evaluation originated by time-dependent travel times.

In order to speed up the search process, a strategy to quickly guide the search towards higher accuracy was designed. Solutions which violated time windows were allowed, and the objective function was minimising the sum of violations of time windows (penalty). Due to the myopic behaviour of LNS, the search quickly reached stagnation. Small time window violations were generated by frequent pairs of customers. In order to avoid stagnation, a tabu list of forbidden pairs of customers was introduced to force the recreation procedure to unexplored trajectories.

At the extend of the knowledge of the authors, this is the first implementation of LNS with allowed time window violations that exploits the destruction and recreation procedures and introduces a tabu list of forbidden pairs of customers for VRP variants with time windows. This strategy allows sequences of customers to be identified that are particularly difficult to accommodate in a solution with a reduced number of vehicles, and allows the algorithm to focus on scheduling these customers without incurring in-time window violations while avoiding stagnation. The local search consists of a modified *Worst-removal* procedure to remove customers that generate penalties along with other destruction procedures before executing a modified *Basic-greedy heuristic* for minimisation of time window penalties (*LNS-Penalty Procedure*). The pseudo code of *Number of vehicles minimisation procedure* is presented as Algorithm 2.

Route reduction procedure follows a descendent–ascent approach, (lines 24 to 26). Initially, a feasible incumbent solution is created with the *IRCI* heuristic proposed by Figliozzi (2012). The vehicle with the minimum number of customers is removed and its customers are allocated to the request bank, (line 2). Each time that the search process reaches a solution with the penalty equal to 0, a feasible schedule, the solution is stored, line 19, and a new vehicle is removed, line 20.

Algorithm 2: *Number of vehicles minimisation procedure*

Start
1.   Incumbent solution ← Construction of Initial Solution with *IRCI*
2.   Incumbent-penalty solution ← Remove one route (Incumbent solution)
3.   Tabu List = { Ø }
4. $h \leftarrow 1$
5.   **Repeat**
6.   **Repeat**
7.   Current solution ← *Shaking-Route Reduction Procedure* (h, Tabu List, Incumbent-penalty solution)
8.   Current solution ← *LNS Penalty Reduction Procedure* (Current solution, Tabu List)
9.   **If** *penalty*(Current solution) > 0
10.   Tabu List ← Tabu List U Elements-Generate-Penalty(Current solution)
11.   **If** *penalty*(Current solution) < *penalty* (Incumbent-penalty solution)
12. $h \leftarrow 1$
13.   Incumbent-penalty solution ← Current solution
14.   **else**
15. $h \leftarrow h + 1$
16.   **EndIf**
17.   **Until** $h = h_{max}$ or *penalty* (Current solution) = 0
18.   **If** *penalty* (Current solution) = 0
19.   Incumbent solution ← Current solution
20.   Incumbent-penalty solution ← Remove one route (Incumbent solution)
21.   Current solution ← Incumbent-penalty solution
22.   Tabu List = { Ø }
23. $h \leftarrow 1$
24.   **Else**
25.   Incumbent-penalty solution ← Current solution
26. $h \leftarrow 1$
27.   **End if**
28.   **Until** stop criteria met
Return Incumbent solution
End

As previously mentioned the local search consists of minimising penalties with LNS procedures described in subsection 2.2, line 8 (*LNS Penalty Reduction Procedure*). Pairs of customers that are in the tabu list are not allowed. In order to prevent stagnating search processes within the LNS neighbourhood movement, *IRCI* is used to diversify the search. All customers in a number of vehicles are removed from the solution and assigned to these vehicles with *IRCI* without violation of time windows. Customers that could not be inserted in these vehicles are inserted in other vehicles using the *Basic-greedy heuristic*.

The proposed *Shaking-Route Reduction Procedure*, line 7, exchanges h customers between routes in order to create new solutions using the exchange operator. Customers to be exchanged are preferably those that generate penalty.

Each time the search reaches a local optima, the pairs of customers that generate penalties in Incumbent-penalty solution are identified and recorded, if they appear in different occasions they are included into the tabu list, line 10.

3.2. Travel time and travel distance minimisation

This procedure relies on the exploration of promising distant regions. In each iteration, a distance region is visited and explored with a fast algorithm. It is established if the region is promising for intensification with a threshold value. Intensification is based on LNS and VNS. The proposed *Travel timed and travel distance minimisation procedure* is presented below as follows:

Algorithm 3: *Travel time and travel distance minimisation procedure*

Start
1. Incumbent solution ← *Number of vehicles minimisation procedure*
2. Current solution ← Incumbent solution
3. **Repeat**
4.   Current solution ← *Shaking procedure* (Current solution)
5.   Current solution ← Educate procedure (Current solution)
6.   **If** *objective* (Current solution) < Threshold value
7.   Current solution ← LNS-VNS Intensification procedure (current solution)
8.   **If** *objective (*Current solution) < *objective* (Incumbent solution)
9.   Incumbent solution ← Current solution
10.   **EndIf**
11.   **EndIf**
12. **Until** stop criteria met
End

This procedure starts with the solution provided by *Number of vehicles minimisation procedure*, line 1. The objective function value in the search can be travel time or travel distance. *Shaking procedure*, line 4, creates a solution in a distant region of the search space, only feasible solutions are allowed. Firstly, a random $vehicle_k$ is selected and all v_i in $vehicle_k$ are inserted in the remaining vehicles when insertions are feasible (no violation of time windows are allowed). Secondly, vehicles are randomly sorted, $S' = \{ vehicle_1, \dots , vehicle_k, \dots , vehicle_m \}$. Thirdly, each v_i in $vehicle_1$ is exchanged, with the exchange operator, in the first feasible insertion in the subsequent vehicles. The third part of the procedure is repeated in all $vehicle_k \in S', k \neq$ m.

Promising solutions are evaluated by executing the *Educate procedure* that consist of 2-opt operator, 2-opt* operator and relocate operator with the length of the sequences to be exchanged bounded by an upper limit of three, line 6. These operators conduct a systematic search by modifying the position of customers within the same route and between routes. Although computationally expensive for a large number of iterations, they are used to identify search regions where a fast reduction in the objective function can be achieved. *Educate procedure* consist of a limited number of iterations.

Identified regions that achieve certain objective function value are selected for intensification, line 7. *LNS-VNS Intensification procedure* is guided by VNS with LNS movements and a "*Shaking*" procedure based on the exchange operator and relocate operator.

4. Benchmark instances

Due to the fact that the TDVRPTW is an extension of the VRPTW, Figliozzi (2012) proposed a modification to the well-known instances for the VRPTW of Solomon (1987) to account for congestion by adding speed profiles. Solomon instances consist of 56 problems with 100 customers and a single depot. Problems are divided in six classes namely: R1, R2, C1, C2, RC1 and RC2. R accounts for random locations, C for clustered locations and RC for a mix of random and clustered locations. Type 1 consist of schedules with tight time windows that allow fewer customers per vehicle than type 2.

Figliozzi (2012) proposed 4 types of speed profiles, with 3 cases for each type, for a total of 12 speed cases. The depot working time $[e_0, l_0]$ is divided into five periods of equal duration. An additional instance with constant speed of 1 is also introduced in order to compare the performance of the proposed solution for the TDVRPTW with available best-known values for the largely studied VRPTW. Speed profiles are as following:

CASES TYPE a (Fast periods between depot opening and closing times)

TD1a = [1.00, 1.60, 1.05, 1.60, 1.00]
TD2a = [1.00, 2.00, 1.50, 2.00, 1.00]
TD3a = [1.00, 2.50, 1.75, 2.50, 1.00]

CASES TYPE b (Higher travel times at the extremes of the working day)

TD1b = [1.60, 1.00, 1.05, 1.00, 1.60]
TD2b = [2.00, 1.00, 1.50, 1.00, 2.00]
TD3b = [2.50, 1.00, 1.75, 1.00, 2.50]

CASES TYPE c (Higher travel speeds are found at the beginning of the working day)

TD1c = [1.60, 1.60, 1.05, 1.00, 1.00]
TD2c = [2.00, 2.00, 1.50, 1.00, 1.00]
TD3c = [2.50, 2.50, 1.75, 1.00, 1.00]

CASES TYPE d (Higher travel speeds at the end of the working day)

TD1d = [1.00, 1.00, 1.05, 1.60, 1.60]
TD2d = [1.00, 1.00, 1.50, 2.00, 2.00]
TD3d = [1.00, 1.00, 1.75, 2.50, 2.50]

5. Implementation and experimental results

Algorithm benchmarking is commonly evaluated in terms of accuracy and speed (Bräysy & Gendreau, 2005a, 2005b; Toth & Vigo, 2001). Accuracy can be easily evaluated when data sets are available. However, different factors have an impact on speed such as hardware (processor, ram), code efficiency and compiler (Figliozzi, 2012). Additionally, it is mentioned in the literature that better results might be obtained by tailoring algorithms accordingly to the test instance. This practice is impractical in industry applications that require fast and reliable solution procedures capable to consistently provide high accurate results (Cordeau et al., 2002; Drexl, 2012; Figliozzi, 2012; Golden et al., 1998).

The proposed LNS approach was coded in Java Eclipse version Juno. It has random elements and results might vary in each run, where multi-core processors offer the possibility to execute multiple threads simultaneously. In this research a computer with processor Intel core i7-2600 3.40GHz and 16 GB of ram was used, three independent threads were run simultaneously and the best result of the three was chosen. The algorithm was run with two different sets of parameters according to termination criteria, which consist of maximum number of iterations, maximum running time, and allowed running time without improvement. The first set of parameters (named F) was set to produce a fast algorithm whereas the second one (named L) produces a more thorough search.

Donati et al. (2008) and Figliozzi (2012) have presented results for metaheuristic approaches for the TDVRPTW using Solomon instances with constant speed in order to compare results with best-known values for the VRPTW, table 1. Best-known values for the minimum number of vehicles for the 56 problems was 405. The result for the proposed algorithm, with set of parameter L, is 408 and best-known values can be achieved by extending running time. However, parameter tuning was set to deal with 672 problems (56 problems x 12 speed cases).

Running the proposed algorithm with sets of parameters F and L provide higher accuracy than previous implementations for the TDVRPTW in the primary objective (average number of vehicles) for instances

R1, RC1 and RC2. Analysis of the secondary objective (distance) shows that the proposed LNS obtained higher accuracy than *IRCI-R&R* (Figliozzi, 2012) in all instances, results within 1% of best-known values can be achieved by increasing running time. Ant colony approach (Donati et al., 2008) obtained higher accuracy in the secondary objective. However, reduced distances might be achieved easily when more vehicles are used, e.g.: problem type R1, reduction of 0.93% from best-known values is achieved with 5.88% more vehicles.

Table 1
VRPTW results for Solomon's 56 problems with 100 customers – Constant speed

Method	R1	Δ	R2	Δ	C1	Δ	C2	Δ	RC1	Δ	RC2	Δ
Average NV												
(1) Best known value	**11.9**		**2.7**		**10.0**		**3.0**		**11.5**		**3.3**	
(2) IRCI – R&R	12.6	5.88%	3.0	11.11%	10.0	0.00%	3.0	0.00%	12.1	5.22%	3.4	4.62%
(3) Ant Colony	12.6	5.88%	3.1	14.81%	10.0	0.00%	3.0	0.00%	12.1	5.22%	3.8	16.92%
(4) LNS F Best 3 runs	12.2	2.52%	3.0	11.11%	10.0	0.00%	3.0	0.00%	11.6	1.09%	3.3	0.00%
(5) LNS L Best 3 runs	12.0	0.84%	2.8	4.38%	10.0	0.00%	3.0	0.00%	11.6	1.09%	3.3	0.00%
Average Distance												
(1) Best known value	**1210.3**		**954.0**		**828.4**		**589.9**		**1384.8**		**1119.2**	
(2) IRCI – R&R	1248.0	3.11%	1124.0	17.82%	841.0	1.52%	626.0	6.12%	1466.0	5.86%	1308.0	16.87%
(3) Ant Colony	1199.0	-0.93%	967.0	1.36%	828.0	-0.05%	590.0	0.02%	1374.0	-0.78%	1156.0	3.29%
(4) LNS F Best 3 runs	1222.3	0.99%	961.7	0.81%	834.0	0.68%	590.3	0.07%	1405.9	1.52%	1170.0	4.54%
(5) LNS L Best 3 runs	1232.2	1.81%	969.6	1.64%	828.6	0.02%	590.3	0.07%	1404.1	1.39%	1160.0	3.65%

(1) Best-known values as reported in Nagata, Bräysy, and Dullaert (2010) (2) Figliozzi (2012) CPU Time 19 min, processor Intel Core Duo 1.2 GHz. (3) Donati et al. (2008) CPU Time 168 min, Pentium IV 2.66 GHz. (4) LNS F (3 threads of 26 min) Intel Core i7 3.4 GHz. (5) LNS L (3 threads of 62 min) Intel Core i7 3.4 GHz.

In the TDVRPTW the proposed primary objective is the minimisation of the number of vehicles, secondary objective might be minimisation of travel distance, travel time or both. Figliozzi (2012) proposed the sum of distance and travel time as secondary objective. Tables (2-5) show benchmarking of *IRCI-R&R* and the proposed LNS. The proposed algorithm is capable of obtaining a reduction in vehicles required of up to 12.96% (cases type b, set of parameter L, instance R2, Table 3) and secondary reduction objective up to 19.60% in travelled time and distance with fewer vehicles (cases type a, set of parameter L, instance RC2, Table 2) in reasonable computational time. In average each problem can be solved in 26.78 seconds using 3 threads with set of parameter S and 65.35 seconds with set L. Table 6 shows the overall sum of number of vehicles, total travelled distance and total travelled time required to solve the 56 problems in each speed profile.

Table 2
TDVRPTW Average results for 3 instances in Cases Type (a) 100 customers

Method	R1	Δ	R2	Δ	C1	Δ	C2	Δ	RC1	Δ	RC2	Δ
(1) Figliozzy IRCI-R&R												
NV	10.8		2.5		10.0		3.0		10.6		3.0	
Distance	1263.3		1243.0		874.3		669.3		1387.3		1444.0	
Travel Time	923.3		875.0		660.3		514.3		1004.3		1012.3	
Second objective	2197.4		2120.5		1544.7		1186.7		2402.3		2459.3	
(2) LNS F												
NV	10.4	-3.53%	2.5	-1.83%	10.0	0.00%	3.0	0.00%	10.2	-4.02%	2.9	-2.03%
Distance	1164.4	-7.83%	1007.4	-18.95%	841.6	-3.74%	589.9	-11.87%	1309.9	-5.58%	1168.6	-19.07%
Travel Time	833.2	-9.76%	674.2	-22.95%	625.9	-5.21%	447.6	-12.97%	924.8	-7.92%	789.2	-22.04%
Second objective	1997.6	-9.09%	1681.7	-20.69%	1467.5	-5.00%	1037.5	-12.57%	2234.7	-6.98%	1957.7	-20.40%
(3) LNS L												
NV	10.3	-4.66%	2.4	-7.19%	10.0	0.00%	3.0	0.00%	10.0	-5.90%	2.8	-7.09%
Distance	1165.6	-7.74%	1013.9	-18.43%	834.7	-4.53%	589.4	-11.94%	1313.0	-5.36%	1178.7	-18.37%
Travel Time	832.6	-9.82%	685.9	-21.61%	619.6	-6.16%	447.2	-13.05%	922.5	-8.14%	795.9	-21.38%
Second objective	2008.5	-8.60%	1702.2	-19.73%	1464.3	-5.20%	1039.7	-12.39%	2245.6	-6.52%	1977.3	-19.60%

(1) Figliozzi (2012) CPU Time 54.1 min, processor Intel Core Duo 1.2 GHz. (2) LNS F (3 threads of 78 min) processor Intel Core i7 3.4 GHz. (3) LNS L (3 threads of 183 min) processor Intel Core i7 3.4 GHz.

Table 3

TDVRPTW Average results for 3 instances in Cases Type (b) 100 customers

Method	R1	Δ	R2	Δ	C1	Δ	C2	Δ	RC1	Δ	RC2	Δ
(1) Figliozzy IRCI-R&R												
NV	11.8		2.8		10.0		3.0		11.5		3.2	
Distance	1277.7		1225.0		880.3		683.7		1441.7		1439.7	
Travel Time	925.7		917.0		655.3		486.0		1035.3		1078.0	
Second objective	2215.1		2144.8		1545.7		1172.7		2488.5		2520.9	
(2) LNS F												
NV	11.2	-4.92%	2.7	-4.26%	10.0	0.00%	3.0	0.00%	10.9	-4.89%	3.0	-6.54%
Distance	1197.7	-6.26%	1004.7	-17.98%	847.3	-3.75%	590.0	-13.70%	1633.2	13.29%	1202.7	-16.46%
Travel Time	853.9	-7.75%	730.6	-20.33%	602.3	-8.09%	432.0	-11.11%	966.4	-6.66%	896.7	-16.82%
Second objective	2051.6	-7.38%	1735.3	-19.09%	1449.6	-6.22%	1022.0	-12.85%	2332.6	-6.26%	2099.4	-16.72%
(3) LNS L												
NV	11.1	-5.68%	2.5	-12.96%	10.0	0.00%	3.0	0.00%	10.8	-6.20%	2.9	-9.14%
Distance	1204.5	-5.72%	1027.8	-16.10%	837.8	-4.83%	589.9	-13.72%	1373.9	-4.70%	1213.1	-15.74%
Travel Time	859.5	-7.15%	752.6	-17.93%	599.3	-8.54%	432.0	-11.11%	968.1	-6.50%	903.7	-16.17%
Second objective	2075.2	-6.32%	1782.9	-16.88%	1447.2	-6.37%	1024.9	-12.60%	2352.7	-5.46%	2119.7	-15.91%

(1) Figliozzi (2012) CPU Time 57.4 min, processor Intel Core Duo 1.2 GHz. **(2)** LNS F (3 threads of 75 min) processor Intel Core i7 3.4 GHz. **(3)** LNS L (3 threads of 186 min) processor Intel Core i7 3.4 GHz.

Table 4

TDVRPTW Average results for 3 instances in Cases Type (c) 100 customers

Method	R1	Δ	R2	Δ	C1	Δ	C2	Δ	RC1	Δ	RC2	Δ
(1) Figliozzy IRCI-R&R												
NV	10.9		2.5		10.0		3.0		10.8		2.9	
Distance	1280.0		1242.0		863.3		668.7		1419.0		1439.0	
Travel Time	916.0		868.7		626.7		502.3		1034.0		1020.0	
Second objective	2206.9		2113.2		1500.0		1174.0		2463.8		2461.9	
(2) LNS F												
NV	10.4	-4.50%	2.5	-1.83%	10.0	0.00%	3.0	0.00%	10.3	-4.57%	2.8	-4.00%
Distance	1171.6	-8.47%	1142.0	-8.05%	836.8	-3.07%	589.3	-11.87%	1335.3	-5.90%	1197.2	-16.80%
Travel Time	826.2	-9.80%	800.0	-7.90%	605.1	-3.44%	445.3	-11.35%	960.9	-7.07%	857.2	-15.96%
Second objective	1997.9	-9.47%	1942.0	-8.10%	1441.9	-3.87%	1034.6	-11.87%	2296.2	-6.80%	2054.3	-16.56%
(3) LNS L												
NV	10.3	-5.37%	2.3	-9.57%	10.0	0.00%	3.0	0.00%	10.1	-6.19%	2.7	-7.14%
Distance	1172.7	-8.39%	1022.4	-17.68%	829.7	-3.90%	589.3	-11.86%	1322.6	-6.79%	1196.3	-16.87%
Travel Time	828.7	-9.53%	710.6	-18.19%	601.2	-4.06%	445.3	-11.35%	954.6	-7.68%	855.9	-16.09%
Second objective	2011.6	-8.85%	1735.3	-17.88%	1440.9	-3.94%	1037.6	-11.61%	2287.3	-7.16%	2054.9	-16.53%

(1) Figliozzi (2012) CPU Time 55.9 min, processor Intel Core Duo 1.2 GHz. **(2)** LNS F (3 threads of 72 min) processor Intel Core i7 3.4 GHz. **(3)** LNS L (3 threads of 174 min) processor Intel Core i7 3.4 GHz.

Table 5

TDVRPTW Average results for 3 instances in Cases Type (d) 100 customers

Method	R1	Δ	R2	Δ	C1	Δ	C2	Δ	RC1	Δ	RC2	Δ
(1) Figliozzy IRCI-R&R												
NV	11.6		2.8		10.0		3.0		11.3		3.3	
Distance	1292.0		1216.3		865.0		678.3		1421.7		1410.0	
Travel Time	976.0		935.3		665.0		502.3		1063.7		1073.3	
Second objective	2279.6		2154.5		1540.0		1183.7		2496.7		2486.6	
(2) LNS F												
NV	11.3	-3.32%	2.7	-3.23%	10.0	0.00%	3.0	0.00%	10.7	-5.51%	3.1	-5.87%
Distance	1195.1	-7.50%	1026.5	-15.61%	833.5	-3.64%	590.1	-13.01%	1373.0	-3.42%	1178.2	-16.44%
Travel Time	901.3	-7.65%	782.6	-16.33%	642.9	-3.32%	437.7	-12.87%	1020.8	-4.03%	890.4	-17.04%
Second objective	2096.3	-8.04%	1809.1	-16.03%	1476.4	-4.13%	1030.0	-12.98%	2393.8	-4.12%	2068.6	-16.81%
(3) LNS L												
NV	11.3	-3.32%	2.6	-6.59%	10.0	0.00%	3.0	0.00%	10.7	-5.51%	3.0	-7.64%
Distance	1184.2	-8.34%	1005.2	-17.36%	831.3	-3.90%	591.4	-12.82%	1367.3	-3.83%	1169.6	-17.05%
Travel Time	894.2	-8.38%	764.1	-18.31%	641.5	-3.53%	438.6	-12.68%	1013.0	-4.76%	883.9	-17.65%
Second objective	2089.6	-8.34%	1771.9	-17.76%	1482.8	-3.71%	1033.1	-12.72%	2391.0	-4.23%	2056.5	-17.30%

(1) Figliozzi (2012) CPU Time 56.8 min, processor Intel Core Duo 1.2 GHz. **(2)** LNS F (3 threads of 81 min) processor Intel Core i7 3.4 GHz. **(3)** LNS L (3 threads of 189 min) processor Intel Core i7 3.4 GHz.

Table 6

Total Number of vehicles, distance and travelled time in all 56 problems (100 customers) in each of the 12 speed profiles (case types)

Case Type	(1) Figliozzy IRCI-R&R			(2) VNS L		
	NV	Distance	Travel Time	NV	Distance	Travel Time
TD1a	402	64875.0	53643.0	387	57439.0	46703.4
TD2a	378	64580.0	45847.0	361	57105.5	39505.4
TD3a	360	64667.0	41198.0	348	57358.7	35105.3
TD1b	420	65044.0	54053.0	403	57950.2	47892.0
TD2b	398	64925.0	46773.0	378	59178.5	41878.0
TD3b	393	65781.0	42837.0	370	59018.2	37480.2
TD1c	402	65304.0	53346.0	387	57842.2	47051.2
TD2c	380	64921.0	45583.0	360	57794.1	40599.2
TD3c	365	64791.0	40985.0	350	57317.0	36004.6
TD1d	417	64858.0	54930.0	401	57639.0	48841.1
TD2d	399	64304.0	47905.0	387	57317.9	42465.6
TD3d	388	65084.0	44466.0	375	58368.9	39472.9
TOTAL	**4702**	**779134.0**	**571566.0**	**4507**	**694329.1**	**502998.9**
Δ				**-4.15%**	**-10.88%**	**-12.00%**

(1) Figliozzi (2012) **(2)** LNS L

5.1. Analysis of results

The proposed algorithm consistently provided improved results for the TDVRPTW. As previously mentioned, route evaluation in the search process is computationally expensive in TDVRP variants. Therefore, selection of neighborhood structures and its adequate tailoring is at the most importance in algorithm design.

Analysis of the computational complexity of different neighborhood structures and performance shows the capability of the proposed LNS tailoring to quickly achieve high accuracy over other procedures. Well-known neighborhood structures involve the deletion of (up to) x arcs of the current solution and the generation of x new arcs to create the subsequent solution, the complexity of neighborhood exploration is $O(n^x)$ (Zachariadis & Kiranoudis, 2010). 2-opt and 2-opt* operators are commonly used in VRP variants with time windows (Bräysy & Gendreau, 2005a), the first one relocates customers within the same vehicle and the second one relocates customers in different vehicles and their complexity of exhaustively examining all possible solutions "naive exploration" is $O(n^2)$, more complex operators with more arc removals are consequently more computationally complex (Zachariadis & Kiranoudis, 2010).

The computational complexity of a LNS procedure that makes use of *Basic-greedy heuristic(U)* for recreation depends on the number of elements in the request bank, the number of routes (vehicles) and the number of customers in the modified route in current solution. After the first insertion, the subsequent customer insertions are only evaluated in the previously modified route with *insertion-cost(v_i, k, p)*. Therefore, computational complexity largely varies according to the number of routes in current solution, being the worst case a solution with one route and quickly reducing computational complexity with more routes.

It is important to highlight that the concept of LNS relies on designing a neighborhood exploration using a group of LNS procedures, that might make use of random elements to diversify the search process, and effectively exploration of neighborhood is wider than well-known neighbourhood structures.

In order to understand the computational complexity of the proposed LNS tailoring and its benefits, a simplified algorithm for travel time minimisation is introduced where different neighborhood structures are used for local search, namely LNS procedures and 2-opt along with 2-opt* procedures. The pseudo code is presented as follows:

Algorithm 4: *Travel time minimisation procedure 2*

Start
 1.   Incumbent solution ← *Construction heuristic*
 2.   Current solution ← Incumbent solution
 3.   **Repeat**
 4.   Current solution ← *Local Search procedure* (Current solution)
 5.   **If** *objective* (Current solution) < *objective* (Incumbent solution)
 6.   Incumbent solution ← Current solution
 7.   **EndIf**
 8.   Current solution ← *Diversification procedure* (Current solution)
 9.   **Until** stop criteria met
End

The tested instances are R101 and R201 with speed case TD1a. The termination criterion is allowed execution time and it was executed 50 times in a single thread for different execution times in order to illustrate the impact of parameter variation. Note that 2-opt* is restricted, the "naive exploration" only performs a fraction of iterations obtaining deteriorated results in solutions with few routes, such as R201. The behaviour of different local search procedures in *Travel time minimisation procedure 2* is shown in Fig. 5. The numbers of routes as a result of Construction heuristic are respectively 21 in R101 and 5 in R201.

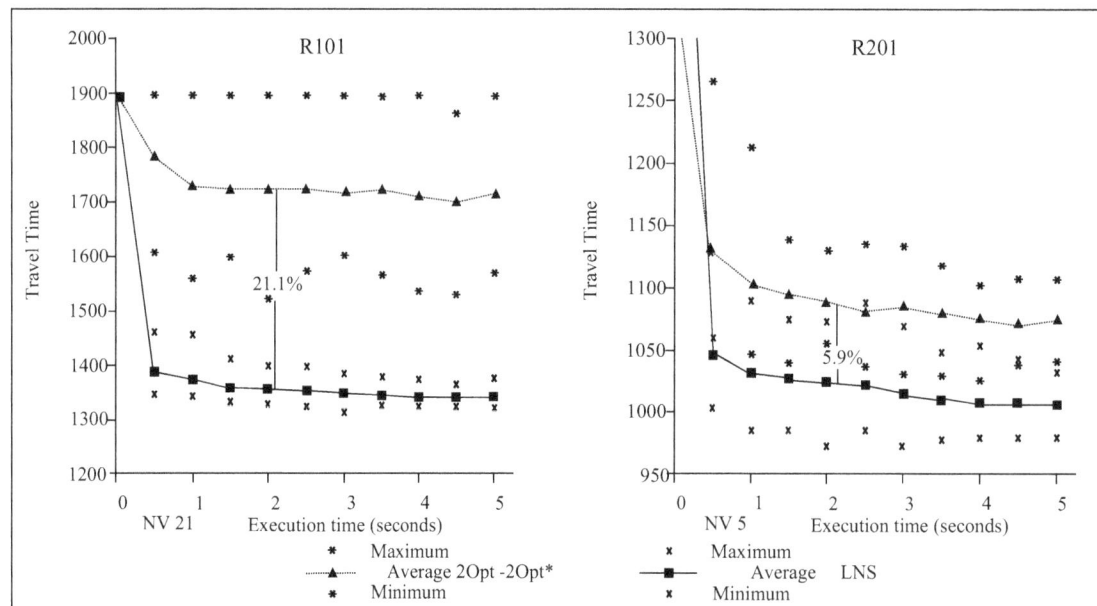

Fig. 5. Behaviour of LNS movements vs. 2-opt and 2-opt* in the presence of time-dependent travel times. Solomon instances R101 and R102 (100 customers) – Figliozzi (2012) speed case TD1a.

Results in execution time of 0.5 seconds illustrate the computational complexity and the accuracy of the proposed LNS, see table 7. 2-opt and 2-opt* were executed a few hundred times whereas LNS complete removals and insertions were executed in average 7,544 times in R101 and 4073 in R102. LNS in instance R101 obtained an average travel time reduction of 22.0% and minimum value reduction of 16.16% over 2-opt and 2opt* local search, in the case of R102 reductions respectively are 7.5% and 5.3%. LNS clearly provides a more accurate local search.

Table 7
Results of executing *travel time minimisation procedure* with different local search procedures at execution time 0.5 s

Instance			2-opt & 2-opt*	LNS	Δ
R101	Best average travel time		1,608.3	1,348.4	16.2%
	Average travel time		1,781.0	1,389.0	22.0%
	Worst travel time		1,893.5	1,460.0	22.9%
	Average Iterations	2-opt	178.9	7,544.2	
		2-opt*	192.7		
R201	Best average travel time		1,059.9	1,003.5	5.3%
	Average travel time		1,132.0	1,046.5	7.5%
	Worst travel time		1,265.3	1,131.2	10.6%
	Average Iterations	2-opt	239.9	4,073.1	
		2-opt*	229.1		

Solomon instances (100 customers) – Figliozzi (2012) speed case TD1a.

Although Figliozzi (2012) implementation for the TDVRPTW was based on the 'ruin-and-recreate' concept, alternative destruction and recreation procedures were proposed rather than LNS procedures. The route improvement procedure consisted of iteratively removing all the customers in selected vehicles in order to rearrange them with a fast heuristic introduced by the author. The criteria to select the vehicles were: a) geographical proximity (distance between any two routes' centre of gravity), b) number of customers in vehicles.

Donati et al. (2008) made use of well-known neighbourhood operators and algorithm tailoring was based on restricting movements taking under consideration customer proximity and the introduction of a variable called "slack time" for each delivery in order to evaluate how long the delivery could be delayed so subsequent visits in the route will not violate time windows in the search process in the presence of time-dependent travel times.

This research consistently provides improved results for the TDVRPTW when compared with previous implementations. It is based on the capabilities of LNS movements to provide a fast and wide exploration of the search space that can quickly reach high accurate results in the presence of time-dependent travel times and time windows. Furthermore, taking advantage of the capabilities of LNS, additionally tailoring of destruction and recreation procedures also achieved high accuracy in the minimisation of required number of vehicles.

6. Conclusions

It is clear from the literature that time-dependent algorithms are necessary to substantially improve vehicle planning and scheduling in congested environments, with existing approaches that do not take congestion into consideration leading to extra time and missed deliveries. However, the added complexities of including time-dependent functions in models requires increased computational process capacity to provide as near optimal results as constant-speed models.

Tailoring algorithms to effectively and efficiently solve VRP variants is proven to be a challenging task. In this research, it is shown how two different strategies are used to solve different elements of the time-dependent vehicle routing problem with hard time windows.

For the minimisation of vehicles it was necessary to have a very specific approach of minimisation of time window violations in order to focus the search in scheduling customers that are particularly difficult to accommodate. Large Neighbourhood Search procedures guided with Variable Neighbourhood Search achieved high accuracy with the proposed algorithm tailoring. It provided a reduction of 4.15% vehicles than previous implementations in the 672 test instances.

Travel time or travel distance minimisation strategy was based on the search of distant regions in order to obtain a robust exploration of the search space. When compared to previous implementations, the algorithm was capable to obtain reductions in some test problems up to 19.60% in travel time and distance with fewer vehicles. It consistently provided improved solutions in the 672 test instances. Although the proposed algorithm makes use of random elements to escape from local optima and can therefore be run on a single processor core if required, parallel computing is also demonstrated here to take advantage of current processor architecture to execute multiple threads and explore different regions of the search space simultaneously without increasing the overall time of the search. Large Neighborhood Search is a strategy that stands out in vehicle routing problem solution algorithms due to its simplicity and wider applicability to solve different variants. The proposed approach shows its capacity to provide planners and drivers with accurate and reliable schedules when congestion is present using current computer architecture in reasonable time with adequate algorithm tailoring.

References

Bräysy, O. (2003). A reactive variable neighborhood search for the vehicle-routing problem with time windows. *INFORMS Journal on Computing, 15*(4), 347-368.

Bräysy, O., & Gendreau, M. (2005a). Vehicle routing problem with time windows, Part I: Route construction and local search algorithms. *Transportation science, 39*(1), 104-118.

Bräysy, O., & Gendreau, M. (2005b). Vehicle routing problem with time windows, Part II: Metaheuristics. *Transportation science, 39*(1), 119-139.

Bräysy, O., & Hasle, G. (2014). Software Tools and Emerging Technologies for Vehicle Routing and Intermodal Transportation. *Vehicle Routing: Problems, Methods, and Applications, 18*, 351.

Chang, Y. S., Lee, Y. J., & Choi, S. B. (2015). More Traffic Congestion in Larger Cities?-Scaling Analysis of the Large 101 US Urban Centers.

Chopra, S., & Meindl, P. (2007). *Supply chain management. Strategy, planning & operation*: Springer.

Cordeau, J.-F., Gendreau, M., Laporte, G., Potvin, J.-Y., & Semet, F. (2002). A guide to vehicle routing heuristics. *Journal of the Operational Research society, 53*(5), 512-522.

Dabia, S., Ropke, S., Van Woensel, T., & De Kok, T. (2013). Branch and price for the time-dependent vehicle routing problem with time windows. *Transportation science, 47*(3), 380-396.

DFT. (2010). Integrated Research Study: HGV Satellite Navigation and Route Planning. Retrieved April 01, 2013, from http://www.freightbestpractice.org.uk/products/3705_7817_integrated-research-study--hgv-satellite-navigation-and-route-planning.aspx.

Donati, A. V., Montemanni, R., Casagrande, N., Rizzoli, A. E., & Gambardella, L. M. (2008). Time dependent vehicle routing problem with a multi ant colony system. *European journal of operational research, 185*(3), 1174-1191.

Drexl, M. (2012). Rich vehicle routing in theory and practice. *Logistics Research, 5*(1-2), 47-63.

Eglese, R., Maden, W., & Slater, A. (2006). A road timetableTM to aid vehicle routing and scheduling. *Computers & operations research, 33*(12), 3508-3519.

Ehmke, J. F., Steinert, A., & Mattfeld, D. C. (2012). Advanced routing for city logistics service providers based on time-dependent travel times. *Journal of Computational Science, 3*(4), 193-205.

Figliozzi, M. A. (2012). The time dependent vehicle routing problem with time windows: Benchmark problems, an efficient solution algorithm, and solution characteristics. *Transportation Research Part E-Logistics and Transportation Review, 48*(3), 616-636. doi: DOI 10.1016/j.tre.2011.11.006

Fleischmann, B., Gietz, M., & Gnutzmann, S. (2004). Time-varying travel times in vehicle routing. *Transportation science, 38*(2), 160-173.

FTA. (2015). Logistics Report 2015.Delivering Safe, Efficient, Sustainable Logistics. In FTA (Ed.), About. Tunbridge Wells: FTA.

Gendreau, M., Ghiani, G., & Guerriero, E. (2015). Time-dependent routing problems: A review. *Computers & operations research, 64*, 189-197.

Golden, B., Wasil, E., Kelly, J., & Chao, I. (1998). Fleet Management and Logistics, chapter The Impact of Metaheuristics on Solving the Vehicle Routing Problem: algorithms, problem sets, and computational results: Kluwer Academic Publishers, Boston.

Haghani, A., & Jung, S. (2005). A dynamic vehicle routing problem with time-dependent travel times. *Computers & operations research, 32*(11), 2959-2986.

Hallamaki, A., Hotokka, P., Brigatti, J., Nakari, P., Bräysy, O., & T, R. (2007). Vehicle Routing Software: A Survey and Case Studies with Finish Data. Technical Report. Finland: University of Jyväskylä.

Hansen, P., & Mladenović, N. (2001). Variable neighborhood search: Principles and applications. *European journal of operational research, 130*(3), 449-467.

Hansen, P., Mladenović, N., & Pérez, J. A. M. (2010). Variable neighbourhood search: methods and applications. *Annals of Operations Research, 175*(1), 367-407.

Ichoua, S., Gendreau, M., & Potvin, J.-Y. (2003). Vehicle dispatching with time-dependent travel times. *European journal of operational research, 144*(2), 379-396.

IMRG. (2012). UK valuing home delivery review 2012. In IMRG (Ed.).

Javelin-Group. (2011). How many stores will we really need? UK non-food retailing in 2020. In L. J. Group. (Ed.).

Jiang, J., Ng, K. M., Poh, K. L., & Teo, K. M. (2014). Vehicle routing problem with a heterogeneous fleet and time windows. *Expert Systems with Applications, 41*(8), 3748-3760.

Kok, A., Hans, E., & Schutten, J. (2012). Vehicle routing under time-dependent travel times: the impact of congestion avoidance. *Computers & operations research, 39*(5), 910-918.

Kritzinger, S., Tricoire, F., Doerner, K. F., & Hartl, R. F. (2011). Variable neighborhood search for the time-dependent vehicle routing problem with soft time windows *Learning and Intelligent Optimization* (pp. 61-75): Springer.

Kytöjoki, J., Nuortio, T., Bräysy, O., & Gendreau, M. (2007). An efficient variable neighborhood search heuristic for very large scale vehicle routing problems. *Computers & operations research, 34*(9), 2743-2757.

Liu, F.-H., & Shen, S.-Y. (1999). The fleet size and mix vehicle routing problem with time windows. *Journal of the Operational Research society*, 721-732.

Malandraki, C. (1989). Time dependent vehicle routing problems: Formulations, solution algorithms and computational experiments.

Malandraki, C., & Daskin, M. S. (1992). Time dependent vehicle routing problems: Formulations, properties and heuristic algorithms. *Transportation science, 26*(3), 185-200.

Mattos Ribeiro, G., & Laporte, G. (2012). An adaptive large neighborhood search heuristic for the cumulative capacitated vehicle routing problem. *Computers & operations research, 39*(3), 728-735.

Mladenović, N., & Hansen, P. (1997). Variable neighborhood search. *Computers & operations research, 24*(11), 1097-1100.

Nagata, Y., Bräysy, O., & Dullaert, W. (2010). A penalty-based edge assembly memetic algorithm for the vehicle routing problem with time windows. *Computers & operations research, 37*(4), 724-737.

Pisinger, D., & Ropke, S. (2007). A general heuristic for vehicle routing problems. *Computers & operations research, 34*(8), 2403-2435.

Polacek, M., Hartl, R. F., Doerner, K., & Reimann, M. (2004). A variable neighborhood search for the multi depot vehicle routing problem with time windows. *Journal of heuristics, 10*(6), 613-627.

Rincon-Garcia, N., Waterson , B. J., & Cherret, T. J. (2015). Requirements from Vehicle Routing Software: Perspectives from literature, developers and the freight industry. *Under Review*.

Ropke, S., & Pisinger, D. (2006). An adaptive large neighborhood search heuristic for the pickup and delivery problem with time windows. *Transportation science, 40*(4), 455-472.

Schrimpf, G., Schneider, J., Stamm-Wilbrandt, H., & Dueck, G. (2000). Record breaking optimization results using the ruin and recreate principle. *Journal of Computational Physics, 159*(2), 139-171.

Shaw, P. (1997). A new local search algorithm providing high quality solutions to vehicle routing problems. *APES Group, Dept of Computer Science, University of Strathclyde, Glasgow, Scotland, UK.*

Shaw, P. (1998). Using constraint programming and local search methods to solve vehicle routing problems *Principles and Practice of Constraint Programming—CP98* (pp. 417-431): Springer.

Solomon, M. M. (1987). Algorithms for the Vehicle Routing and Scheduling Problems with Time Window Constraints *Operations Research, 35*(2), 254-265.

Sörensen, K. (2015). Metaheuristics—the metaphor exposed. *International Transactions in Operational Research, 22*(1), 3-18.

Toth, P., & Vigo, D. (2001). *The vehicle routing problem*: Siam.

US-Department-Of-Transport. (2003). Final Report -Traffic Congestion and Reliability: Linking Solutions to Problems.

Vidal, T., Crainic, T. G., Gendreau, M., & Prins, C. (2013). A hybrid genetic algorithm with adaptive diversity management for a large class of vehicle routing problems with time-windows. *Computers & operations research, 40*(1), 475-489.

Visser, J., Nemoto, T., & Browne, M. (2014). Home delivery and the impacts on urban freight transport: A review. *Procedia-social and behavioral sciences, 125*, 15-27.

Zachariadis, E. E., & Kiranoudis, C. T. (2010). A strategy for reducing the computational complexity of local search-based methods for the vehicle routing problem. *Computers & operations research, 37*(12), 2089-2105.

Appendix A: Travel time function

Travel time between any two given customers i,j depends on the travel speed function, specific data format and departure time from customer i, the following algorithm proposed by Ichoua et al. (2003) (with notation of Figliozzi (2012)), returns arrival time a_j to customer j when departing from customer i at b_i:

Data:
 T=$T_1,T_2, ...,T_p$ where each period T_k has an associated constant travel speed s_k, an initial time $t_{\underline{k}}$ and a final time $t_{\overline{k}}$

Start
  if $a_i < e_i$ **then**
    $b_i \leftarrow e_i + g_i$
  else
    $b_i \leftarrow a_i + g_i$
  endif
  find k **where** $t_{\underline{k}} \le b_i \le t_{\overline{k}}$
  $a_j \leftarrow b_i+d_{ij}/s_k$
  $d \leftarrow d_{ij}, t \leftarrow b_i$
  while $a_j > t_{\overline{k}}$ **do**
    $d \leftarrow d - (t_{\overline{k}} -t) s_k$
    $t \leftarrow t_{\overline{k}}$
    $a_j \leftarrow t + d_{ij}/s_{k+1}$
    $k \leftarrow k+1$
  end while
  Return: a_j, arrival time at customer j
End

4

A mathematical model for the product mixing and lot-sizing problem by considering stochastic demand

Dionicio Neira Rodado[a], John Willmer Escobar[b*], Rafael Guillermo García-Cáceres[c] and Fabricio Andrés Niebles Atencio[a]

[a]Departamento de Ingenieria Industrial, Universidad de la Costa. Calle 58 # 55 - 66. Barranquilla, Colombia
[b]Departamento de Ingeniería Civil e Industrial, Pontificia Universidad Javeriana. Calle 18 No 118-250, Cali (Colombia)
[c]Profesor Investigador, Universidad Antonio Nariño, Bogotá, Colombia. Colombia

CHRONICLE	ABSTRACT
Keywords: *Lot Sizing* *Product-mix planning* *Stochastic demand* *EVA* *Sample average approximation*	The product-mix planning and the lot size decisions are some of the most fundamental research themes for the operations research community. The fact that markets have become more unpredictable has increaed the importance of these issues, rapidly. Currently, directors need to work with product-mix planning and lot size decision models by introducing stochastic variables related to the demands, lead times, etc. However, some real mathematical models involving stochastic variables are not capable of obtaining good solutions within short commuting times. Several heuristics and metaheuristics have been developed to deal with lot decisions problems, in order to obtain high quality results within short commuting times. Nevertheless, the search for an efficient model by considering product mix and deal size with stochastic demand is a prominent research area. This paper aims to develop a general model for the product-mix, and lot size decision within a stochastic demand environment, by introducing the Economic Value Added (EVA) as the objective function of a product portfolio selection. The proposed stochastic model has been solved by using a Sample Average Approximation (SAA) scheme. The proposed model obtains high quality results within acceptable computing times.

1. Introduction

It is a huge challenge for companies to reach strategic decisions related to the production system. Some of the most difficult decisions are the definition of the mix of products, the determination of the lot size for each item, and the scheduling for a monthly view. Currently, these decisions are relevant, especially when the life cycle of products is short and customers ask for more customized products. Indeed, the companies must satisfy the demand of the customers of specific products with eminent quality and short delivery times. The decision making process becomes very difficult due to the fact that companies usually do not have the tools to solve the problem of lot sizing and product-mix planning by considering the maximization of the profit of the companies. This decision affects both, the productivity and the

* Corresponding author
E-mail: jwescobar@javerianacali.edu.co (J. W. Escobar)

profitability of them. The problem of product mixing and lot sizing has been commonly addressed by considering the maximization of profit and the maximization of incomes. On the other hand, some authors have considered the minimization of costs. The decision of lot sizing a mix product is a key success for the worldwide companies. This decision involves the participation of different areas of the company such as sales, supplies and production and marketing. In particular, each area tries to impose its arguments in order to increase or decrease the lot size and to include or eliminate products from the companies´ portfolio achieving greater or lesser amount of finished products or raw materials. We have proposed a Stochastic Mixed Integer Linear Programming Model (SILP) for the product mixing and lot-sizing problem. The decisions of the stochastic model are defined in two steps by considering variability of the demand. It is first proposed as the objective function of the SILP model, the maximization of EVA, and the latter is defined as the economic value added generated by the company. The solution strategy used for the SILP model solution is known as Sample Average Approximation (SAA). This methodology has been proposed by Kleywegt et al. (2002) and uses a scheme of sample averages by Monte Carlo Simulation for stochastic linear programming problems.

The main contribution of this paper is the mathematical construction of the proposed SMILP model by considering the EVA as the objective function. In addition, the paper extends the literature of mathematical modeling applied to the problem of lot sizing and product mixing by considering variability of demand. In particular, we seek to assess the applicability and effectiveness of a stochastic linear model for the considered problem. In Section 2, the literature review of topics related to the paper with stochastic elements is introduced. In Section 3, the proposed mathematical model is presented. The solutions strategy SAA is presented in Section 4. Finally, computational results and conclusions are given in Section 5.

2. Literature Review

2.1 Economic Value Added

Although Marshall (1980) stated the principles of EVA almost 100 years before, this financial measure began to be widely carried out just until the decade of 1980. This financial measure tries to define a relationship between the cost of the investment in the company and its minimum expected profit. The main goal of any company is to generate economic wealth in order to give out the profits, increment the value of shares and reinvest in economic projects. Indeed, the generation of economic value, will result in the wellbeing of society, shareholders and employees. A company generates economic value only if the return over the investment is greater than the cost of capital. On the other hand, if the return on investment is less than the cost of capital, the company is its destroying value. Indeed, the Economic Value Added (EVA) is the product obtained from the difference between profitability of assets and the monetary value of capital required to possess those assets. The EVA is obtained by deducting from the operational income the operational expenses, value of taxes, and the costs of opportunity of capital investment (Lin & Qiao, 2010; Singh & Mehta, 2012). Therefore, the productivity of all elements used to develop the entrepreneurial activity is considered in this measure. In other words, EVA is the result obtained once all the expenses and a minimum profitability expected by the shareholders have been covered (Lin & Qiao, 2010). The EVA can be calculated as follows:

$$EVA = NOPAT - (WACC * NOA) \tag{1}$$
$$NOA = FOA + NOWC \tag{2}$$

where,

- $NOPAT$ corresponds to the Net Operating Profits After Taxes,
- NOA corresponds to the Net Operating Assets,

- *WACC* corresponds to Weighted Average Cost of Capital,
- *FOA* corresponds to the Fixed Operating Assets,
- *NOWC* corresponds to the Net Operating Working Capital.

The Eq. (1) calculates the difference between the return rate of capital and the cost of Weighted Average Capital Cost (WACC), multiplied by the economic value of the invested capital of the concern. NOWC depends on the mean accounts receivable, the mean accounts payable, and the mean inventory. From the Eq. (1), it is possible to infer that a company could have earnings, but they may not be adequate to obtain a positive value of EVA. According to Ray (2001), the importance of EVA in comparison with the generally accepted accounting principles stays in the fact that EVA shows in a better way the relationship between productivity and the financial wellbeing of the company.

2.2 Product-mix planning and Lot Sizing

Many researches have studied deeply the lot-sizing problem. Rizk and Martel (2001) proposed a literature review of related problems with lot-sizing decisions. In this work, it is remarked that nearly all the proposed models are based on the Economic Order Quantity (EOQ) principle or based on the Wagner and Within Model (Dynamic lot sizing). According to those authors, the lot-sizing problem (LP) can be split in two main categories: the Single Facility LP (SLP) and the Multifacility LP (MLP). In addition, both categories could be divided in the following subcategories: single item and multi-item problems, uncapacitated and capacitated problems, and single stage and multistage problems.

Single Facility LP (SLP) problems are the most studied in the literature. In this type of problem, the transportation cost is not considered. In general, this type of problem considers a multi machine or multi-stage facility system. The interdependency between the different facilities in the system are often not considered due to the predominant assumption that the general multifacility problem could be solved by optimizing the costs of each facility independently, and due to the difficulties of solving big size problems in real time (Rizk & Martel, 2001, 2006). Within SLP category, it is possible to find several variants dealing with Uncapacitated Single Item Problem (Rizk & Martel, 2001), Capacitated Single Item Problems (Haksever & Moussourakis, 2008), Continuous Setup Lot-Sizing Problem (Karimi et al. 2003), Discrete Lot-Sizing and Scheduling Problem (Absi & Kedad-Sidhoum, 2008) and (Van Hoesel, et al. 1994), Uncapacitated Multi-item Problems (Van Eijs et al. 1992) and Capacitated Multi-item Problems (Fleischmann, 1990). The Multi-Level Lot-sizing Problem (MLP) considers products that are manufactured through several levels, possibly involving several production/stocking points. The objective of the MLP is to determine a multi-stage production/procurement schedule, which minimizes the total cost, while known demand is fulfilled. There are two types of demand in the MLP: independent demand and dependent demand. Independent demand is activated outside the firm. Dependent demand is triggered by the production/supply required to fulfill the independent demand. Within MLP category, several variants have been proposed related to Uncapacitated Single Item Problems (Rizk & Martel, 2001), Capacitated Single Item Problems (Rizk & Martel, 2001) and Capacitated Multi-Item Problems (Mabin & Gibson, 1998). The different variations of the lot sizing problems use deterministic demand to determine the optimum. The solution for these types of problems could be obtained by different techniques such as branch and bound schemes, ant colony optimization among others. Usually, the objective function is expressed as the minimization of costs or the maximization of earnings or incomes (Alejandro et al. 2013)

2.3 Methodologies for the solution of LP problems

Over the final 50 years, researchers have found many applications of optimization in the study of production planning and scheduling, product mix, lot sizing, and location and transportation problems. Earliest approaches have considered deterministic parameters avoiding the appropriate representation of realistic situations. The evolution of processors has been a relevant fact, which help researchers work with more complex models in order to represent real problems more accurately by including random parameters within mathematical models. The inclusion of variability within LP models has included higher complexity. The complexity grew exponentially, and gave an impulse to new schemes, that served to obtain faster solutions for complex problems. These new methodologies included nature-based heuristics, such as Ant Colony Optimization (ACO), Particle Swarm Optimization (PSO), Artificial Neural Networks (ANN), and Genetic Algorithms (GA). These methodologies are characterized by using one, although random, structured search approximation (Niebles-Atencio & Neira-Rodado, 2016).

The different types of techniques to solve stochastic programming for LP problems are listed as follows: Programming with recourse, stochastic linear programming, stochastic integer programming, stochastic non-linear programming, Robust Stochastic Programming, Probabilistic Programming, Fuzzy Mathematical Programming, Flexible programming, Possibility programming, and Stochastic Dynamic Programming (Sahinidis, 2003; Ahmed et al. 2004; Su & Wong, 2008). It is also important to point that techniques such as Sample Average Approximation (SAA) are very useful to find solutions with stochastic variables (Santoso et al., 2005; Escobar, 2012; Escobar et al., 2012; Escobar, 2009; Escobar et al., 2013; Paz et al., 2015; Mafla & Escobar, 2015).

The proposed SAA scheme used on this paper is based on the work proposed by Kleywegt et al. (2002) and Paz et al. (2015). First, a limited sample of the lot sizing problem and configurations of the mix of products is generated; in which each of these decisions is determined from multiple random generation by using random demand scenarios and corresponding stochastic model solutions. For each scenario, the indicators proposed in Mafla and Escobar, (20015) are calculated to verify the stopping criterion, and then the procedure is repeated until the criterion is met, ensuring the selection of the best configuration.

3. Proposed Mathematical Model

The proposed mathematical model for the product mixing and lot sizing planning has been inspired on a company with a multistage process. Each stage of the process has different number of machines and each machine receives a different operating cost and a different speed. The statistics needed for the problem modeling such as speed of machines, standard processing times, set up times, percentage of defects, maximum sales standard deviation, are recognized. The submitted information corresponds to a four-stage production process company. This company trades its products to retail stores. The company operates with an MTS (make to stock) policy, and makes a monthly production program by placing production orders considering the projected sales and the finished inventory per item.

The following assumptions are considered for the proposed model:

- The shortage is permitted,
- Multi-product and multi-stage production process have been considered,
- Working process is not considered,
- Each product has a different proportion of defective points,
- Each process has a different ratio of defective items,
- The defective items imply a reprocessing time,
- No constraints of space have been considered related to raw material,
- The processing times in each stage and each machine are considered as deterministic,
- The set up times are considered deterministic,
- The demand for each item follows a normal density distribution function.

We have considered a possible portfolio of 64 items. All the info related to the items is known (the expected sales, the standard deviation of the sales, the maximum possible sales, and the monthly stationary index). In addition, all the information related to the process is recognized (the processing times of each item at each machine and the set up time of each machine for each particular item). Moreover, it is relevant to state that in each process, some machines are faster than others. Each of these machines has a different production rate. Indeed, each machine has a different activation cost and is able to develop a different amount of items. This situation obviously affects the financial outcomes of the company. This is a very significant issue for the formulation of the problem due to that it commonly forces to consider a great amount of variables of the type Q_{im} (quantity of product i to be processed on machine m). In order to reduce the size of the problem, the proposed formulation contains average costs and speeds in each stage, which are conformed to some speed and cost correction factors depending on the machine assigned to process the items. In addition, we have considered that the company produces Ready-to-assemble furniture (RTA). Indeed, each furniture is made of a several number of pieces, which could be not assembling in the manufacturing plant. Therefore, it is assumed that a similar amount of work can be assigned to each machine in each place of the process. In the proposed model, the defective items are generated and collected in each stage, and then are reprocessed only at the conclusion of the process. The problem of the product mix planning and lot size decisions is addressed as a problem of maximization of the Economic Value Added of the company (EVA) by seeing the different stages of the process, the annual demand of the different items manufactured by the company, the total costs of the company, and the schedule of the different items. The problem of the product mix and lot size was tackled in two phases. The first phase deals with the product mix problem. The proposed model analyzes the products produced by the company and the costs of each asset. This approach leads to the maximization of the EVA of the company. The decisions of the first phase consider the selection of the products that the company must be produced and the assets (machines) that the company must use to produce the selected items. This first phase of the stochastic model was stated as the maximization of the EVA of the company. In the second phase, the selected products must be scheduled according to monthly plans and employing the selected assets in the first phase. The second phase problem is tackled as the maximization of the company's earnings. The expected results of the second phase must be the amount of product to be produced for each month.

Notation of the proposed Model

Sets

ITEM	Portfolio of products of company, indexed by i
PROCESS	Productive process of company, indexed by j
MACHINES	Available machine in each process, indexed by m
FAMILIES	Groups of items with similar characteristics (tables, libraries, etc) indexed by f

Parameters

TAX	Tax fare [%]
UC_i	Cost per unit of the product i [\$ / unit]
SP_i	Sale price for product i [\$ / unit]
WACC	Weighted average cost of capital [%]
OPC_j	Overall time available in each process in the time horizon T [hrs]
st_{ij}	Standard time of the product i in the process j [time unit]
FOA_{mj}	Value of the asset (machine) m of the process j [\$]
M_j	Number of machines available in process j
dff_{ij}	Mean proportion of defectives generated in process j of item i
FC	Fixed costs of the company [\$]
AC	Administration costs of the company [\$]
PS_i	Potential sales of item i [unit]

Variables

SUT_i	Units of item i considered in the final mix of the portfolio.
AR_i	Accounts receivable associated with the sales of item i [$]
AP_i	Accounts payable associated with the purchase of raw material for item i [$]
$\bar{I}_i$	Mean inventory of raw materials for item i [$]
o_{jm}	Binary variable equal to 1 if the machine m of process j is included on the assets needs to maximize the EVA, 0, otherwise.
SU_{if}	Sold units of item i included in family f
y_{jm}	Activation variable (binary) for the cost FOA_{mj}

The first stage maximization formulation is stated taking into account equation (1) and (2).

The objective function of the first stage is formulated as follows:

$$Z(max) = \left(\sum_{i=1}^{n}(SP_i * SUT_i) - (FC + AC + \sum_{i=1}^{n}(UC_i * SU_i)\right) * (1 - tax\ fare) \tag{3}$$

$$- \left(\sum_{i=1}^{n} AR_i - \sum_{i=1}^{n} AP_i + \left(\sum_{i=1}^{n} \bar{I}_i\right) + \sum_{j=1}^{4}\sum_{m=1}^{M_j} FOA_{mj}\right) * WACC$$

The amount of the accounts receivable and accounts payable depends on the policies of the companies. In particular, we have considered a policy with ar days for the accounts receivable and ap days policy for the accounts payable. ar and ap are given parameters.

The average inventory of raw materials could be considered constant in time and it is determined with respect to the raw materials requirements. In the first stage, we assume that the inventory of finished product is zero, i.e. once a product is manufactured it is immediately delivered to the customer. The accounts receivable is a function of the sales and the rotation (rotation is determined as 6 times, i.e. 360 days / 60 days), assuming that all the sales have been performed by credit. The accounts payable is a function of the expenditures and the rotation, assuming that only materials expenditures have been performed by credit. Another important assumption at this point is that the fixed costs and management costs not vary or its variation is considered negligible. In addition, the total production cost considers a correction factor related with the amount of defectives that the process generates of each item. Additionally, the portion of defective items is not the same for all the items because of their complexity level, which is also the reason to work with different processes and set up times for each particular item.

The mathematical model is subject to the following constraints:

- Capacity of the different processes:

For each process, a capacity constrain is defined by the used machines. For example, in the case of the first process (sawing process) the set of constraint must be defined as:

$$\sum_i (st_{i1} * \sum_f SU_{if})/\prod_f (1 - deff_{i1}) \leq (1,5 * o_{11} + 1,5 * o_{12} + 0,5 * o_{13} + 0,5 * o_{14}) * \frac{OPC_1}{4} \tag{4}$$

This set of constraints ensure that the sum of the time needed for making the amount of the selected items must to be less than or equal to the available time of this process for each year by considering only working days. This set of constrains consider the defectives items. The right hand side of the inequality shows the overall process capacity (OPC), which is divided by the number of machines of the process (4 machines available of the sawing process) to find the capacity of one machine per year. The value of 1,5 that multiplies the variable o_{11} indicates that the machine is 50% faster than the average speed of the process. In this case, machine number 1 of the sawing process, is 50% faster than the mean of the sawing process.

- Proportion of families of portfolio products:

This group of constraints ensures must have participation less than 30% of the sales of the company. This constraint considers that the company has six product families (tables, miscellaneous, entertainment centers, computer centers, armoires, and libraries), and is interested in having a market participation in as many categories as possible. In the case of the libraries:

$$\sum_{f=1} \sum_i SU_{if} \leq 0,3 * \sum_f \sum_i SU_{if} \tag{5}$$

In order to optimize the process is necessary to have a proportion between the armory families and tables of 1:1.5

$$1.5 * \sum_{f=2} \sum_i SU_{if} \geq \sum_{f=3} \sum_i SU_{if} \tag{6}$$

The company has established the maximum sales of each item by using forecast marketing techniques. These potential sales per item give another group of constraints.

$$SU_i \leq PS_i \tag{7}$$

- Activation of shop floor assets.

In the case of machine 1 of sawing process these constraints is formulated as follows:

$$o_{11} \leq 10 * y_{11} \tag{8}$$

At the end of the first stage, the shop floor assets of the company are used and the products and quantities to be produced has been determined. Then, the lot size of these products has to be calculated in order to meet the monthly demand. This decision considers that the company works with a monthly production plan. The general formulation of the second stage (lot size decision) is the following. We have described the additional sets, parameters and variables from the first stage:

Additional sets

ITEM	Portfolio of the selected items (in the first stage) of company, indexed by i
PERIODS	Planning period times, indexed by t

Additional parameters

$Demand_{it}$	Quantity of SU_i to be satisfied by the market in the period t [unit / time]
SC_i	Shortage cost of item i [\$ / unit]
CC_i	Carrying cost of item i [\$ / unit]
$index_{it}$	Stationary index for the sales of item i in period t
SU_i	Quantity of item i that the company sells for the time horizon T. This decision was taken in the first stage. (It becomes a parameter in the second stage)
$FOPC_j$	Final overall capacity of process j [units of time] (With the machines selected in stage 1)
$s0_i$	Initial shortage of item i in period 1[units]
$bl1_i$	Initial inventory of item i in period 1[units]
Storage	Storage capacity of the finished goods warehouse.

Additional variables

FS_{it}	Final shortage of item i in period t [units]
Q_{it}	Quantity of item i that the company schedules in period t [units]
FI_{it}	Final inventory of item i in period t [units]
$setup_{ij}$	Set up time of item i in process j [units of time]
w_{it}	Binary variable equal to 1 if the process set up time for item i in the period t, 0

otherwise.

The objective function is stated as a maximization of the annual profit of the company as follows,

$$Annual\ Profit = \sum_i \sum_t \left([(FS_{i\,t-1} + Demand_{i\,t} - FS_{i\,t}) * SP_i] - \frac{(Q_{i\,t} * UC_i)}{\prod_s (1 - deff_{i\,s})} - (FS_{i\,t} * SC_i) - (FI_i * CC_i) \right) \quad (9)$$

where:

$$Demand_{i\,t} = index_{i\,t} * \frac{SU_i}{12} \qquad \forall\ i, t \quad (10)$$

The incomes and expenditures depend of the items, the monthly demand, the batch size, the level of inventory, and the shortage for each period. Therefore, the monthly profit of item i is obtained by the subtraction of the expenditures of the period from the incomes of the period. The annual profit is obtained by adding the monthly profit of the items; and the total profit is calculated by the sum of the annual profits of each item. The key factors of the second stage have been determined in the previous stage. At the second stage, the variables of decision are related to with the machines to be used, the amount of available time in each process and the annual demand for each item. The demand is converted into monthly horizon affecting its mean value by a stationary index. The proposed model considers shortage; therefore the available units for each period are not enough to satisfy the monthly requirements. In case that in the period t the available product is less than the monthly requirements, the sales in that particular period be equal to the available products. The units of item i sold in month t ($Sales_{it}$) are:

$$Sales_{it} = FS_{i\,t-1} + Demand_{i\,t} - FS_{i\,t} \quad (11)$$

We have defined the $Sales_{it}$ by considering that shortage is allowed. Indeed, the available product for each period t is given by:

$$Q_{i\,t} + FI_{i\,t-1} \quad (12)$$

Moreover, the monthly requirements are given by:

$$FS_{i\,t-1} + Demand_{i\,t} \quad (13)$$

The sales for each period are equal to the available products. In this particular case the value of $FS_{i\,t}$ indicates the additional units that the company must sell, i.e. the lack of inventory. Additionally, if the available units are greater than the requirements, the units sold are the same as the requirements, and the value of $FS_{i\,t}$ is zero. This type of formulation is important also in cases where the products have different sale prices at different periods of the year. This objective function of the second stage is subject to the following constraints:

- Capacity of the different processes:

$$\sum_i \left((setup_{ij} * w_{it}) + (st_{ij} * Q_{i\,t} / \prod (1 - deff_{ij})) \right) \leq \frac{FOPC_j}{12}; \quad \forall\ j \quad (14)$$

This set of constraints control the capacity of each process. In particular, it is only possible to use a maximum amount of time. The machines selected for each process at the first stage give this available time. If a lot of the product i is manufactured for this period, this constraint guarantees that the set up time consumes part of the available time of the stage.

- Production balance:

For the first period this constraints could be expressed as:

$$bl1_i + Q_{i\,1} + FS_{i\,1} = Demand_{i\,1} + FI_{i\,1} + s0_i \quad \forall i \tag{15}$$

In general, for the other periods it is expressed as:

$$FI_{i\,t-1} + Q_{i\,t} + FS_{i\,t} = Demand_{i\,t} + FI_{i\,t} + FS_{i\,t-1} \quad \forall i, t \tag{16}$$

For the period 12, shortage is allowed:

$$FS_{i\,12} \leq 0,2 * Demand_{i\,12} \quad \forall i \tag{17}$$

In particular, the way to reduce the overall capacity of the different processes by the setup time affects the processes capacity at the second stage. Indeed, the sum of Q_{it} is less than the expected sales. There are two ways to handle this:

- Manage the information of the first stage as proportions and include a constraint of the second stage that forces the model to meet this proportion for each product of the portfolio.
- Allow a certain amount of shortage for all items in the last period. Therefore, the model considers the decisions of which product (or products) contributes less to the profit generation and assigns a value of shortage to those products. In particular, for the products that make a greater contribution to profit generation, the sum of Q_{it} meet the expected sales, i.e. the sum of Q_{it}, is less than the expected sales for that item.

- Activation of set up time of item i in period t:
$$Q_{i\,t} \leq 1000000 * w_{it} \quad \forall i, t \tag{18}$$

This constraint controls the assignation of the set up time.

- Space constraint:

$$\sum_i FI_{i\,t} \leq storage \quad \forall t \tag{19}$$

This constraint assures that the space needed to keep the inventory in each period t fits the space available of the finished goods warehouse.

4. Solutions strategy: Sample Average Approximation (SAA)

The proposed stochastic model has been solved by using the Sample Average Approximation (SAA) technique. According to Paz et al. (2015), the SAA algorithm is summarized in 4 steps:

i.   Generate S independent samples of size N each, for $s = 1, ..., S$ for each sample solve the corresponding SAA problem.

$$max_{w \in W} \left\{ c^T w + \frac{1}{N} \sum_{n=1}^{N} Q(w, \xi_s^{\ n}) \right\} \tag{20}$$

For each s, it is possible to obtain the optimum value and the corresponding optimal solution, v_N and $\hat{y}_N$.

ii.   Calculate the following statistical indicators:

$$\bar{v}_{N,S} = \frac{1}{S} \sum_{s=1}^{S} v_N^s \tag{21}$$

$$\sigma_{\bar{v},N,S}^2 = \frac{1}{(S-1)S} \sum_{s=1}^{S} (v_N^s - \bar{v}_{N,S})^2 \tag{22}$$

The value of $\bar{v}_{N,S}$ provides a statistical lower bound of the real optimal($v *$) of the original problem, while Eq. (22) is an estimator of its variance.

iii.   Select a feasible solution $\bar{w} \in W$ of the real problem, using as example the best of solutions $\hat{w}_N^s$ calculated in Eq. (20). Calculate the objective function value of ($\bar{w}$), using Eq. (23).

$$\tilde{f}_{N'}(\bar{w}) = c\bar{w} + \frac{1}{N'} \sum_{n=1}^{N'} Q(\bar{w}, \xi^n) \tag{23}$$

In Eq. (23), $(\xi_j^1, \ldots, \xi_j^{N'})$ is a sample of size N' which is independent of the samples used in step 1. In general, it is usual to take a value N' much greater than N. Since the samples are independent and identically distributed, the variance of Eq. (23) can be expressed as follows:

$$\sigma_{N'}^2(\bar{w}) = \frac{1}{(N'-1)N'} \sum_{j=1}^{N'} \left[c\bar{w} + Q(\bar{w}, \xi^n) - \tilde{f}_{N'}(\bar{w}) \right]^2 \tag{24}$$

In this case, since the problem to be solved is maximization, it is natural to choose $\bar{w}$ with the highest estimated objective function value $\tilde{f}_{N'}(\bar{w})$.

iv.   Calculate an estimate of the optimality gap of solution $\bar{w}$ by using the results obtained in steps 2 and 3 as follows:

$$gap_{S,M,N'}(\bar{w}) = \tilde{f}_{N'}(\bar{w}) - \bar{v}_{N,S} \tag{25}$$

The estimated gap variance is calculated as follows:

$$\sigma_{gap}^2 = \sigma_{N'}^2(\bar{w}) + \sigma_{\bar{v},N,S}^2 \tag{26}$$

The SAA algorithm developed in this study is based on the steps proposed by Paz et al. [25]. This optimization algorithm must to solve initially a sample of M problems of the deterministic model each one with S demand scenarios generated by Monte Carlo Simulation. The indicators of the SAA is calculated for the supply network design for $N' = 3S$. If the stop criteria ($gap \leq 5\%$) is not met, the algorithm is repeated with $M = 2M$ until find an optimal solution.

5. Results Analysis

The optimization process was performed generating 600 scenarios for the first stage and 300 scenarios for the second stage. A branch bound algorithm has been performed for each scenario. These scenarios have been grouped in samples of 20. Then the number of samples performed was 30. The proposed stochastic model has been solved using CPLEX 12.1. After performing some preliminary tests, it became clear that the ideal runtime for CPLEX is 1 hour, due to the change of the objective function was of 0.01%, when increasing the time limit from 1 hour to 2 hours, and from 2 hours to 7 hours of runtime the objective function remained constant.

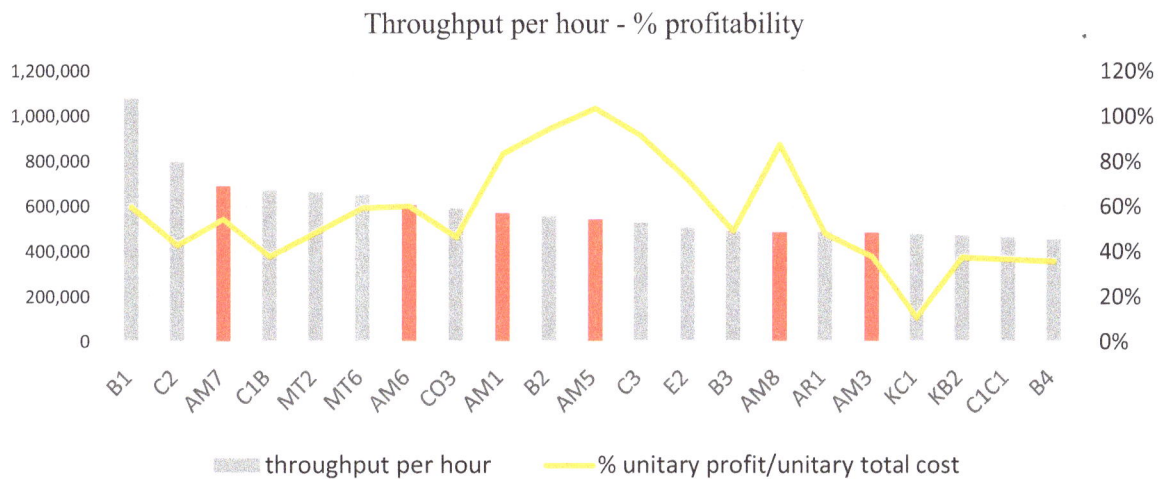

Fig. 1A. Items throughput and profitability, ordered in throughput descending trend

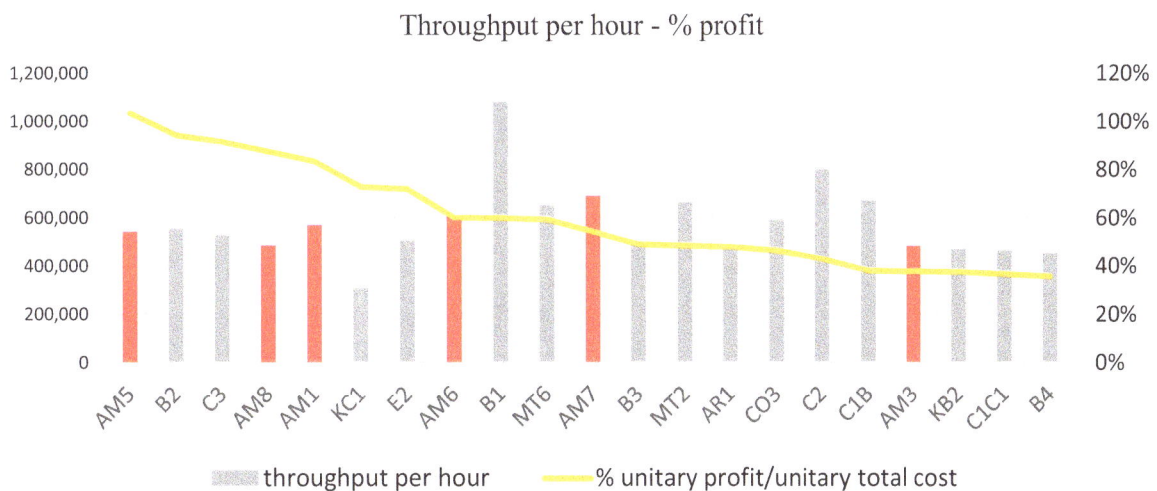

Fig. 1B. Items throughput and profit, ordered in profitability descending trend

The first stage of the optimization indicates the items that contribute to the maximization of the EVA. In addition, the results of the first model allow analyzing the rest of the portfolio of products taking into account the references, which reduce less the EVA of the company.

Nevertheless, the scenario with the higher value of the objective function activates and chooses the machines and items the company should produce. These items and machines are used in the stage 2 (lot sizing). It is very important to analyze the items chosen in this first stage. Commonly the criteria used for the lot size determination are lower cost, higher throughput and higher unitary profitability (% unitary profit/unitary total cost). In this work, the proposed objective function couldn´t found any dominance of these criteria.

Fig. 1A and Fig. 1B show the throughput of products and the profitability for some items respectively. Fig. 1A shows the items in a descending trend of throughput per hour, while Fig. 1B shows the same items of Fig. 1A considering unitary profitability (% unitary profit/unitary total cost) ordering them in a descending trend. This is an important result because shows a different way to select the mix of products for the company. On Figs. 1A and 1B items included in the armoire product family are highlighted in red. On Fig. 1A it can be observed that the armoires with the higher throughput are AM6, AM7 and AM1. And the armoires with the higher profitability (Fig. 1B) are AM5, AM8 and AM1. Nevertheless the selected armoires in the product selection are AM6, AM7, and AM8; because these are the products that contribute in a better way to EVA generation. Similar behavior can be observed among the other product families.

In particular, we have used the EVA as criterion, which is not commonly used. The EVA considers a holistic point of view considering costs, productivity and profitability. The constraint that forces the company not concentrate its production and sales in a few items, also forces to the model to activate some other wildcard items depending on the variability for each scenario.

Table 1
Results of the proposed stochastic model. **Source:** Owner

	Results of the Proposed Model			
	Final Inventory	Batch Size	Number of batches	Shortage
	Mean Value (Monthly)	Mean Value	(Year)	Mean (Monthly)
AM6	252.16	2179.6	9.84	3.75
AM7	308.63	1863.78	8.56	3.58
AM8	205.22	990.6	7.23	18.79
B1	699.04	2507.43	4.35	56.77
B2	362.81	1979.69	5.64	95.77
B3	708.36	3416.51	3.98	238.84
B4	552.89	3180.83	4.06	269.75
C2	480.47	2324.05	2.67	132.05
CO3	578	3754.08	4.24	348.72
CT1	337.43	1920.88	3.35	195.05
CV2	87.76	1010.73	4.08	265.46
E2	57.73	1184.86	11.01	91.09
MT1	214.68	5245.3	5.53	1537.4
MT2	312.09	1705.57	7.85	57.29
MT4	327.71	1615.93	2.01	146.96
MT6	367.33	2516.46	5.56	242.41
KB2	661.38	3650.08	4.44	264.03

Table 1 shows a summary for the results of the 300 scenarios performed. On this table a comparison between selected items AM6 and CT1 can be performed. AM6 is scheduled an average of 9.84 times in a year. On the other hand CT1 is scheduled only 3.35 times per year. This happens because AM6 is a very important item for EVA generation and additionally it has a high sale price and high unitary cost. Scheduling this valuable items many times the solver tries to minimize their inventory and shortage cost. This issue can be noticed also when checking the mean shortage and the mean inventory of the two references. CT1 is not so important for EVA generation (wildcard item) therefore falling into carrying and shortage cost costs in these references is not so critical than falling into these costs with EVA relevant items such as AM6. The proposed model can be applied both for MTS (make to stock) and MTO (make to order) systems by manipulating the shortage cost, the carrying cost, and variables such as shortage, final inventory and warehouse available space. Another important issue is that the maximization of the EVA as the objective function, allows the decision maker to involve costs, earnings, profits and productivity together.

6. Concluding remarks and future works

In this study, we propose a Stochastic Mixed Integer Linear Programming Model (SILP) for the product mixing and lot-sizing problem. The decisions of the stochastic model are defined in two steps by considering variability of the demand. It is proposed as objective function of the SILP model, the maximization of EVA, the latter is defined as the economic value added generated by the company. The solution strategy used for the SILP model solution is known as Sample Average Approximation (SAA). The results of the application of this methodology show that the proposed approach is an effective decision support tool of product mix and lot sizing problems by considering the expected economic contribution of products and variability of the demand. In addition, we have demonstrated that the methodology SAA provides near-optimal solutions for linear stochastic programming problems with small sample sizes on production problems. Dynamic aspects of the problem, which help to make decisions according to the demand seasonality and the variability of the prices, could be considered as future research work. In addition, decisions of scheduling of products and decisions of space constraints for working process must be taking into account. Finally, applicability of other methods for solving stochastic linear optimization models and other real cases such as companies with different production systems could be considered as extensions of this paper.

References

Absi, N., & Kedad-Sidhoum, S. (2008). The multi-item capacitated lot-sizing problem with setup times and shortage costs. *European Journal of Operational Research, 185*(3), 1351-1374.

Ahmed, S., Tawarmalani, M., & Sahinidis, N. V. (2004). A finite branch-and-bound algorithm for two-stage stochastic integer programs. *Mathematical Programming, 100*(2), 355-377.

Alejandro, G. H. J.W Escobar, & Álvaro, F. C. (2013). A Multi-Product Lot-Sizing Model for a Manufacturing Company. *Ingeniería Investigación y Tecnología, 14*(3), 413-419.

Azaron, A., Tang, O., & Tavakkoli-Moghaddam, R. (2009). Dynamic lot sizing problem with continuous-time Markovian production cost. *International Journal of Production Economics, 120*(2), 607-612.

Escobar J.W. (2012). Rediseño de una red de distribución con variabilidad de demanda usando la metodología de escenarios. *Revista Facultad de Ingeniería UPTC, 21*(32), 9-19.

Escobar, J. W. (2009). Modelación y Optimización de diseño de redes de distribución de productos de consumo masivo con elementos estocásticos. *Proceedings of XIV Latin American Summer Workshop on Operations Research (ELAVIO), El Fuerte, México.*

Escobar J.W., Bravo, J.J., & Vidal, C.J. (2012). Optimización de redes de distribución de productos de consumo masivo en condiciones de riesgo. *Proceedings of XXXIII Congreso Nacional de Estadística e Investigación Operativa (SEIO), Madrid, Spain.*

Escobar, J. W., Bravo, J. J., & Vidal, C. J. (2013). Optimización de una red de distribución con parámetros estocásticos usando la metodología de aproximación por promedios muéstrales. *Ingeniería y Desarrollo, 31*(1), 135 –160.

Fleischmann, B. (1990). The discrete lot-sizing and scheduling problem. *European Journal of Operational Research, 44*(3), 337-348.

Haksever, C., & Moussourakis, J. (2008). Determining order quantities in multi-product inventory systems subject to multiple constraints and incremental discounts. *European Journal of Operational Research, 184*(3), 930-945.

Karimi, B., Ghomi, S. F., & Wilson, J. M. (2003). The capacitated lot-sizing problem: a review of models and algorithms. *Omega, 31*(5), 365-378.

Kleywegt, A. J., Shapiro, A., & Homem-de-Mello, T. (2002). The sample average approximation method for stochastic discrete optimization. *SIAM Journal on Optimization, 12*(2), 479-502.

Lin, C. H. E. N., & QIAO, Z. L. (2010). What influence the Company's Economic Value Added? Empirical Evidence from China's Securities Market. *Management Science and Engineering, 2*(1), 67-76.

Mabin, V. J., & Gibson, J. (1998). Synergies from spreadsheet LP used with the theory of constraints—a case study. *Journal of the Operational Research Society, 49*(9), 918-927.

Mafla I., Escobar J.W., (2015). Rediseño de una red de distribución para un grupo de empresas que pertenecen a un holding multinacional considerando variabilidad de la demanda. *Revista de la Facultad de Ingeniería U.C.V., 30*(1), 37 – 48.

Marshall, A. (1890). *Principles of Political Economy*. Maxmillan, New York.

Niebles-Atencio, F. N., & Niera-Rodado, D. N. (2016). A Sule´s method initiated genetic algorithm for solving QAP formulation in facility layout design: A real world application. *Journal of Theoretical and Applied Information Technology, 84*(2), 157.

Paz, J. C., Orozco, J. A., Salinas, J. M., Buriticá, N. C., & Escobar, J. W. (2015). Redesign of a supply network by considering stochastic demand. *International Journal of Industrial Engineering Computations, 6*(4), 521 – 538.

Ray, R. (2001). Economic value added: Theory, evidence, a missing link. *Review of Business, 22*(1/2), 66.

Rizk, N., & Alain, M. (2001). Supply chain flow planning methods: a review of the lot-sizing literature. Canada: Universite Laval Canada and Centor. Centre de recherche sur les technologies de l'organisation réseau (CENTOR)

Rizk, N., Martel, A., & Ramudhin, A. (2006). A Lagrangean relaxation algorithm for multi-item lot-sizing problems with joint piecewise linear resource costs. *International Journal of Production Economics, 102*(2), 344-357.

Santoso, T., Ahmed, S., Goetschalckx, M., & Shapiro, A. (2005). A stochastic programming approach for supply chain network design under uncertainty. *European Journal of Operational Research, 167*(1), 96-115.

Singh, T., & Mehta, S. (2012). EVA VS Traditional Accounting Measures: A Pre-Recession Case Study of Selected IT Companies. *International Journal of Marketing and Technology, 2*(6), 95-120.

Su, C. T., & Wong, J. T. (2008). Design of a replenishment system for a stochastic dynamic production/forecast lot-sizing problem under bullwhip effect. *Expert Systems with Applications, 34*(1), 173-180.

Van Eijs, M. J. G., Heuts, R. M. J., & Kleijnen, J. P. C. (1992). Analysis and comparison of two strategies for multi-item inventory systems with joint replenishment costs. *European Journal of Operational Research, 59*(3), 405-412.

Van Hoesel, S., Kuik, R., Salomon, M., & Van Wassenhove, L. N. (1994). The single-item discrete lot sizing and scheduling problem: optimization by linear and dynamic programming. *Discrete Applied Mathematics, 48*(3), 289-303.

5

Comments on "A note on multi-objective improved teaching-learning based optimization algorithm (MO-ITLBO)"

Dhiraj P. Rai[a]*

[a]*Sardar Vallabhbhai National Institute of Technology, Surat – 395007, India*

CHRONICLE	ABSTRACT
	A note published by Chinta et al. (2016) [Chinta, S., Kommadath, R. & Kotecha, P. (2016) A note on multi-objective improved teaching–learning based optimization algorithm (MO-ITLBO). *Information Science*, 373, 337-350.] reported some impediments in implementation of MO-ITLBO algorithm. However, it is observed that their comments are based on incorrect understanding of TLBO, ITLBO and MO-ITLBO algorithms. Their raised issues are thoroughly addressed in this paper and it is proved that MO-ITLBO algorithm has no lacunae.
Keywords: *Multi-objective optimization* *Teaching-learning based* *optimization* *MO-ITLBO*	

1. Introduction

The real world optimization problems consist of many objectives that need to be optimized, simultaneously. Unlike single objective optimization problems (SOOPs) there is no single solution to multi-objective optimization problems (MOOPs). Therefore, in MOOPs the attempt is always towards finding the set of Pareto-optimal solutions. Obtaining the Pareto-optimal set of solutions is theoretically a challenging task and, therefore, has attracted the attention of many optimization researchers.

A number of population based metaheuristic multi-objective optimization algorithms have been proposed by researchers (Zhou et al., 2011). However, a major issue in implementation of population based metaheuristic optimization algorithms is the tuning of common control parameters and algorithm-specific parameters which increases the burden on the user. Teaching-learning-based optimization (TLBO) algorithm was proposed by Rao et al. (2011) as a algorithm-specific parameter-less algorithm. The TLBO algorithm has been applied widely by the researchers in various disciplines of engineering (Rao, 2016a; Rao, 2016b).

* Corresponding author
E-mail: dhiraj.p.rai@gmail.com (D. P. Rai)

Different multi-objective versions of TLBO algorithm have been also developed by researchers (Li et al., 2014; Zou et al., 2013; Medina et al., 2014; Yu et al., 2015; Sultana & Roy, 2014, Rao et al., 2016). Rao and Patel (2014) applied the multi-objective improved teaching-learning based optimization (MO-ITLBO) algorithm to solve the multi-objective benchmark functions of CEC 2009 and showed its effectiveness. Later, Patel and Savsani (2016) proposed the same work but with the addition of Friedman's rank test.

Recently, Chinta et al. (2016) published a note on the MO-ITLBO algorithm by Rao and Patel (2014) published in *International Journal of Industrial Engineering Computations* and Patel and Savsani (2016) in *Information Sciences*. The authors feel that there are impediments in the implementation of MO-ITLBO algorithm. However, it is observed that the issues raised by Chinta et al. (2016) are evolved out of misunderstandings regarding the working of MO-ITLBO algorithm. Chinta et al. (2016) raised many questions on the work of Rao and Patel (2014) and Patel and Savsani (2016). However, it seems that neither Rao nor Patel was involved in the review work of Chinta et al. (2016) nor their opinions were sought. This paper aims to address the note of Chinta et al. (2016) thoroughly and to prove that MO-ITLBO algorithm has no lacunae.

2. Comments on the note of Chinta et al. (2016)

The comments on the note of Chinta et al. (2016) on various steps of MO-ITLBO algorithm such as selection of teacher; assigning learners to teachers; teacher phase; learner phase; external archive; adaptive teaching factor and exploration factor; removal of duplicate solution; updating population members; computational time; number of variables; random seed; runs and plots; code and notations and schematic diagram of MO-ITLBO algorithm are addressed in the following sub-sections.

2.1 Selection of teachers

This step includes determination of (i) the chief teacher and (ii) the teacher for each group. Chinta et al. (2016) reported difficulties in the implementation of these two steps due to misunderstanding of the working of MO-ITLBO algorithm. The replies to the note of Chinta et al. (2016) on the working of MO-ITLBO algorithm are as follows.

2.1.1 Determination of chief teacher

In MOOPs, as a solution improves in terms of one objective it may deteriorate in terms of the other objectives, if the objectives are mutually conflicting in nature. In the case of MO-ITLBO algorithm, a user can select any solution which is best in terms of any objective chosen randomly out of the multiple objectives as the 'chief teacher' of the class. If a particular solution is the best with respect to one objective, either of the three cases are possible (1) the solution may be best with respect to other objective(s) also or (2) the solution may be poor with respect to other objective(s) or (3) the solution may be the best with respect to some objectives and poor with respect to some other objectives. In either of the three cases the particular solution is capable of improving the result of the class in terms of whichever objective it is superior. Therefore, the 'chief teacher' can be easily decided by considering the value of objective function of the solutions in the population in any one of the objective chosen randomly out of the multiple objectives.

It is implicit that this approach of deciding the 'chief teacher' by considering the value of the objective function of the solutions in any one of the objective chosen randomly out of the multiple objectives ensures good diversity, as all the solutions in the class get a fair chance of improving in different objectives in every iteration. The same concept was used by Rao and Patel (2014) and Patel and Savsani (2016) for selection of teacher, and therefore, it is not required to explicitly specify a separate strategy for selection of teacher in the MO-ITLBO algorithm. The statement of Chinta et al. (2016) that *"Under these circumstances, the article does not provide any strategy to determine the best solution which is crucial for the working of the algorithm"* is not meaningful. In general, it is adequate to say that one of the solutions which is the best with respect to any of the objectives chosen randomly out of the multiple objectives can be treated as the 'chief teacher' of the class.

Furthermore, the strategy 1A.1 suggested by Chinta et al. (2016) is completely different from the approach used by Rao and Patel (2014) and Patel and Savsani (2016). The strategy 1A.2 proposed by Chinta et al. (2016) for the selection of 'chief teacher', is not new and is the same as the approach used Rao and Patel (2014) and Patel and Savsani (2016). In strategy 1A.3 Chinta et al. (2016) mentioned that *"One of the anonymous reviewers of this article has suggested another strategy....".* This statement shows that strategy 1A.3 is suggested by an anonymous reviewer and is not an original idea of Chinta et al. (2016). Also strategy 1A.3 is completely different from the approach used by Rao and Patel (2014) and Patel and Savsani (2016).

2.1.2 Determination of group of teachers

Rao and Patel (2014) had given a pseudo-code for 'determination of group teachers' in Fig. 1. In Fig. 1, the term $f(.)$ is written in general sense and $f(.)$ may represent any objective out of the multiple objectives. However, to maintain consistency, it is implicit that $f(.)$ in Fig. 1 of Rao and Patel (2014) represents the same objective which was previously chosen randomly out of the multiple objectives for determination of the 'chief teacher' in section *2.1.1*. In this context, the phrase 'closeness of the solutions' can only mean closeness of objective function value of the solutions corresponding to the particular objective which is represented by $f(.)$. There is no need of any explicit explanation on 'closeness of the solutions'. Therefore the statement of Chinta et al. (2016) that *"the article does not explicitly provide any strategy to determine 'closeness' in the context of multiple objectives"* is incorrect.

Assuming that the user of MO-ITLBO is well aware about the preliminaries of multi-objective optimization, Rao and Patel (2014) and Patel and Savsani (2016) had implicitly specified the strategy for closeness measure. Furthermore, Chinta et al. (2016) described two strategies (1B.1 and 1B.2) for forming groups and selection of teachers for the groups. These two approaches suggested by Chinta et al. (2016) in their paper are completely different from the approach of Rao and Patel (2014) and Patel and Savsani (2016).

2.2 Assigning learners to teachers

In the pseudo-code given by Rao and Patel (2014) in Fig. 1 for selection of teachers of the groups and distributing students to different groups the term $f(.)$ was written in general sense and $f(.)$ may represent any objective out of the multiple objectives. However, to maintain consistency, it is implicit that $f(.)$ in Fig. 1 of Rao and Patel (2014) represents the same objective which was previously chosen randomly out of the multiple objectives for determination of the 'chief teacher' in section *2.1.1*.

Now, based on the objective function value $f(.)$ the selection of teachers for the groups and distribution of solutions to these groups can be done following the pseudo-code given by Rao and Patel (2014). Therefore, pseudo-code provided by Rao and Patel (2014) is self-explanatory and there is no need to provide any explicit strategy for selection of teachers for groups and assignment of students to various groups in the context of MOOPs. Thus, the statement of Chinta et al. (2016) that *"Despite the critical nature of this step, the article does not provide any details about the strategy used for assignment of students to various groups in the context of MOOP"* is incorrect. The approach of Rao and Patel (2014) and Patel and Savsani (2016) for selecting the teachers of the groups and assigning solutions to these groups is correct and undisputable in the context of MOOPs.

2.3 Teacher phase

The teacher phase of MO-ITLBO algorithm involves (i) Calculating the mean result of each group of learners in each subject (ii) Utilizing the adaptive teaching factor for calculating the difference mean of that group (iii) updating the learners according to *learning through tutorial phase*. The comments of Chinta et al. (2016) regarding the implementation of *teacher phase* of MO-ITLBO algorithm are addressed as follows.

2.3.1 Modifying student

In the MO-ITLBO algorithm the solutions are modified according to the '*learning through tutorial*' phase using Eq. (4a) and Eq. (4b) given by Rao and Patel (2014) or Eq. (6a) and Eq. (6b) given by Patel and Savsani (2016). In the case of two competing solutions $X_{j,i}^{h}$ and $X_{j,i}^{k}$ a decision is required to be made as to which solution will share knowledge with the other. In single objective optimization the solution which has a better objective function value shares knowledge with the solution with poor objective function value. The same is reflected in the above equation. The above equation can also be applied to MOOPs, as it is, because in the case of MOOPs also if a solution $X_{j,i}^{h}$ is better than a solution $X_{j,i}^{k}$ in terms of any one objective $f(.)$, irrespective of the other objective(s), then solution $X_{j,i}^{h}$ is capable of sharing knowledge with solution $X_{j,i}^{k}$ because solution $X_{j,i}^{h}$ may improve solution $X_{j,i}^{k}$ in terms of $f(.)$ irrespective of whether solution $X_{j,i}^{h}$ is superior, inferior or equally good with respect to solution $X_{j,i}^{h}$ in terms of the remaining objectives.

The Eq. (4a) and Eq. (4b) are written by Rao and Patel (2014) in general terms and term $f(.)$ means any one objective chosen randomly out of the multiple objectives. However, the solutions $X_{j,i}^{h}$ and $X_{j,i}^{k}$ may be compared by considering the objective function value of any one objective and it is implicit that it should be chosen randomly from among the multiple objectives to ensure diversity. It is clear that the Eq. (4a) and Eq. (4b) of Rao and Patel (2014) are very much applicable to the multi-objective optimization scenario. Therefore, the statement of Chinta et al. (2016) that *"In context of MOOP, it is possible that the student h and the student k are non-dominated and thus it is not clear as to which of the above mathematical expression would be applicable in such a scenario"* is not correct. Furthermore, Chinta et al. (2016) suggested two strategies (3A.1 and 3A.2) for modifying the solutions in teacher phase. However, both the strategies (3A.1 and 3A.2) suggested by Chinta et al. (2016) in their paper are very different from the approach used by Rao and Patel (2014) and Patel and Savsani (2016).

2.3.2 Updating student

Chinta et al. (2016) stated that *"One of the anonymous reviewers have pointed out that in multi-objective perspective, 'if the result has improved' can be interpreted as a solution that is 'totally dominating', i.e., a solution which is better in all objectives. Hence it is necessary to clarify whether 'if the results have improved' corresponds to a non-dominated solution or a totally dominating solution"*.

If X^{mod} is the modified solution and X^{k} is the old solution, then the phrase '*if the result has improved*' can be interpreted as X^{mod} has improved at least in terms of one objective. From the basic concept of dominance it can be inferred either X^{mod} totally dominates X^{old}, or X^{mod} and X^{old} both are non-dominated with respect to each other. In such case, following the philosophy of TLBO algorithm, it is obvious that, if X^{mod} totally dominates X^{old} then X^{mod} replaces X^{old}. Furthermore, Rao and Patel (2014) mentioned that *"The MO-ITLBO algorithm uses a fixed size archive to maintain the good solutions obtained in every iteration. The ε- dominance method is used to maintain the archive (Deb et al. (2005))"*. Therefore it is clear that, in a scenario where X^{mod} and X^{old} both are non-dominated, then *"population acceptance procedure*" and "*archive acceptance procedure*" suggested by Deb et al. (2005) is followed by Rao and Patel (2014) and Patel and Savsani (2016). As the paper (Deb et al., 2005) was well cited by Rao and Patel (2014) and Patel and Savsani (2016) and there is no ambiguity in this regard.

The points mentioned in the above discussion are basic and explicit discussion on such points is not necessary. Therefore, the statement of Chinta et al. (2016) that *"However, the article does not discuss as to how MO-ITLBO handles a scenario in which X^{mod} and X^k are non-dominated solutions"* is not meaningful.

2.3.3 Adaptive factor

In the basic TLBO algorithm the teaching factor (T_F) is chosen heuristically either 1 or 2 which mimics the scenario that a student is either unable to learn anything from the teacher or learns everything from the teacher. In actual teaching-learning scenario the teaching factor will not be always at its extreme values (i.e. 1 or 2) but varies in between (1 and 2) also in a continuous manner. This means that the teaching factor can also be a decimal fraction between 1 and 2. The students may learn in any proportion from the teacher. Therefore, in the ITLBO algorithm (Rao & Patel, 2013) and MO-ITLBO algorithm (Rao & Patel, 2014; Patel & Savsani, 2016) the value of teaching factor is fixed adaptively, in every iteration, using Eq. (1a) and Eq. (1b).

$$\left(T_F\right)_{s,i} = \left(\frac{f(X^k)}{T_s}\right)_i \qquad if\ T_s \neq 0 \qquad (1a)$$

$$\left(T_F\right)_i = 1 \qquad if\ T_s = 0 \qquad (1b)$$

where $f(X^k)$ is the result of any learner k associated with the group 's' and T_s is the result of the teacher of the same group at iteration i.

In a multi-objective scenario, it is implicit that in the MO-ITLBO algorithm, the objective function value $f(X^k)$ corresponding to the objective which was previously chosen for identifying the teachers of the groups must be used to determine the teaching factor T_F to maintain consistency. Therefore, there is no need to separately specify the objective function value that should be used to determine the teaching factor in multi-objective scenarios. Thus, the statement of Chinta et al. (2016) that *"The article does not clearly specify the objective function value that should be used to determine the teaching factor"* has no meaning.

2.3.4 Updating teacher

It is very clear from the Fig. 2 and Appendix-A provided by Rao and Patel (2014) that selection of 'chief teacher' and selection of 'other teachers' (i.e. teachers for the groups) is performed before the commencement of the teacher phase. After the solutions are modified in the teacher phase all the groups are updated by considering the newly modified values of the solutions, this step is clearly shown in Appendix A provided by Rao and Patel (2014). Since all the groups are updated after the teacher phase it is obvious that the teachers for the updated groups have to be freshly selected. The approach used by Rao and Patel (2014) for selection of teachers for the groups is implicit and has already been discussed in section *2.1.2*. Therefore, the statements of Chinta et al. (2016) that *"The article does not clearly describe whether the modified member (who might be superior to all others in the group at the end of the teacher phase) becomes the new teacher for the learner phase of the group or does the current teacher continue to be the teacher for the learner phase"* is not meaningful.

2.4 Learner phase

In the MO-ITLBO algorithm the solutions are modified according to the '*Self-motivated learning*' phase using the following equation given by Rao and Patel (2014) and Patel and Savsani (2016).

$$X'^p_{j,i} = \left\lfloor X^p_{j,i} + r_i \left(X^p_{j,i} - X^q_{j,i} \right) \right\rceil + \left\rfloor r_i \left(X^s_{j,i} - E_F X^p_{j,i} \right) \right\rceil, \qquad if \; f(X^p) < f(X^q) \tag{2a}$$

$$X'^p_{j,i} = \left[X^p_{j,i} + r_i \left(X^q_{j,i} - X^p_{j,i} \right) \right] + \left[r_i \left(X^s_{j,i} - E_F X^p_{j,i} \right) \right]. \qquad if \; f(X^q) < f(X^p) \tag{2b}$$

It is obvious that Eq. (2a) and Eq. (2b) mentioned above are applicable even if solution p and solution q are non-dominated with respect to each other because if solution p is non-dominated with respect to solution q, then it implies that solution p is better than solution q at least in terms of one objective, assuming that solution p and solution q are not the duplicate solutions. Therefore, solution p can share its knowledge with solution q with respect to the particular objective(s) in which it is better than solution q which may improve solution q in those particular objective(s). The same is true if solution q is non-dominated with respect to solution p. Thus, there is no need to explicitly specify any strategy to handle multi-objective optimization scenarios. Therefore, the statement of Chinta et al. (2016) that *"However, the article does not provide any strategy to handle such circumstances"* is not meaningful.

Chinta et al. (2016) provided two strategies (4.1 and 4.2) to handle MOOPs. However, both the strategies suggested by Chinta et al. (2016) are very different from the approach used by Rao and Patel (2014) and Patel and Savsani (2016). Furthermore, Chinta et al. (2016) did not provide any logical explanation as to how strategies (4.1) and (4.2) can effectively improve the quality of the Pareto-front and the performance of the algorithm.

It has already been proved in section *2.3.2* that the strategy suggested by Rao and Patel (2014) and Patel and Savsani (2016) to handle newly modified solution is effective in the multi-objective optimization scenarios and does not suffer from any drawback. Therefore, the statement of Chinta et al. (2016) that *"However, this strategy also suffers from the same drawbacks that had been explained for updating the students in the teacher phase"* is incorrect.

2.5 External archive

It is clear that a user can select the value of ε based on their own discretion following the guideline given by Deb et al. (2005). In their work, the aim of Rao and Patel (2014) and Patel and Savsani (2016) was to show the effectiveness of MO-ITLBO algorithm in solving MOOPs and encourage other researchers to apply MO-ITLBO algorithm to their respective MOOPs. A value of ε which is suitable for the MOOPs considered by Rao and Patel (2014) and Patel and Savsani (2016) may not be suitable for other MOOPs. Researchers can set the value of ε by means of trial and error and try to identify the value of ε which best suits their MOOP.

Chinta et al. (2016) could have used the same procedure in order to set the value of ε in solving the MOOPs of CEC 2009. Finally, it can be said that the absence of exact value of ε used by Rao and Patel (2014) and Patel and Savsani (2016) does not stop the users from applying MO-ITLBO to their own MOOPs because the methodology to select the value of ε is well documented in the literature and the relevant paper was also cited by Rao and Patel (2014) and Patel and Savsani (2016).

2.6 Adaptive teaching factor and exploration factor

MO-ITLBO algorithm is a metaheuristic optimization algorithm. The algorithm automatically selects the value of E_F randomly either 1 or 2. Selection of E_F in MO-ITLBO algorithm is a heuristic step and requires no user intervention. Therefore, in order to preserve the heuristic nature of the algorithm it is not reasonable to convert each and every heuristic step into an adaptive step. Thus, Rao and Patel (2014) and Patel and Savsani (2016) allowed the algorithm to randomly choose the value of E_F either 1 or 2. Therefore, the statement of Chinta et al. (2016) that *"...the articles do not provide any reason for fixing the value of the exploration factor using $(E_F = round (1+r_i))$"* is not meaningful.

Furthermore, Rao and Patel (2014) and Patel and Savsani (2016) provided one application by selecting the E_F randomly either 1 or 2. However, if someone wants to determine the value of E_F adaptively, then he/she may do so and check its effect on the performance of the algorithm.

2.7   Removal of duplicate solutions

Rao and Patel (2014) clearly stated that *"The total number of function evaluations in the proposed algorithm is = {(2 × population size × number of generations) + (function evaluations required for the duplicate elimination)}"*.

It is absolutely clear from the above equation that a complete duplicate removal process was employed by Rao and Patel (2014). Further, Rao and Patel (2014) explicitly took into account the number of function evaluations spent by MO-ITLBO algorithm for duplicate elimination in the above equation given for calculating the maximum objective function evaluations required by MO-ITLBO algorithm. In the opinion of Mernik et al. (2015) such an approach for deciding the 'function evaluations required for duplicate removal' is imprecise, seems to be biased and invalid. However, previous researchers had used algorithms such as GA, SA, PSO, ACO, ABC, DE, ES, NSGA, SPEA, NSGA-II, VEGA, etc. but did not mentioned about function evaluations required for duplicate removal by those algorithms for solving optimization problems. In fact, the previous researchers had not even considered the concept of function evaluations required for duplicate elimination. Actually, there may not be any such need (Rao, 2016b).

Mernik et al. (2015) attempted to invalidate the approach followed by Rao and Patel (2012) for computing the function evaluations required for duplicate removal. However, Mernik et al. (2015) themselves did not compute any function evaluations required for duplicate elimination while describing certain misconceptions in comparing different versions of ABC algorithm. In general, the function evaluations required by well-known algorithms is calculated as (population size × no. of generations), however, the function evaluations required by TLBO algorithm is calculated as (2 × population size × no. of generations).

Therefore, the statement of Chinta et al. (2016) that *"As the duplicate removal step requires evaluation of objective function which in turn governs the termination of the algorithm, it is necessary to explicitly know whether the duplicate removal step is included in MO-ITLBO algorithm"* raises some questions regarding their understanding of the working of metaheuristic optimization algorithms, in general, and their understanding of MO-ITLBO algorithm, in particular.

2.8   Updating population members

The flowchart for MO-ITLBO in Fig. 2 given by Rao and Patel (2014) and the demonstration of TLBO and ITLBO algorithms in Appendix A given by Rao and Patel (2014) clearly depicts that the complete class of learners first undergoes teacher phase and subsequently the entire class of learners undergoes learner phase, therefore the reporting of MO-ITLBO algorithm in Rao and Patel (2014) and Patel and Savsani (2016) and actual implementation of MO-ITLBO algorithm are consistent and free from any discrepancies. Thus the statement of Chinta et al. (2016) that *"This issue also exists in MO-ITLBO algorithm and it is not explicitly specified if a member of the group undergoes teacher phase followed by the learner phase or if the entire group completes the teacher phase before undergoing the learner phase"* is not correct.

2.9 Computational time

The steps such as number of teachers concept, adaptive teaching factor, tutorial training and self-motivated learning are introduced in the ITLBO and MO-ITLBO algorithms in order to enhance the exploration and exploitation capabilities of the algorithm (Rao and Patel, 2014) as compared to the basic TLBO algorithm.

Undoubtedly, the inclusion of these new steps will increase the computational effort and computational time required by the algorithm per iteration. However, with the improved exploration and exploitation capability the ITLBO algorithm and MO-ITLBO algorithm may require fewer number of iterations to search the space and may converge at the global optima within a few number of function evaluations. This may reduce the overall computational effort and the overall computational time required by the ITLBO algorithm and MO-ITLBO algorithm. Thus the statement of Chinta et al. (2016) that *".... the amount of computational time required in ITLBO algorithm could be higher than the base TLBO"* may not be true.

In general, the optimization researchers calculate the number of objective function evaluations required by any algorithm as $N_G \times N_c$, where N_G is the number of generations and N_c is the class size. The number of function evaluations required by TLBO algorithm is calculated as $2 \times N_G \times N_c$ (Rao, 2016b). However, Chinta et al. (2016) calculated the objective function evaluations required by TLBO and ITLBO algorithms as $N_c \times (1+2 \times N_G)$ which is unusual, if not incorrect.

2.10 Number of variables

It was clearly mentioned by Rao and Patel (2014) that the computational challenge provided by CEC 2009 competition had been considered in their work. CEC 2009 competition requires the participants to demonstrate the performance of their algorithm with 30 variables. Rao and Patel (2014) had already mentioned that *"The detailed mathematical formulations of the considered test functions are given in Zhang et al. (2009)"*. In that mathematical expression it is clearly mentioned that n = 30 (i.e. number of variables equal to 30).

Furthermore, Rao and Patel (2014) and Patel and Savsani (2016) compared the results of MO-ITLBO algorithm with a number of other algorithms which had considered the same challenge provided by CEC 2009. Therefore, it is clear that the number of variables used by Rao and Patel (2014) and Patel and Savsani (2016) was the same as that defined by CEC 2009. Thus the statement of Chinta et al. (2016) that *"The article does not seem to specify the number of variables used in the benchmark problems"* is incorrect.

2.11 Random seed

In the metaheuristic optimization algorithms it is unhealthy to use the same set of random numbers or a predefined set of random numbers in every simulation run of the algorithm. Using a constant random seed results in selection of random numbers from the same random number series and the same sequence of random numbers is followed. This may increase the tendency of the algorithm to converge at the same Pareto-front in different independent runs.

It is obvious that a researcher would not tend to use a fixed or predefined value of random seed. The random seed itself should be chosen randomly in every simulation run of the algorithm in order to allow the algorithm to work liberally and converge at the true Pareto-front in a probabilistic manner. Therefore, there must not be any user intervention as far as selection of random seed is concerned.

Hence, in order to allow liberal functioning of MO-ITLBO algorithm, Rao and Patel (2014) and Patel and Savsani (2016) might not have used a fixed or predefined set of random seed. Thus the statements of Chinta et al. (2016) that *"Thus it would have been beneficial if the article had specified the seed (and their selection for multiple runs) and the algorithm that was used to generate the random numbers"* is not meaningful.

2.12   Runs and Plots

It is quite obvious that Fig. 3 and Fig. 4 of Rao and Patel (2014) are for the run in which the MO-ITLBO algorithm had shown its best performance and there is no need of misinterpretation in this regard. The previous researchers had also attempted the same problems and reported the Pareto fronts without mentioning the particular run number. It was implicit that those plots were for the run in which those algorithms had shown their best performance. Thus, the statement of Chinta et al. (2016) that *"However, it is not clear as to which of the 30 runs have been plotted...."* is not meaningful.

2.13 Code

A demonstration of TLBO and ITLBO algorithms on Rastrigin function was given by Rao and Patel (2014) as Appendix A in their paper. Readers can make use of the demonstration to develop their own code suiting to their application. Furthermore, many new algorithms and the improved versions of existing algorithms have been proposed by various researchers and are available in the literature. The codes of all the versions of those algorithms are not public. It is not necessary that the code of every version is to be made public. Once the pseudo-code, flowchart and/or steps of the algorithm are explained in the paper then the peers are expected to understand the same instead of looking or searching for the public code.

2.14   Notations and schematic diagram of MO-ITLBO algorithm

Fig. 3 to Fig. 6 provided by Rao and Patel (2014) are only to enable the readers to visualize the true Pareto-front and the Pareto-front obtained by MO-ITLBO algorithm. The true performance of the MO-ITLBO algorithm was shown based on the mean and standard deviation of the values of IGD measure obtained over 30 runs of MO-ITLBO algorithm. The performance of MO-ITLBO algorithm was not judged based on Fig. 3 to Fig 6. Therefore, detailing the font size of the symbols used in Fig. 3 to Fig. 6 is a trivial issue.

3. Additional Comments

- It is observed that Chinta et al. (2016) commented on several trivial issues in the work of Rao and Patel (2014) and Patel and Savsani (2016). However, Chinta et al. (2016) themselves committed basic mistakes in their paper. For example, in their work, Chinta et al. (2016) incorrectly cited Rao and Patel (2014) as Rao et al. (2014); and Patel and Savsani (2016) as Patel et al. (2016) throughout their paper. Thus, it makes unworthy of looking into the trivialities in the works of Rao and Patel (2014) and Patel and Savsani (2016).

- It must be noted that the results shown by Chinta et al. (2016) in Fig. 1 to Fig. 4 are obtained by using their self-proposed variants (variant 1 and variant 2) of the improved TLBO algorithm and the results are not of MO-ITLBO algorithm proposed by Rao and Patel (2014) and Patel and Savsani (2016). Rao and Patel (2014) had shown the results of application of MO-ITLBO algorithm for the unconstrained functions and constrained functions of CEC 2009 and proved the effectiveness of the algorithm. The two variants imagined by Chinta et al. (2016) are inferior to the MO-ITLBO algorithm and there is no point in questioning the effectiveness of MO-ITLBO based on the imagined versions. It is surprising to see the statement of Chinta et al. (2016) that *"Nevertheless, this exercise has been reported on the insistence of an anonymous reviewer to emphasize on the lacunae of the information provided in the MO-ITLBO and its impact"*. Rao and Patel (2014) need not be blamed if someone has not properly understood the working of MO-ITLBO algorithm. This means that Chinta et al. (2016) had not understood the MO-ITLBO algorithm and they had imagined two variants and tried those variants on CEC 2009 functions and had reported inferior results. Had Chinta et al. (2016) applied MO-ITLBO algorithm with correct understanding then the results, shown by Rao and Patel (2014) and Patel and Savsani (2016), would have been obtained by the authors. It is not meaningful to propose two variants (due to misunderstanding about the working about MO-ITLBO algorithm) and then to comment about MO-ITLBO algorithm.

- Just for the sake of publication the authors must not resort to pessimistically persuade the readers. It is worth contemplating that, many papers may not be explicit, however, it is not justifiable to speculate upon or write a note on all such works. Furthermore, the interest of the reviewers in encouraging such works is also questionable. Chinta et al. (2016) had mentioned that *"Nevertheless, this exercise has been reported on the insistence of an anonymous reviewer to emphasize on the lacunae of the information provided in the MO-ITLBO algorithm….".* It is unfortunate that reviewers, instead of reviewing the paper assigned to them, resort to the practice of insisting the authors to propose these variants and to emphasize on "lacunae" of the information provided in MO-ITLBO. It reflects that the authors had resorted to write the extended note due to the insistence of an anonymous reviewer.

- Chinta et al. (2016) had presented the results of the two variants assumed by them (because of misunderstanding about the MO-ITLBO algorithm) for only unconstrained benchmark functions UF1 to UF10 of CEC 2009. But they had not reported any results of the constrained functions CF1 to CF10 of CEC 2009 that were reported in Rao and Patel (2014) and Patel and Savsani (2016). The reasons for not attempting CF1 to CF10 were not provided by Chinta et al. (2016).

- Many multi-objective versions of TLBO algorithm were reported by the researchers (Li et al., 2014; Zou et al., 2013; Medina et al., 2014; Yu et al., 2015; Sultana and Roy, 2014, Rao et al., 2016) and many explanations were implicit in those versions and those versions had provided good results. The codes of all those algorithms are not public. It is not necessary that the code of every version is to be made public. Once the pseudo-code, flowchart and/or steps of the algorithm are explained in the paper, the peers are expected to understand the same instead of looking or searching for the public code.

- Although many multi-objective versions of TLBO algorithm are available it is surprising that Chinta et al. (2016) chose to write a note only on MO-ITLBO algorithm and proposed two variants of MO-ITLBO algorithm and reported dissatisfactory performance of the two variants on multi-objective benchmark functions of CEC 2009.

- It is not clear whether Chinta et al. (2016) had contacted Rao and/or Patel before jumping to write a note on their work. The authors should have contacted Rao and/or Patel in the case of difficulty in understanding the working of MO-ITLBO algorithm for clarification of doubts. Furthermore, it is not clear whether Rao and/or Patel were given any reviewing opportunity to present their side.

4.   Conclusions

The MO-ITLBO algorithm was proposed by Rao and Patel (2014) and has shown superior performance in solving multi-objective constrained and unconstrained benchmark problems of CEC 2009 as compared to the other optimization algorithms. The "issues" raised by Chinta et al. (2016) do not have any meaningful base and are unfounded. Chinta et al. (2016) have simply speculated on the works of Rao and Patel (2014) and Patel and Savsani (2016). Such tendency to write notes just due to misunderstanding of the concepts may be discouraged by the Journals. Furthermore, the reviewers need to play a constructive role while reviewing the papers instead of insisting the authors to search and report the "lacunae".

Acknowledgement

The authors would like to thank the anonymous referees for constructive comments on earlier version of this paper.

References

Chinta, S., Kommadath, R. & Kotecha, P. (2016). A note on multi-objective improved teaching–learning based optimization algorithm (MO-ITLBO). *Information Science*, 373, 337-350.

Deb, K., Mohan, M., & Mishra, S. (2005). Evaluating the ε-Domination Based Multi-Objective Evolutionary Algorithm for a Quick Computation of Pareto-Optimal Solutions. *Evolutionary computation*, *13*(4), 501-525.

Li, D., Zhang, C., Shao, X., & Lin, W. (2014). A multi-objective TLBO algorithm for balancing two-sided assembly line with multiple constraints. *Journal of Intelligent Manufacturing*, 27(4), 725-739.

Medina, M.A., Das, S., Coello, C.A.C. & Ramírez, J.M. (2014). Decomposition-based modern metaheuristic algorithms for multi- objective optimal power flow—A comparative study. *Engineering Applications of Artificial Intelligence*, 32, 10–20.

Mernik, M., Liu, S.H., Karaboga, D. & Crepinsek, M. (2015). On clarifying misconceptions when comparing variants of the Artificial Bee Colony Algorithm by offering a new implementation. *Information Sciences*, *291*, 115–127.

Patel, V. & Savsani, V.J. (2016). A multi-objective improved teaching–learning based optimization algorithm (MO-ITLBO). *Information Science, 357*, 182–200.

Rao, R.V., Savsani, V.J. & Vakharia, D.P. (2011). Teaching-learning-based optimization: A novel method for constrained mechanical design optimization problems. *Computer Aided Design, 43*(3), 303-315.

Rao, R.V., & Patel, V. (2012). An elitist teaching-learning-based optimization algorithm for solving complex constrained optimization problems. *International Journal of Industrial Engineering Computations*, 3(4), 535-560.

Rao, R. V., & Patel, V. (2013). An improved teaching-learning-based optimization algorithm for solving unconstrained optimization problems. *Scientia Iranica, 20*(3), 710-720.

Rao, R.V., & Patel, V. (2014). A multi-objective improved teaching-learning based optimization algorithm for unconstrained and constrained optimization problems. *International Journal of Industrial Engineering Computations, 5*(1), 1-22.

Rao, R.V. (2016a). Review of applications of TLBO algorithm and a tutorial for beginners to solve the unconstrained and constrained optimization problems. *Decision Science Letters, 5*(1), 1-30.

Rao, R.V. (2016b). *Teaching–learning-based optimization (TLBO) algorithm and its engineering applications*. Switzerland: Springer International Publishing.

Rao, R.V., Rai, D.P. & Balic, J. (2016) Multi-objective optimization of machining and micro-machining processes using non-dominated sorting teaching–learning-based optimization algorithm. *Journal of Intelligent Manufacturing*. DOI 10.1007/s10845-016-1210-5

Sultana, S., & Roy, P. K. (2014). Multi-objective quasi-oppositional teaching learning based optimization for optimal location of distributed generator in radial distribution systems. *Electrical Power and Energy Systems*, 63, 534–535.

Yu, K., Wang, X., & Wang, Z. (2015). Self-adaptive multi-objective teaching–learning-based optimization and its application in ethylene cracking furnace operation optimization. *Chemometrics and Intelligent Laboratory Systems, 146*, 198–210.

Zhang, Q., Zhou, A., Zhao, S., Suganthan, P.N., Liu, W. & Tiwari, S. (2009). Multi-objective optimization test instances for the congress on evolutionary computation (CEC 2009) special session & competition. *Working Report CES-887*. University of Essex, UK.

Zhou, A., Qu, B.Y., Li, H., Zhao, S.Z., Suganthan, P.N. & Zhang Q. (2011). Multi-objective evolutionary algorithms: a survey of the state-of-the-art. *Swarm & Evolutionary Computation*, *1*(1), 32–49.

Zou, F., Wang, L., Hei, X., Chen, D. & Wang, B. (2013). Multi-objective optimization using teaching-learning-based optimization algorithm. *Engineering Applications of Artificial Intelligence*, *26*, 1291–1300.

6

Modeling and optimization of surface roughness and tool vibration in CNC turning of Aluminum alloy using hybrid RSM-WPCA methodology

Priyabrata Sahoo[a]*, **Ashwani Pratap**[b] **and Asish Bandyopadhyay**[c]

[a, c]*Mechanical Engineering Department, Jadavpur University, Kolkata, India, 700032*
[b]*Mechanical Engineering Department, Indian Institute of Technology Patna, India, 801103*

CHRONICLE	ABSTRACT
Keywords: *CNC turning* *Surface roughness* *Tool vibration* *RSM* *WPCA* *ANOVA*	This paper suggests an advanced hybrid multi output optimization technique by applying weighted principal component analysis (WPCA) incorporated with response surface methodology (RSM). This investigation has been carried out through a case study in CNC turning of Aluminum alloy 63400 for surface roughness (Ra) and tool vibration (db) optimization. Primarily, input parameters such as spindle speed (N), feed rate (S) and depth of cut (t) are designed for experiment by using RSM Box-Behnken methodology. The aluminum alloy workpieces are machined by using coated carbide tool (inserts) in dry environment. Secondly, the empirical model for the responses as the functions of cutting parameters are obtained through RSM technique and the adequacy of the models have been checked using analysis of variance (ANOVA). Finally, the process parameters are optimized using WPCA technique. The confirmatory experiment has been performed using optimized result and it reveals that multiple response performance index (MPI) value was increased by 0.2908 from initial setting. The increases in MPI value indicates that the aforesaid optimization methodology is suitably acceptable for multi response optimization for turning process.

1. Introduction

Turning is one of the primitive and widely applicable machining processes. To cope with minimum cost and flexibility in machining process, automation to the conventional turning process has been adopted. In general, the machining study emphasizes on work material composition and its mechanical properties, cutting tool properties and all the input parameter settings that influence output quality characteristics as well as process efficiency. Process efficiency can be significantly improved by optimization of process parameters which identifies the zone of critical process parameter leading to desired responses by ensuring low cost of production (Montgomery, 1997).Aluminum and its alloys have been widely used in biomedical, pharmaceutical, food processing, health care, automotive and space research industries for various applications. They exhibit strong, light weight, higher ductility as well as excellent oxidation and

* Corresponding author
E-mail: sahoopriyabrata89@gmail.com (P. Sahoo)

corrosion resistance. However, machining of aluminium is associated with some difficulties due to its softness. Due to the soft and sticky nature of aluminium, specific geometries and characteristics of carbide inserts are generally used for efficient machining and to avoid built up edge formation (Roy et al., 2009).

Surface roughness has a viable importance in machining process as it is regarded as a measure of quality index of a product (Wang et al., 2010). It measures the closely spaced irregularities smaller than waviness in the surface texture. However, achieving a desired surface finish is a major issue with respect to functional behaviour of a machined part. The production cost and performance of machined parts are greatly influenced by surface finish as it affects various factors such as friction, geometrical tolerance, ease of holding, transition fit, thermal and electrical conductivity and many more (Makadia & Nanavati, 2013). In turning process, the surface finish is influenced by various parameters such as feed rate, depth of cut, cutting speed, nose radius of cutting tool, cutting environment, tool wear, machine vibrations, mechanical and material properties of work material and tool. A small deviation in any of the aforementioned parameters may affect the machined surface significantly (Knight & Boothroyd, 2005). The un-avoidable dynamic interactions between the cutting tool and work piece cause vibration and hence the chatter noise. This chatter has adverse effect on various parameters including surface roughness, dimensional accuracy, tool wear, etc. (Siddhupura & Paurobally, 2012). According to Tobias (1961), mechanical vibration in turning process is classified into three categories viz. natural, force and self-excited. These vibrations mainly occur due to the lack of rigidity of the machine tool. Among them, self-excited vibration or chatter is the most harmful and uncontrollable because of its complex nature. Therefore, there is a need to analyze and optimize such responses which has adverse effect on machining process to make the process more efficient and cost effective. Although, process modelling and optimization is not a new approach, still complicacy persists in modelling of surface roughness and vibration as these responses are affected by various parameters up to different ranges. Various modelling, optimization and simulation approaches have been adopted by researchers for optimization of surface roughness and vibration so far.

Palanikumar (2010) carried out statistical modelling using RSM to investigate the effect of process parameters on surface roughness and delamination factor in turning operation of glass fiber reinforced composite. RSM central composite design matrix was employed for the experiment. The adequacy of the model was verified at 95% confidence level within limit of input parameters being considered. Asilturk and Naseli (2012) conducted multi objective optimization along with empirical models for surface roughness parameters optimization with respect to cutting parameters (cutting speed, feed rate, depth of cut) in CNC turning of AISI 304 stainless steel. It was observed from the analysis that feed was the most significant factor influencing surface roughness. Gupta et al. (2011) applied Taguchi-Fuzzy multi response optimisation for output responses like surface roughness, power consumption, cutting force and tool life with respect to cutting parameters along with cutting conditions (dry, wet, cryogenic) and nose radius in CNC turning of tool steel. They fuzzified the signal to noise ratio (S/N) of output parameters to make comprehensive output measure (COM) which was being used as response for single objective optimization.

Surface roughness not only varies according to cutting parameters but also with the progress of tool wear. So, a signal which presents the interaction effect between cutting tool and workpiece should be studied when surface roughness is the main concern. Kirby et al. (2004) developed a mathematical model in terms of feed rate and vibration signal along cutting, feed and thrust direction during investigation on surface roughness. Risbood et al. (2003) developed a prediction equation using ANN taking radial vibration of tool holder as a feedback signal. Rudrapati et al. (2016) studied the effect of process parameters of cylindrical grinding process on the responses like workpiece vibration and surface roughness using RSM methodology. Process parameters were optimized for the desired responses using multi objective genetic algorithm and predicted model was verified using confirmatory test. Routara et al. (2010) attempted a new multi response optimization for cylindrical grinding parameters by incorporating Taguchi philosophy with weighted principal component analysis (WPCA).

Most of the optimization methods are associated with some assumptions, conflicts and constraints. Usual conflict faced by the researchers is the implementation of weightage to find out the optimal settings. From the literature, it is evident that optimization problems neglect the correlation of responses but this is not valid in real application for all cases.

To meet such challenges, a new hybrid optimization technique has been implemented in the present work. The study has been carried out experimentally. Data analysis and modelling has been attempted using RSM and ANOVA. Effect of process parameters (spindle speed, feed rate, depth of cut) on responses such as surface roughness and tool vibration has been analyzed using 3-D contour plot. Optimization of the output responses for evaluating the best combinations of input parameters has been carried out by using WPCA technique. In this method the co-related responses are converted into uncorrelated indices termed as principal components considering accountability proportion (AP) value as weightage. Finally the individual principal components have been merged to get a single composite principal component regarded as multi response performance index (MPI). MPI values have been utilised as index for single objective optimization.

2. Analysis and optimisation methodology

2.1 Response Surface Methodology (RSM)

RSM is a collection of statistical and empirical methods which is very useful for developing, improving, and optimizing processes parameters (Phadke, 1995). RSM is extensively used in the situation where various input variables significantly influence some output responses or quality index of the process. Sometimes the input variables are also called independent variables. RSM is mostly used to analyze the influence of input parameters on the specific responses (Routara et al., 2012). This method can also be used to define the contribution of the independent variables (input parameters) on the desired responses (output parameters). The vital purpose of developing regression models is to co-relate the responses with the independent variables which lead to optimization of the process parameters. Commonly used mathematical model for the response y and independent variables $\xi_1, \xi_2 \ldots \xi_k$ can be represented as

$$y = f(\xi_1, \xi_2, \ldots \xi_k) + \varepsilon,$$

(1)

where, ε is termed as a statistical error, which is normally distributed by response y with mean zero and variance σ^2. Then,

$$E(y) = \eta = E[f(\xi_1, \xi_2, \ldots, \xi_k)] + E(\varepsilon) = f(\xi_1, \xi_2, \ldots, \xi_k).$$

(2)

The variables $\xi_1, \xi_2, \ldots, \xi_k$ in Eq. (2) are called as natural variables, as they are expressed in the natural units of measurement. However it is more convenient to use coded variables ($x_1, x_2, x_3 \ldots$) which are dimensionless. The response function (η) can be written as

$$\eta = f(x_1, x_2, \ldots, x_k).$$

(3)

It is evident from the literature that second orders mathematical model is mostly used due to flexibility, wide variety of functional forms and use of significant least square method. Second order quadratic model can be expressed as

$$\eta = \beta_0 + \sum_{j=1}^{k} \beta_j x_j + \sum_{j=1}^{k} \beta_{jj} x_j^2 + \sum_{i<}\sum_{j=2}^{k} \beta_{ij} x_i x_j,$$

(4)

where, β_js are regression coefficient and x_js are coded form of independent variables.

2.2 Weighted Principal Component Analysis (WPCA)

To optimize the multi response objective, Su and Tong (1997) proposed a new method called as principal component analysis. Approach of this method is preceded through different steps (1) normalization of each response, (2) Checking of co-relation between different normalized responses, (3) conversion of the normalized responses to un-correlated index regarded as principal components (Liao, 2006). The principal components are calculated as

$$P_j = \sum_{i=1}^{r} a_{ji} Y_i, \text{ for } j = 1,\ldots\ldots,k \tag{5}$$

where, Y_i is the normalized value of ith responses (i= 1, 2,.....,r). For computation of PCA, k (k $\leq$ r) number of components can be obtained for the explanation of variance of the responses. The coefficient a_{ji} is termed as eigen vector, where $\sum_{i=1}^{r} a_{ji}^2 = 1$. The selected principal component is used as an index to optimize the multi response problem. However, two shortcomings are related with PCA method. First, if the principal component value is greater than one and second, as the principal components are explained by total variation, multi response performance index can't be replaced by multi response solutions. In order to overcome these challenges, weighted principal component analysis (WPCA) is incorporated (Datta et al., 2009). As each principal component has its own variance which may not be same, in this study variance of every principal component is used as the weight to compute multi response performance index (MPI). MPI can be calculated as

$$MPI = \sum_{j=1}^{k} W_j P_j, \tag{6}$$

where, W_j is regarded as weight of the corresponding principal component. MPI value defines the response for the single objective optimization and it is regarded as the quality index. Hence, greater is the MPI value, more is the quality.

3. Experimental Details

3.1 Design of Experiment (DOE)

Under the umbrella of DOE, most of the researchers design their experimental plans to reduce time and resources without compromising with quality. The result obtained from such experiments is quite easy to analyze as the experimental errors are minimized (Rudrapati et al., 2016). In this study, RSM Box-Behnken design has been applied for experimentation considering three levels of the three input parameters (N, S, t). The input machining parameters and their levels are shown in Table 1. The experimental design matrix developed by Box-Behnken design is given in Table 2. It represents all combinations of factors consisting of twelve points at the centre of the edge of the face of the cube and one point at centre of the cube with three replications. Total 15 numbers of combinations are formulated for the experimentation.

Table 1
Levels of process parameters

Coded levels	Spindle speed, N (rpm)	Feed rate, S (mm/min)	Depth of cut, t (mm)
-1	600	25.0	0.2
0	650	37.5	0.3
+1	700	50.0	0.4

Table 2
Experimental design matrix

Standard order	Spindle speed (N)		Feed rate (S)		Depth of Cut (t)	
	Code (A)	Actual (rpm)	Code (B)	Actual (mm/min)	Code (C)	Actual (mm)
1	-1	600	0	37.5	+1	0.4
2	0	650	0	37.5	0	0.3
3	-1	600	0	37.5	-1	0.2
4	-1	600	+1	50.0	0	0.3
5	0	650	-1	25.0	+1	0.4
6	+1	700	+1	50.0	0	0.3
7	+1	700	-1	25.0	0	0.3
8	0	650	+1	50.0	-1	0.2
9	0	650	+1	50.0	+1	0.4
10	-1	600	-1	25.0	0	0.3
11	0	650	-1	25.0	-1	0.2
12	0	650	0	37.5	0	0.3
13	+1	700	0	37.5	-1	0.2
14	+1	700	0	37.5	+1	0.4
15	0	650	0	37.5	0	0.3

3.2 Machine and material

The machine tool used for this experiment is a HMT CNC lathe having FANUC control system. The CNC lathe is equipped with 3200 RPM maximum spindle speed, 1-699 mm/min feed rate and 0.3 KW output power. For generation of machined surface, CNC part programmes have been created using G-codes. For the better performance in machining, CVD coated tool are used in the experiments. The tool used for machining is K-10, WIDIA inserts having hardness 1570 HV, density 14.5 g/cc, transverse rupture strength 3800 N/mm^2 and tool specifications: -6°- -6°-11°-8°-36°-90°-0.8 (mm). The experiments are carried out on 15 numbers of same composition aluminum alloys bar having dimension $\Phi22 \times 60$ mm (Fig. 1). The mechanical properties of the work piece specimen are as follows: hardness: 124 HV, ultimate tensile strength: 1.8-26 kgf/mm^2, melting Point: 660° C, density: 2600-2800 kg/mm^3 and its chemical composition is Cu-0.01%, Mg-0.64%, Si-0.70%, Fe-0.17%, Mn-0.11%, Zn-0.06%, Ti-0.03%, Cr-0.01%, remaining Al.

Fig. 1. Photographic view of machined workpieces

3.3 Vibration and Roughness Measurement

Tool vibration has been measured online during machining by placing uniaxial piezoelectric accelerometer (Make: syscon Instrument Pvt. Ltd., Bangalore, Model: 353b31) on the tool holder, as shown in Fig. 2. The output of the accelerometer is fed to the digital real-time oscilloscope with FFT module (Make: Tektronix, Model: TDS 2MM), which gives the mechanical signal in terms of electrical signal. From literature (Feng, 2001), it is evident that amplitude parameter of vibration has significant effect on tool vibration so the amplitude parameter (db) is selected as a response. Root mean square (RMS) value of the amplitude parameter has been calculated which is used as vibration response. Initially, trial run are carried out by placing the accelerometer in radial (X), axial (Z) and cutting (Y) directions to observe the significant effect of tool vibration. Initial run reveals that tool vibration has significant effect in Z-direction, therefore vibration data in Z-direction has been acquired for the case study. Surface roughness measurement is carried out by using contact type stylus profilometer Talysurf (Make:Taylor Hobson, Surtronic 3+, UK). The photographic view of the profilometer set up is shown in Fig. 3. Profilometer is equipped with a diamond stylus of 5 µm tip radius. The arithmetic average amplitude parameter (R_a) is widely used in industry; the same parameter has been considered in this study. Roughness of each specimen is measured thrice and average values of the calculated R_a values are recorded. Obtained results from tool vibration and surface roughness measurements are shown in Table 3.

Fig. 2. Photographic view of position of accelerometer on tool holder

Fig. 3. Photographic view of stylus type profilometer, Talysurf

Table 3

Experimental data of surface roughness (Ra) and tool vibrations (V) corresponding to Box-Behnken design of experiment

Experiment No.	Process parameters			Observed values	
	Spindle speed (rpm)	Feed rate (mm/min)	Depth of cut (mm)	Surface Roughness, R_a (µm)	Vibration amplitude V (db)
1	600	37.5	0.4	0.749	74.506
2	650	37.5	0.3	0.905	81.304
3	600	37.5	0.2	0.704	73.2626
4	600	50.0	0.3	1.167	78.156
5	650	25.0	0.4	0.653	78.904
6	700	50.0	0.3	1.163	79.811
7	700	25.0	0.3	0.589	77.986
8	650	50.0	0.2	1.168	79.798
9	650	50.0	0.4	1.200	83.440
10	600	25.0	0.3	0.596	69.580
11	650	25.0	0.2	0.592	76.947
12	650	37.5	0.3	0.893	81.579
13	700	37.5	0.2	0.682	74.409
14	700	37.5	0.4	0.723	76.703
15	650	37.5	0.3	0.886	81.597

4. Results and Analysis

To analyze the measured responses for the input parameters, checking for adequacy of the model is essential. The model adequacy can be checked by incorporating verification of regression model, model coefficient, and lack of fit value (Senthilkumar, 2014). To satisfy all these purposes, ANOVA is the appropriate choice and the same has been applied in this study. To find out the best combination of input parameters for better surface finish and less vibration, a multi output optimisation technique is also performed.

4.1 Analysis of Surface Roughness and Tool Vibration

The main purpose of data analysis is to find the significant individual and interaction effects of independent variables on the dependent responses (Rudrapati et al., 2016). ANOVA analysis has been performed using the software package Design Expert 7.0. There is various type of regression model such as linear, quadratic, interaction. Quadratic model is chosen for both surface roughness and tool vibration as it is recommended by fit summary. The results of the ANOVA analysis are presented in Table 4 (Surface Roughness) and Table.5 (Tool Vibration). The value of R^2 for both surface roughness and tool vibration are more than 90%, which reveals that the model has very good correlation with the experimental data. The adjusted R^2 value for surface roughness and vibration are 99.84% and 89.41% which indicates good correlation. Both of the Surface roughness model and Vibration model are statically significant as p-values are less than 0.05, considering 95% confidence level. For surface roughness, lack of fit term is not significant which is desirable; however in case of vibration, lack of fit is significant. Therefore further investigation is necessary for tool vibration analysis. From Table 4, it can be concluded that individual effect of feed rate (S) is most significant for surface roughness as it has the highest F-value (Lowest P value). Along with the feed rate, the individual effect of spindle speed (N), depth of cut (t) and the second order effect of spindle speed (N^2), feed rate (S^2) and depth of cut (t^2) are significant terms for the model as the corresponding p-values are less than 0.05. The interaction effect of spindle speed-feed rate (N-S), spindle speed-depth of cut (N-t) and feed rate-depth of cut (S-t) are not found to be significant as p values of the terms are less than 0.05. ANOVA analysis for tool vibration is presented in Table 5. From Table 5, it revealed that second order term of spindle speed (N^2) has the highest significance in the model for tool vibration as it is having lowest p-value. Apart from this, individual effect of spindle speed (N), feed rate (S), depth of cut (t), second order effect of depth of cut (t^2) and interaction effect of spindle speed-depth of cut (N-t) are significant terms for this model. Rest of the terms are not significant. Although, for both the cases models are statistically significant but to improve the performance of the models, non significant terms are eliminated.

Table 4
ANOVA table for surface roughness

Source	DOF	Sum Square	Mean square	F- value	P-value
Model	9	0.74	0.082	974.92	<0.0001 *significant*
N	1	4.351E-004	4.351E-004	5.15	0.0725
S	1	0.64	0.64	7610.71	< 0.0001
T	1	4.005E-003	4.005E-003	47.41	0.0010
N^2	1	0.039	0.039	457.68	< 0.0001
S^2	1	0.028	0.028	326.38	< 0.0001
t^2	1	0.022	0.022	264.76	< 0.0001
N*S	1	2.250E-006	2.250E-006	0.027	0.8768
N*t	1	4.000E-006	4.000E-006	0.047	0.8363
S*t	1	2.102E-004	2.102E-004	2.49	0.1755
Res. error	5	4.224E-004	8.448E-005	-	-
Lack of fit	3	2.377E-004	7.925E-005	0.86	0.5778 *not significant*
Pure error	2	1.847E-004	9.233E-005	-	-
Total	14	0.74	-	-	-

R^2= 0.9994, R^2 adjusted = 0.9984

Table 5
ANOVA table for tool vibration

Source	DOF	Sum Square	Mean square	F- value	P-value
Model	9	187.21	20.80	14.13	0.0047*
N	1	22.46	22.46	15.25	0.0113
S	1	39.55	39.55	26.86	0.0035
T	1	10.43	10.43	7.09	0.0448
N^2	1	95.33	95.33	64.75	0.0005
S^2	1	3.103E-003	3.103E-003	2.108E-003	0.9652
t^2	1	10.57	10.57	7.18	0.0438
N*S	1	11.39	11.39	7.74	0.0388
N*t	1	0.28	0.28	0.19	0.6831
S*t	1	0.71	0.71	0.48	0.5184
Res. Error	5	7.36	1.47	-	-
Lack of fit	3	7.31	2.44	90.33	0.0110*
Pure error	2	0.054	0.027	-	-
Total	14	194.57	-	-	-

$R^2 = 0.9622$, R^2 adjusted = 0.8941   *Significant

After elimination of the non significant terms, improved ANOVA analysis is shown in Table 6 and Table 7, for surface roughness and tool vibration, respectively. It can be concluded that for both surface roughness and tool vibration, R^2 and adjusted R^2 values are more than 90%. Lack of fit term is non significant for both the cases which was previously significant for tool vibration analysis. It proves backward elimination method is viable to improve the model performance.

Table 6
ANOVA table for surface roughness (after backward elimination)

Source	DOF	Sum Square	Mean square	F- value	P-value
Model	6	0.74	0.12	1546.51	< 0.0001*
N	1	4.351E-004	4.351E-004	5.45	0.0479
S	1	0.64	0.64	8050.85	< 0.0001
T	1	4.005E-003	4.005E-003	50.15	0.0001
N^2	1	0.039	0.039	484.15	< 0.0001
S^2	1	0.028	0.028	345.25	< 0.0001
t^2	1	0.022	0.022	280.08	< 0.0001
Res. error	8	6.389E-004	7.986E-005	-	-
Lack of fit	6	4.542E-004	7.571E-005	0.82	0.6406 not significant
Pure error	2	1.847E-004	9.233E-005	-	-
Total	14	0.74	-	-	-

$R^2 = 0.9991$, R^2 adjusted = 0.9985   *Significant

Table 7
ANOVA table for tool vibration (after backward elimination)

Source	DOF	Sum Square	Mean square	F- value	P-value
Model	6	186.22	31.04	29.74	< 0.0001 significant
N	1	22.46	22.46	21.52	0.0017
S	1	39.55	39.55	37.89	0.0003
T	1	10.43	10.43	10.00	0.0134
N^2	1	95.81	95.81	91.79	< 0.0001
t^2	1	10.61	10.61	10.16	0.0128
N*S	1	11.39	11.39	10.92	0.0108
Res. Error	8	8.35	1.04	-	-
Lack of fit	6	8.30	1.38	51.28	0.0693 Not Significant
Pure error	2	0.054	0.027	-	-
Total	14	194.57	-	-	-

$R^2 = 0.9571$, R^2 adjusted = 0.9249

After elimination of the non-significant terms from the model, the final regression equation for surface roughness and tool vibration are given by:

$$R_a = -17.14417 + 0.053066 \times N - 0.018800 \times S + 4.89375 \times t - 4.09333E\text{-}005 \times N^2 + 5.53067E\text{-}004 \times S^2 - 7.78333 \times t^2, \quad (7)$$

$$V = -889.76191 + 2.77578 \times N + 1.93314 \times S + 112.8129 \times t - 2.70040E\text{-}003 \times N \times S - 2.03154E\text{-}003 \times N^2 - 168.98615 \times t^2. \quad (8)$$

To predict tool vibration and surface roughness these developed regression models can be used. Fig. 4 and 5 depicts the predicted and actual values and it shows that patterns are obvious. Fig. 6 and 7 illustrates the normal probability of residuals. It reveals that all the residual points are falling in a straight line which means, the errors are normally distributed. This indicates that the developed regression model is acceptable under given experimental range.

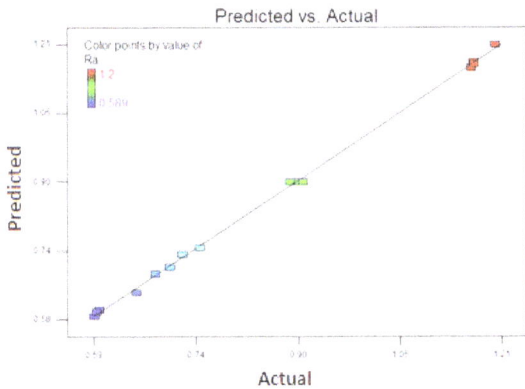

Fig. 4. Predicted vs. actual response plot of surface roughness

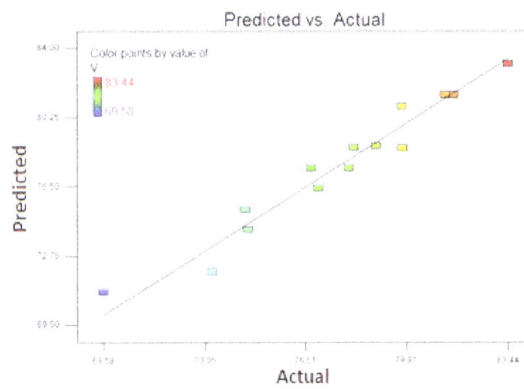

Fig. 5. Predicted vs. actual response plot of tool vibration

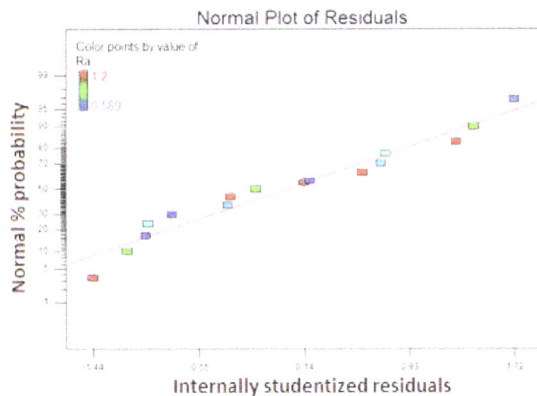

Fig. 6. Normal probability plot of residuals for surface roughness

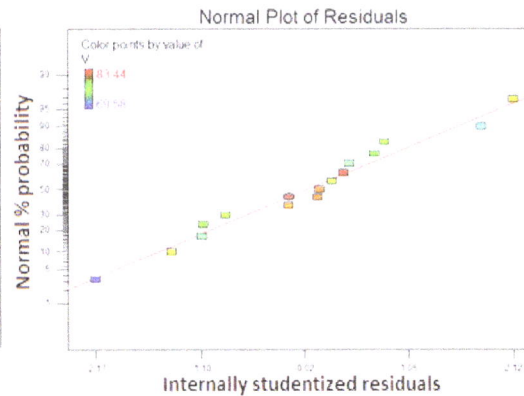

Fig. 7. Normal probability plot of residuals for tool vibration

3-D response surface plots are drawn based on the regression equations 7 and 8 for the better understanding of interaction effects of independent variables on responses. As every model comprises of three independent variables, one variable has been kept constant at its centre level for each plot.

Fig. 8 and Fig. 9 depict 3-D surface plot for surface roughness, whereas Fig. 10 and Fig. 11 represent 3-D surface plot for tool vibration. Except the effect of feed rate for tool vibration all the independent variables show curvature trend for the responses and hence variation is non-linear. Fig. 8 and Fig. 9 are drawn on the basis of Eq. (7). It can be observed that surface roughness increases with increase in rotational speed, depth of cut and feed rate. However, the effect of rotational speed with surface roughness violates the general trends; vibration may be regarded as one of the cause for such violation at higher speeds. 3-D surface plots for tool vibration are drawn on the basis of Eq. (8). It can be concluded from the plots that tool vibration increases for all three input parameters.

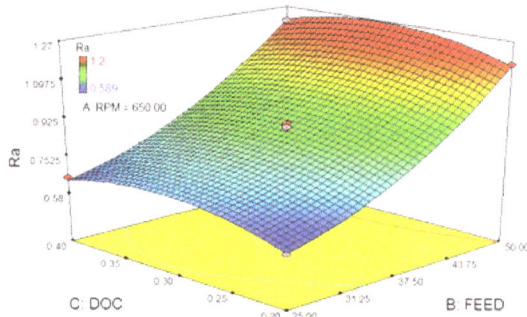

Fig. 8. Effect of depth of cut (t) and feed rate (s) on surface roughness (3-D plot)

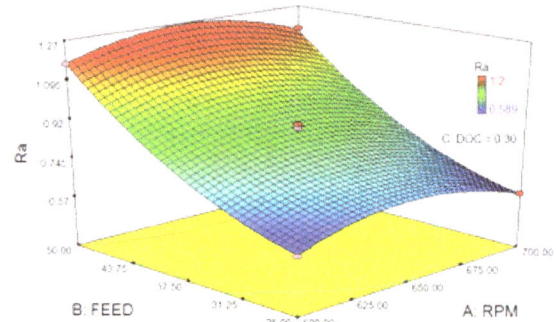

Fig. 9. Effect of feed rate (s) and spindle speed (N) on surface roughness (3-D plot)

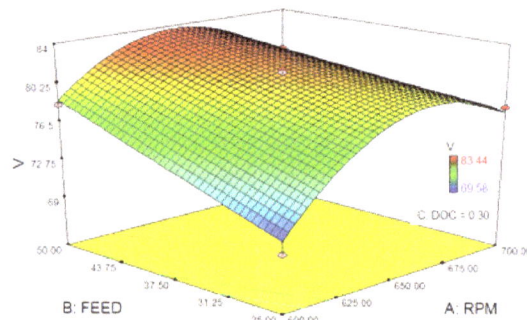

Fig. 10. Effect of spindle speed (N) and feed rate (s) on tool vibration (3-D plot)

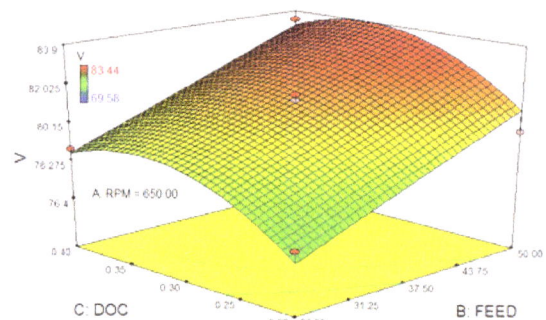

Fig. 11. Effect of feed rate (s) and depth of cut (t) on tool vibration (3-D plot)

4.2 Multi objective optimization

Initially, the experimental data of surface roughness and tool vibration, as shown in Table 3, are normalized. For normalization of all the responses, lower the better criteria have been implemented by using the Eq. (9), which is given as:

$$X_i^{\times}(k) = \frac{\min X_i(k)}{X_i(k)}, \tag{9}$$

where, i = 1,2,…….., m; k = 1,2,………,n. m is the number of experimental runs and n is the number of quality characteristics. The normalized responses range is kept between 0 to 1 and these are presented in Table 8. After normalization the next step is to check the correlation between the responses. The Pearson's correlation coefficient for the responses is shown in Table 9. It can be observed that the coefficient is a non-zero value which indicates that the responses are correlated. To eliminate correlation between the responses, PCA has been applied using the MINITAB. The details of the PCA (Eigen value, eigen vector, accountability proportion) are shown in Table 10. Uncorrelated indexes broadly known as principal components (P_1, P_2) are obtained from the correlated responses using the equation 5. To calculate MPI value accountability proportion (AP) has been used as individual weight of the principal components. The value is calculated using equation 6 and obtained MPI data are shown in Table 11. MPI value is treated as single objective function for optimization in order to minimize surface roughness and tool vibration. Higher MPI value gives better results, From the Table 11 it seems that optimum combination can be obtained nearer to the highest MPI value i.e. 1.1958. The factorial combination that maximizes MPI can be treated as optimal parametric combination or most favourable environment ensuring low surface roughness and tool vibration. This has been performed using main effect plot of mean for MPI as shown in Table 12 and Fig. 12.

The suggested optimum level from main effect plot for MPI is $N_1 S_1 t_1$. Table 13 shows the results of the confirmation experiments using the optimal machining parameters. From Table 13, it is clear that R_a value is decreased from 0.893 μm to 0.541 μm and vibration value decreased from 80.579 db to 70.392 db. MPI value increased from 0.9362 to 1.227 from the initial setting. From Table 11, it is observed that highest MPI value is 1.1958 having combinations of spindle speed 600 rpm, feed 25 mm/min and depth of cut 0.3 mm and responses Ra value 0.596 μm and V value 69.58 db which is higher than that of confirmation experiment shown in Table 12. As a whole, it is found that the multiple performance characteristics in the CNC turning process are improved through this study.

Table 8
Normalized responses for surface roughness and vibrations

Sl. No.	Ra	V
Ideal	1.000	1.000
1	0.7864	0.9833
2	0.6508	0.9011
3	0.8366	1.0000
4	0.5047	0.9374
5	0.9020	0.9285
6	0.5064	0.9180
7	1.0000	0.9394
8	0.5043	0.9181
9	0.4908	0.8780
10	0.9883	1.0529
11	0.9949	0.9521
12	0.6596	0.8981
13	0.8636	0.9846
14	0.8147	0.9551
15	0.6648	0.8979

Table 9
Correlation checking

Correlation between responses	Pearson correlation coefficient	Comment
Ra and V	0.646	Co-related

Table 10
Principal component analysis

	Φ_1	Φ_2
Eigen value	1.6461	0.3539
Eigen vector	$\begin{vmatrix} 0.707 \\ 0.707 \end{vmatrix}$	$\begin{vmatrix} -0.707 \\ 0.707 \end{vmatrix}$
AP	0.823	0.177
CAP	0823	1.000

Table 11
Principal components and MPI

Sl. N0.	Individual principal components		MPI
	P_1	P_2	
Ideal	1.414	0.000	1.1637
1	1.2512	0.1392	1.0544
2	1.0972	0.1769	0.9343
3	1.2985	0.1155	1.0891
4	1.0196	0.3059	0.8932
5	1.2942	0.0187	1.0684
6	1.0071	0.2909	0.8803
7	1.3712	-0.0428	1.1209
8	1.0056	0.2926	0.8794
9	0.9678	0.2737	0.8449
10	1.4431	0.0457	**1.1958**
11	1.3766	-0.0303	1.1276
12	1.1012	0.1686	0.9362
13	1.3067	0.0855	1.0905
14	1.2513	0.0993	1.0474
15	1.1048	0.1648	0.9384

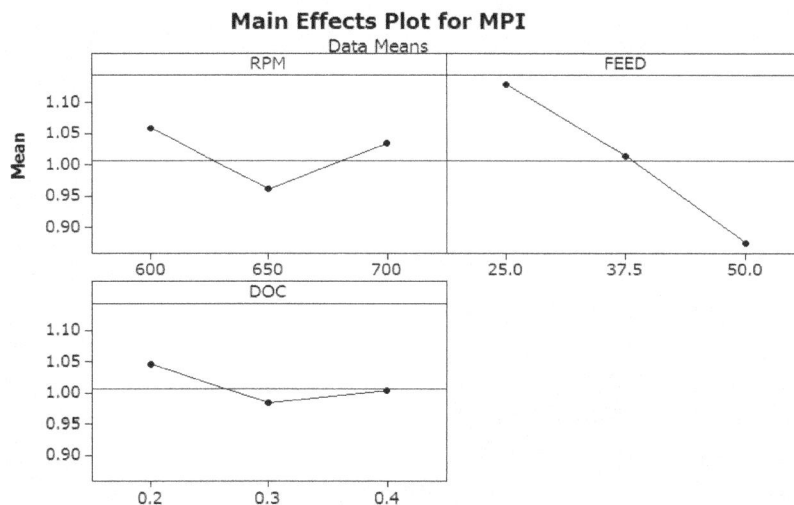

Fig. 12. Main effects plot for MPI

Table 12
Analysis of means of MPI

Level	Spindle speed(N)	Feed rate(s)	Depth of cut(t)
1	**1.0581**	**1.1281**	**1.0466**
2	0.9613	1.0129	0.9855
3	1.0347	0.8744	1.0037
Delta	0.0968	0.2537	0.0611
Rank	2	1	3

Table 13
Comparison of machining performances using initial and optimal cutting parameters

	Initial cutting parameters	optimum cutting parameters
Spindle speed, Feed rate & Depth of cut	650 RPM, 37.5 mm/min, 0 .3 mm	600 RPM, 25 mm/min, 0.2 mm
R_a value (µm)	0.893	0.541
V value (db)	81.579	70.392
MPI value	0.9362	1.227
MPI value increased by 0.2908		

5. Conclusions

The aforementioned study presented an integrated application of RSM and WPCA technique for the modelling and optimization of process parameters to achieve good surface finish and lower tool vibration in CNC turning process. The outcomes of the study can be summarised as:

- ANOVA analysis reveals that feed rate and second order term of spindle speed is most influencing parameter for surface roughness and tool vibration respectively.
- It can be observed that the interaction effects are not significant for the model in case of surface roughness however the interaction effect of spindle speed and feed are significant in case of tool vibration.
- 3 D surface plots of the responses vs. the process parameters indicate that the interaction effects exist. The combined effect of spindle speed and feed rate on surface roughness is however found to be much less than those of a) spindle speed and depth of cut and b) feed rate and depth of cut. Also, the combined effect of a) spindle speed and depth of cut and b) feed rate and depth of cut on vibration is found to be of lesser degree than those of spindle speed and feed rate.
- The response surface methodology used in the recent study has been proven as an effective tool for the analysis of the CNC turning process.
- Application of WPCA is an improved multi output optimisation technique by considering AP as individual weight of the principal components which merely fulfil the weightage criteria for the researchers. MPI value has been regarded as the index for single objective optimisation.
- From the main effect plot for MPI, optimal conditions for both the responses are: spindle speed 600 RPM, feed rate25 mm/min, depth of cut 0.2 mm.
- Confirmatory experiment has been performed at optimum condition and observed surface roughness is 0.541 µm and tool vibration is 70.392 db. MPI value is increased nby 0.2908 from the initial setting which adds a strong acceptance of WPCA technique.
- The discussed approach can be suggested for modelling and multi output optimisation of various machining processes.

References

Asiltürk, I., & Neşeli, S. (2012). Multi response optimisation of CNC turning parameters via Taguchi method-based response surface analysis. *Measurement, 45*(4), 785-794.

Datta, S., Nandi, G., Bandyopadhyay, A., & Pal, P. K. (2009). Application of PCA-based hybrid Taguchi method for correlated multicriteria optimization of submerged arc weld: a case study. *The International Journal of Advanced Manufacturing Technology, 45*(3-4), 276-286.

Feng, C. X. (2001). An experimental study of the impact of turning parameters on surface roughness. In *Proceedings of the industrial engineering research conference* (Vol. 2036).

Gupta, A., Singh, H., & Aggarwal, A. (2011). Taguchi-fuzzy multi output optimization (MOO) in high speed CNC turning of AISI P-20 tool steel. *Expert Systems with Applications, 38*(6), 6822-6828.

Kirby, E. D., Zhang, Z., & Chen, J. C. (2004). Development of an accelerometer-based surface roughness prediction system in turning operations using multiple regression techniques. *Journal of Industrial Technology*, *20*(4), 1-8.

Knight, W. A., & Boothroyd, G. (2005). *Fundamentals of metal machining and machine tools* (Vol. 198). CRC Press.

Liao, H. C. (2006). Multi-response optimization using weighted principal component. *The International Journal of Advanced Manufacturing Technology*, *27*(7-8), 720-725.

Makadia, A. J., & Nanavati, J. I. (2013). Optimisation of machining parameters for turning operations based on response surface methodology. *Measurement*, *46*(4), 1521-1529.

Montgomery, D. C. (1997) *Design and analysis of experiments*, 4th ed., John wiley & Sons Inc., New York.

Phadke, M. S. (1995). *Quality engineering using robust design*. Prentice Hall PTR.

Palanikumar, K. (2010). Modeling and analysis of delamination factor and surface roughness in drilling GFRP composites. *Materials and Manufacturing Processes*, *25*(10), 1059-1067.

Risbood, K. A., Dixit, U. S., & Sahasrabudhe, A. D. (2003). Prediction of surface roughness and dimensional deviation by measuring cutting forces and vibrations in turning process. *Journal of Materials Processing Technology*, *132*(1), 203-214.

Roy, P., Sarangi, S. K., Ghosh, A., & Chattopadhyay, A. K. (2009). Machinability study of pure aluminium and Al–12% Si alloys against uncoated and coated carbide inserts. *International Journal of Refractory Metals and Hard Materials*, *27*(3), 535-544.

Rudrapati, R., Pal, P. K., & Bandyopadhyay, A. (2016). Modeling and optimization of machining parameters in cylindrical grinding process. *The International Journal of Advanced Manufacturing Technology*, *82*(9-12), 2167-2182.

Routara, B. C., Mohanty, S. D., Datta, S., Bandyopadhyay, A., & Mahapatra, S. S. (2010). Combined quality loss (CQL) concept in WPCA-based Taguchi philosophy for optimization of multiple surface quality characteristics of UNS C34000 brass in cylindrical grinding. *The International Journal of Advanced Manufacturing Technology*, *51*(1-4), 135-143.

Routara, B. C., Sahoo, A. K., Parida, A. K., & Padhi, P. C. (2012). Response surface methodology and genetic algorithm used to optimize the cutting condition for surface roughness parameters in CNC turning. *Procedia Engineering*, *38*, 1893-1904.

Senthilkumar, N., Tamizharasan, T., & Anandakrishnan, V. (2014). Experimental investigation and performance analysis of cemented carbide inserts of different geometries using Taguchi based grey relational analysis. *Measurement*, *58*, 520-536.

Siddhpura, M., & Paurobally, R. (2012). A review of chatter vibration research in turning. *International Journal of Machine tools and manufacture*, *61*, 27-47.

Su, C. T., & Tong, L. I. (1997). Multi-response robust design by principal component analysis. *Total Quality Management*, *8*(6), 409-416.

Tobias, S. A. (1961). Machine tool vibration research. *International Journal of Machine Tool Design and Research*, *1*(1), 1-14.

Wang, Z., Meng, H., & Fu, J. (2010). Novel method for evaluating surface roughness by grey dynamic filtering. *Measurement*, *43*(1), 78-82.

7

Solving machine loading problem of flexible manufacturing systems using a modified discrete firefly algorithm

Eleonora Bottani[a], Piera Centobelli[b]*, Roberto Cerchione[c], Lucia Del Gaudio[d] and Teresa Murino[e]

[a]Department of Industrial Engineering, University of Parma, viale G.P.Usberti 181/A, 43124 – Parma, Italy
[b,c]Department of Industrial Engineering, University of Naples Federico II, P.le Tecchio 80, 80125 – Naples, Italy
[d,e]Department of Chemical, Materials and Industrial Production Engineering, P.le Tecchio 80, 80125 – Naples, Italy

CHRONICLE

Keywords:
Discrete Firefly Algorithm
Flexible Manufacturing System
Machine Allocation Problem
Swarm-based Optimization

ABSTRACT

This paper proposes a modified discrete firefly algorithm (DFA) applied to the machine loading problem of the flexible manufacturing systems (FMSs) starting from the mathematical formulation adopted by Swarnkar & Tiwari (2004). The aim of the problem is to identify the optimal jobs sequence that simultaneously maximizes the throughput and minimizes the system unbalance according to given technological constraints (e.g. available tool slots and machining time). The results of the algorithm proposed have been compared with the existing and most recent swarm-based approaches available in the open literature using as benchmark the set of ten problems proposed by Mukhopadhyay et al. (1992). The algorithm shows results that are comparable and sometimes even better than most of the other approaches considering both the quality of the results provided and the computational times obtained.

1. Introduction

Manufacturing systems are changed through the years from high variety and low volume production, typical of job-shop production systems, to dedicated manufacturing lines characterized by low variety and high volume supporting the economies of scale. The FMS paradigm was developed in the 1980s, as a primary effort to include at the shop-floor level the ability to tackle with mass customization and to face the problem of mid-volume and mid-variety of the production requests (ElMaraghy, 2005).

Operatively in FMS pre- and post- release decisions have to be faced. Pre-release decisions consist of the FMS operational planning decision to be carried out before the process starts, e.g. the pre-allocation of jobs to machines, as well as tools, whereas post-release decisions refer to scheduling problems (Kumar, Murthy, & Chandrashekara, 2012). Pre-release decision problems may be very complex and according to Stecke (1983) they may include the following five objectives: machine grouping, selection of the part

* Corresponding author
E-mail: piera.centobelli@unina.it (P. Centobelli)

type, calculation of the production rate, allocation of resource, and machine loading. Among them, the last two problems, i.e. allocation of resource and machine loading, are considered crucially important (Kumar et al., 2006). More specifically, Biswas and Mahapatra (2008) state that the overall performance of the FMS is mainly influenced by machine loading (ML) problem.

Machine loading problem covers many objectives. However, the goal is always to enhance the overall system performance (e.g. minimize the idle time of the machines, maximize the machine utilization) by defining jobs allocation of each part type to be produced to a given number of machines also satisfying technological constraints. Therefore, due to the different and simultaneous factors to be set, the problem of machine loading belongs to the NP-hard problems (Srinivas et al., 2004).

The ML problem was first formulated with the aim to maximize the throughput of the system over a given set of jobs using a branch and backtrack strategy (Stecke & Solberg, 1981). Since then, several approaches based on mathematical models (Mgwatu, 2011), heuristics (Tiwari et al., 2007), meta-heuristics (Sandhyarani & Mahapatra, 2009), simulation approaches (Lee & Jung, 1989) as well as a combination of them (Abazari et al., 2012), have been formulated to solve this optimization problem and reduce computational burden.

In recent years among metaheuristics, particularly interesting are the results obtained using the swarm-based approaches (Pandey, 2011; Yusof et al., 2014; Santuka et al., 2015). In summary it emerges that the ML problem in FMSs is a topic deeply analysed nevertheless currently under investigation.

With these premises, we propose a modified discrete firefly algorithm to be applied in the context of ML problem of FMSs in order to simultaneously reduce the system unbalance, that is the union of the under- and over-utilized times with regard to the set of machines of the system, and to maximize the throughput, which is the maximization of the production rate and the system efficiency, according to given technological constraints (i.e. tool slots available and total machining time available).

The paper is structured as follows. After this introduction, the mathematical formulation of ML in FMS is provided in Section 2. Section 3 illustrates the DFA adapted to the ML problem. In Section 4 the computational results are reported. At the end, conclusions and the future directions of research are discussed in Section 5.

2. Machine loading problem: mathematical formulation

In this article we adopt the mathematical formulation of ML problem developed by Swarnkar and Tiwari (2004) and the systematic review approach defined by Cerchione & Esposito (2016) to identify the main contributions on the topic.

The machine loading problem consists in a set of J jobs to be assigned to a set of M machines characterized by a positive processing time and fixed tool slots. Each job is considered as a set of operations with a predefinite processing time and fixed tool slot, and it can be realized on a specific set of machines. The aim is to identify an individual jobs sequence that simultaneously maximizes the throughput (TP) and minimizes the system unbalance (SU) according to given technological constraints.

The first objective to be minimized (F_1) is the system unbalance, formulated as follows:

$$\min F_1 \sum_{j=1}^{M}(UT_j - OT_j) = \max F_1\left[-\sum_{j=1}^{M}(UT_j - OT_j)\right] = \frac{MXH - \sum_{j=1}^{M}(UT_j - OT_j)}{MXH} \qquad (1)$$

where UT and OT are respectively the under-and over- utilized time.

The second objective to be maximized (F_2) is throughput expressed as:

$$\max F_2 = \frac{\sum_{i=1}^{N}(Bs_i \times x_i)}{\sum_{i=1}^{N} Bs_i} \tag{2}$$

Where Bs_i is the batch size and x_i is a binary variable equal to 1 if the *i-th* part is selected and 0 otherwise.

Thus, the weighted sum criteria has been adopted (Moncayo-Martinez & Zhang, 2011) to convert the multi-objective problem into a single-objective one assigning a weight (w) to each target function. Therefore the overall objective function is formulated as:

$$\max F = \frac{w_1 \times F_1 + w_2 \times F_2}{w_1 + w_2} = \left[w_1 \frac{M \times H - \sum_{j=1}^{M}(UT_j - OT_j)}{M \times H} + w_2 \frac{\sum_{i=1}^{N}(Bs_i \times x_i)}{\sum_{i=1}^{N} Bs_i} \right] / (w_1 + w_2) \tag{3}$$

The problem is subjected to the following assumptions and constraints (Yogeswaran et al., 2009):

1. System unbalance. It is equal to the union of the UT and OT on the set of available machines after the complete allocation of jobs. Therefore, this value should be higher or ideally equal to 0 in case of complete utilization of the system.
2. Unique job routing. Each operation has to start and finish on the same machine.
3. Tool slots. The tools available on the selected machine should be higher - at least equal - than the number of tools necessary to process the operation of the jobs on that machine.
4. Non-splitting of job. The operations on a job should be completed before processing another one.
5. Sharing of tool slots is not allowed, i.e. the number of tool slots after any job assignment should be positive or zero.
6. Parts and tools necessary to process the jobs are readily available in proximity of the machine stations. Therefore the setup times as well as the material handling operations are negligible.

In addition each operation may be essential or optional. This latter may be executed on a certain group of machines with equal or unequal processing time and tool slots whereas the first type may be executed exclusively on a single machine which requires a given number of tool slots. Therefore, the flexibility of the system, as well as its optimization, consists in the allocation of operations to the available machines in order to enhance the overall system performance.

3. Modelling framework

3.1 Firefly algorithm

Firefly Algorithm was firstly formulated at Cambridge University by Yang (2009) and it is categorised into the class of the nature-inspired metaheuristics. It is inspired by the behaviour of fireflies, as well as their flashing feature, and can be applied to different engineering optimization problems. The flashing lights are the result of a biochemical process which allows fireflies to communicate with each other. In fact, in order to live in harmony, individuals within a group need to interact and adapt their behaviour to the whole aim of the groups. Group members cannot behave as if they are solitary. The collective decisions of fireflies represent the main biological basis for developing the firefly algorithm.

The main fireflies characteristics are formalized as follows:

- All fireflies are unisex hence every firefly will be attracted from other fireflies regardless of the sex.
- Their attractiveness is proportional to the brightness, and they both decrease as their distance increases.
- The light intensity of a firefly is affected or determined by the landscape of the fitness function.

The FA is mainly founded on four basic concepts: distance, intensity, attractiveness, and movement.

Distance (r): The distance between two generic fireflies (e.g. i and j) is expressed by the Cartesian distance as follows:

$$r_{ij} = \|p_i - p_j\| = \sqrt{\sum_{k=1}^{n}(p_{ik} - p_{jk})^2} \tag{4}$$

Intensity I(r): In the traditional FA the light intensity perceived of each firefly depends on the distance r and it is proportional to the value of the objective function. In line with the formulation provided by Yang (2009), it can be expressed as:

$$I = I_0\, e^{-\gamma r^2} \tag{5}$$

where:
I_0 is the original light intensity
γ is the coefficient of light absorption.

Attractiveness (β): This represents a relative measure of the light perceived by beholders and other fireflies. Therefore the formulation of the attractiveness is expressed as:

$$\beta = \beta_0 e^{-\gamma r^2} \tag{6}$$

where β_0 represents the attractiveness when the distance r is equal to zero.

Movement: Finally, the movement of the generic firefly i, that is attracted by a brighter one, is expressed as:

$$p_i(t+1) = p_i(t) + \beta_0\, e^{-\gamma r_{ij}^2}(p_i(t) - p_j(t))^2 + \alpha\epsilon_i \tag{7}$$

In this equation $p_i(t)$ represents the position of the firefly at the time t; $\beta_0\, e^{-\gamma r_{ij}^2}(p_i(t) - p_j(t))^2$ represents the attraction between fireflies; $\alpha\epsilon_i$ represents the randomness of the process where the vector ϵ_i includes random numbers extracted from a normal distribution and α is a random parameter.

Summarising, the firefly algorithm is governed by three different parameters: the first (α) concerns the randomness; the second (β) regards the attractiveness, and the third (γ) is the absorption coefficient. This latter influences the attractiveness and affects the convergence process of the firefly algorithm as well as its evolution.

3.2. Modified Discrete Firefly algorithm

Several discrete formulations of the FA have been applied to the discrete optimization problems in the context of operations management (Sayadi et al., 2010; Lu & Wang, 2016; Osaba et al., 2016; Zhou et al., 2015).

In this paragraph a firefly algorithm has been adopted for the ML problem of the FMS. In the FMS considered several jobs, characterized by one or more operations, must be executed with different processing requirements (i.e. processing time and tool slots necessary to execute such operations) on a given number of multifunctional machines. More specifically, in each problem jobs and operations have to be loaded on four machines. Furthermore it is assumed that each machine has 5 tool slots and 480 minutes of available time.

In our approach we defined two vectors for each firefly and they represent a candidate solution to the problem respecting the problem constraints discussed in section 2: the machine assignment vector (M_string) containing the machines assigned for each operations and, the operation scheduling vector

(O_string) reporting the operations sequencing on the machines. The methodology developed is based on the following steps:

<u>Step 1.</u> The population is randomly initialized, i.e. each firefly of the initial population has been generated using the casual permutation of both the part types to be allocated and the available machine on which these part types can be allocated (M_string and O_string) (Karthikeyan, 2014). With reference to the first problem formulated by Mukopadhyay, Midha, & Krishna (1992), an example of firefly representation is reported in Table 1.

Table 1
M_string and O_string representation

Machine assignment vector																		
Position																		
1	2	3	4	5	6	7	8	9	10	11	12	13	14	15	16	17	18	19
Operations (O_{ik})																		
O_{11}	O_{21}	O_{22}	O_{23}	O_{31}	O_{32}	O_{41}	O_{42}	O_{51}	O_{52}	O_{61}	O_{62}	O_{63}	O_{71}	O_{72}	O_{73}	O_{81}	O_{82}	O_{83}
M_string																		
3	4	4	2	1	3	3	4	2	2	4	4	2	2	2	4	2	2	1

Operation scheduling vector																		
Sequence																		
	6		4	1	3		2			8			7			5		
Operation (O_{ik})																		
O_{61}	O_{62}	O_{63}	O_{41}	O_{42}	O_{11}	O_{31}	O_{32}	O_{21}	O_{22}	O_{23}	O_{81}	O_{82}	O_{83}	O_{71}	O_{72}	O_{73}	O_{51}	O_{52}
O_string																		
11	12	13	7	8	1	5	6	2	3	4	17	18	19	14	15	16	9	10

After that, the objective function of each firefly that belongs to the population is calculated in terms of SU and TP according to (3).

<u>Step 2</u>. After evaluating all the objective function value of each firefly of the population, the best one is selected in order to update the solution.

The updating is given by the firefly movement towards the brighter one, i.e. the best firefly of the population (Pbest). In order to calculate this movement, we must first measure the distance between the best firefly and the any other element of the population. Two methods has been adopted to evaluate the distance between two fireflies, i.e. the hamming distance and the position distance. The hamming distance, used in M_string, is expressed as the set of non-corresponding elements in two vectors. The position distance, used in O_string, is the number of minimal switches necessary to convert one string into another one. These distances allow to update the job sequences of the initial population, according to the attractiveness equation, and consequently the objective function values.

Considering as an example the best firefly and a generic firefly reported in table 2, the position distance for O_string is computed as follows:

Job 1. In both fireflies it is in the third position therefore in this case the position distance is 0.

Job 2. In the best firefly it occupies the fourth position while in the generic firefly the second one. The position distance for job 2 is 2.

Repeating the same process for the eight jobs, the position distance r is equal to 24.

Table 2
Position distance illustration

	Jobs							
Best firefly	4	3	1	2	8	5	6	7
Generic firefly	8	2	1	7	6	3	4	5

The attractiveness is expressed using the following equation:

$$\beta = \frac{\beta_0}{1 + \gamma r^2} \tag{8}$$

Once computed the β values, a randomized sequences uniformly distributed in [0,1] is generated. If the random value results lower or equal to β, the corresponding insertion in the machine assignment vector and pair-wise exchange in operation scheduling vector is performed on the elements of the current firefly.

Thereafter a random integer number is generated between 1 and the dimension of M_string. The corresponding position of the machine in M_string is replaced with the one with the shortest processing time. As regards O_string, two integer randomly generated select the two operations to be switched.

Step 3. For each firefly of the population an updated value of the objective function is calculated. In order to explore the solution space each firefly replaces the previous one not only if it is better than the previous solution. However the best solution of the updating replaces the previous global best firefly only if it is better than the previous one.

Step 4. Repeat the previous steps until the maximum number of generation is reached, i.e. 50.

4. Computational experiments

The algorithm was coded in MATLAB R2015b and was run on a personal computer with an Intel® Core™ i5 processor and 4GB RAM in order to evaluate the performance of the proposed modified DFA algorithm.

In this section our computational results are presented to illustrate the benefits of the DFA developed. Several runs and trials were conducted to define the final set of the parameters to be used:

Population size: 20
Attractiveness of firefly β_0: 1.0
Light absorption coefficient γ: 0.1
Randomization parameter α: 1.0
Maximum number of generations Gen_max: 50

In the following section the obtained results are reported.

The algorithm is tested on the benchmark problems available in Mukopadhyay, Midha, & Krishna (1992) and the comparative results between the proposed approach and the existing swarm-based ones reported in the literature are summarized in table 3. The comparative study was performed considering the heuristics with the same hypothesis of the proposed approach and developed by Yusof et al. (2014), Biswas and Mahapatra (2009), Biswas and Mahapatra (2008), Prakash et al. (2008), Prakash et al. (2008), Ponnambalam and Kiat (2008), Prakash et al. (2007), Tiwari et al. (1997), and Mukopadhyay et al. (1992).

The outcomes reported in Table 3 show that the algorithm outperforms the heuristics developed by Tiwari et al. (2007) and Mukopadhyay et al. (1992), and, except for the problem number 2 in which the SU and TP values obtained are respectively 9 and 46, reach the same results of the others swarm approaches.

In conclusion, the algorithm shows results that are comparable, and sometimes even better than, most of the other approaches in terms of quality of solutions obtained and computational times. Indeed for the most complex problem, i.e. the first one, the algorithm reaches the best solution in 8.7 seconds.

Table 3

Comparison of the proposed modified DFA with other-heuristic solutions available in the open literature for the problem set

Pr. N°	Proposed approach		Yusof et al., 2014 AIS		Biswas & Mahapatra, 2009 PSO		Biswas & Mahapatra, 2008 PSO		Prakash et al., 2008 AHACO		Prakash et al., 2008 MIA		Ponnambalam & Kiat, 2008 PSO		Prakash et al., 2007 ACO		Tiwari et al., 1997 AIS		Heuristic		Mukopadhyay et al., 1992 Heuristic	
	SU	TP	SU	TP	SU	TP	SU	TP	SU	TP	SU	TP	SU	TP	SU	TP	SU	TP	SU	TP	SU	TP
1	14	48	0	52	3	49	3	49	14	48	14	48	14	48	14	48	14	48	76	42	122	42
2	9	46	15	64	69	50	69	50	22	63	124	63	22	63	22	63	124	63	234	63	202	63
3	28	73	28	73	32	48	32	48	25	73	72	69	28	73	25	73	72	69	152	69	286	79
4	819	51	819	51	819	51	819	51	819	51	819	51	819	51	819	51	819	51	819	51	819	51
5	264	61	69	61	9	64	9	64	264	61	264	61	264	61	264	61	264	61	264	61	364	76
6	7	64	7	64	35	62	35	62	37	64	28	61	37	64	37	64	28	61	314	63	365	62
7	231	63	21	63	39	66	39	66	147	66	231	63	231	63	147	66	231	63	996	48	147	66
8	63	48	56	54	52	50	52	50	63	48	13	44	63	48	63	48	13	44	158	43	459	36
9	309	88	56	88	309	88	309	88	309	88	309	88	309	88	309	88	309	88	309	88	309	88
10	122	56	205	67	5	58	5	58	122	56	122	56	122	56	122	56	122	56	76	42	320	56

5. Conclusions

This paper proposes to apply a discrete firefly algorithm to the machine loading problem of the flexible manufacturing systems. Despite the machine loading problem in the context of flexible manufacturing is a topic deeply investigated in the literature and already tackled by researchers with several approaches, nevertheless the potentiality of the swarm-based approach makes it interesting and challenging.

Among them, the recently developed firefly algorithm has been selected due to its interesting adaptability features (i.e. for the calculation of distance and attractiveness between fireflies) and therefore suitable for our problem.

The present work has successfully developed an efficient heuristic based on firefly algorithm with the aim to formulate an objective function which takes into account both the throughput and the system unbalance. Iterative experiments had been conducted in order to achieve optimal or semi-optimal solutions and the findings have been analyzed and compared with the previous contributions on the topic.

The analysis of the machine loading problem concerns the challenge of locating the available resources (machines) to load the part types; the constraints of the problem are concurrently considered such as the number of operations for each part type, the allowable machines that may be allocated for the operation, and the machine time available.

The results highlight that the proposed approach offers results that are comparable to the best results of the other swarm algorithms, and outperforms the other heuristics considered for evaluation.

This paper addresses an interesting topics and it will be still the basis of new studies and research works.

Indeed, the proposed approach has been applied on datasets acquired in literature, so an extension of this work may be to apply it in a real life industrial problem; in addition, as future development the DFA procedure proposed may be tested on various similar optimization problems, particularly to the problems that regard the sequencing or the allocation of resources. In addition, this research can be also exploited in multi-objective loading and scheduling problem with the addiction of more flexible attributes. Finally further experiments may be conducted using a more complex procedure to generate the initial population as well as implement a strategy to avoid trapping into local optima.

References

Abazari, A., Solimanpur, M., & Sattari, H. (2012). Optimum loading of machines in a flexible manufacturing system using a mixed-integer linear mathematical programming model and genetic algorithm. *Computers and Industrial Engineering, 62*(2), 469-478.

Biswas, S., & Mahapatra, S. S. (2008). Modified particle swarm optimization for solving machine-loading problems in flexible manufacturing systems. *The International Journal of Advanced Manufacturing Technology, 39*(9-10), 931-942.

Biswas, S., & Mahapatra, S. S. (2009). An improved metaheuristic approach for solving the machine loading problem in flexible manufacturing systems. *International Journal of Services and Operations Management, 5*(1), 76-93.

Cerchione, R., & Esposito, E. (2016). A Systematic Review of Supply Chain Knowledge Management Research: State of the Art and Research Opportunities. *International Journal of Production Economics, 182*, 276-292.

ElMaraghy, H. (2005). Flexible and reconfigurable manufacturing systems paradigms. *International Journal of Flexible Manufacturing Systems, 17*(4), 261–276.

Karthikeyan, S., Asokan, P., & Nickolas, S. (2014). A hybrid discrete firefly algorithm for multi-objective flexible job shop scheduling problem with limited resource constraints. *The International Journal of Advanced Manufacturing Technology, 72*(9-12), 1567-1579.

Kumar, A., Prakash, Tiwari, M., Shankar, R., & Baveja, A. (2006). Solving machine-loading problem of a flexible manufacturing system with constraint-based genetic algorithm. *European Journal of Operational Research, 175*(2), 1043-1069.

Kumar, V. M., Murthy, A. N. N., & Chandrashekara, K. (2012). A hybrid algorithm optimization approach for machine loading problem in flexible manufacturing system. *Journal of Industrial Engineering International, 8*(1), 1-10.

Lee, S. M., & Jung, H. J. (1989). A multi-objective production planning model in a flexible manufacturing environment. *The International Journal of Production Research, 27*(11), 1981-1992.

Lu, S., & Wang, X. (2016). Modeling the fuzzy cold storage problem and its solution by a discrete firefly algorithm. *Journal of Intelligent and Fuzzy Systems, 31*(4), 2431-2440.

Mgwatu, M. (2011). Integration of part selection, machine loading and machining optimisation decisions for balanced workload in flexible manufacturing system. *International Journal of Industrial Engineering Computations, 2*(4), 913-930.

Moncayo-Martinez, L., & Zhang, D. (2011). Multi objective ant colony optimisation:a meta-heuristic approach to Supply Chain design. *International Journal of Production Economics, 131*(1), 407-420.

Mukopadhyay, S., Midha, S., & Krishna, V. (1992). A heuristic procedure for loading problem in flexible manufacturing system. *International Journal of Production Research, 30*(9), 2213-2228.

Osaba, E., Yang, X.-S., Diaz, F., Onieva, E., Masegosa, A., & Perallos, A. (2016). A discrete firefly algorithm to solve a rich vehicle routing problem modelling a newspaper distribution system with recycling policy. *Soft Computing*, 1-14.

Pandey, M. (2011). Operation allocation in flexible manufacturing system using immune algorithm. In M. T. Harding, *Evolutionary Computing in Advanced Manufacturing* (p. 95-121). Hoboken, NJ, USA.

Ponnambalam, S. G., & Kiat, L. S. (2008). Solving machine loading problem in flexible manufacturing systems using particle swarm optimization. *International Journal of Mechanical, Aerospace, Industrial, Mechatronic and Manufacturing Engineering, 2*(3), 242-247.

Prakash, A., Khilwani, N., Tiwari, M., & Cohen, Y. (2008). Modified immune algorithm for job selection and operation allocation problem in flexible manufacturing system. *Advances in Engineering Software, 39*(3), 219-232.

Prakash, A., Shankar, R., Shukla, N., & Tiwari, M. (2007). Solving machine loading problem of FMS: An artificial intelligence (AI) based random search optimization approach. In *Handbook of Computational Intelligence in Manufacturing and Production Management* (p. 19-43).

Prakash, A., Tiwari, M., & Shankar, R. (2008). Optimal job sequence determination and operation machine allocation in flexible manufacturing systems: An approach using adaptive hierarchical ant colony algorithm. *Journal of Intelligent Manufacturing, 19*(2), 161–173.

Sandhyarani, B., & Mahapatra, S. (2009). An improved metaheuristic approach for solving the machine loading problem in flexible manufacturing systems. *International Journal of Services and Operations Management, 5*(1), 76-93.

Santuka, R., Mahapatra, S., Dhal, P., & Mishra, A. (2015). An improved particle swarm optimization approach for solving machine loading problem in flexible manufacturing system. *Journal of Advanced Manufacturing Systems, 14*(3), 167-187.

Sayadi, M., Ramezanian, R., & Ghaffari-Nasab, N. (2010). A discrete firefly meta-heuristic with local search for makespan minimization in permutation flow shop scheduling problems. *International Journal of Industrial Engineering Computations, 1*(1), 1-10.

Srinivas, Tiwari, M., & Allada, V. (2004). Solving the machine loading problem in a flexible manufacturing system using a combinatorial auction-based approach. *International Journal of Production Research, 42*(9), 1879-1893.

Stecke, K. (1983). Formulation and solution of nonlinear integer production planning problems for flexible manufacturing systems. *Management Science, 29*, 273-288.

Stecke, K. E., & Solberg, J. J. (1981). Loading and control policies for a flexible manufacturing system. *The International Journal of Production Research, 19*(5), 481-490.

Swarnkar, R., & Tiwari, M. K. (2004). Modeling machine loading problem of FMSs and its solution methodology using a hybrid tabu search and simulated annealing-based heuristic approach. *Robotics and Computer-Integrated Manufacturing, 20*(3), 199-209.

Tiwari, M., Hazarika, B., Vidyarthi, N., Jaggi, P., & Mukopadhyay, S. (1997). A heuristic solution approach to the machine loading problem of FMS and its petri net model. *International Journal of Production Research, 35*(8), 2269-2284.

Tiwari, M., Saha, J., & Mukhopadhyay, S. (2007). Heuristic solution approaches for combined-job sequencing and machine loading problem in flexible manufacturing systems. *The International Journal of Advanced Manufacturing Technology, 31*(7-8), 716-730.

Yang, X. (2009). Firefly algorithms for multimodal optimization. In Stochastic algorithms:. In *Stochastic Algorithms: Foundations and Applications* (p. 169-178). Springer Berlin Heidelberg.

Yogeswaran, M., Ponnambalam, S., & Tiwari, M. (2009). An efficient hybrid evolutionary heuristic using genetic algorithm and simulated annealing algorithm to solve machine loading problem in FMS. *International Journal of Production Research, 47*(19), 5421-5448.

Yusof, U., Budiarto, R., & Deris, S. (2014). A hybrid of bio-inspired and musical-harmony approach for machine loading optimization in flexible manufacturing system. *International Journal of Innovative Computing, Information and Control, 10*(6), 2325-2344.

Yusof, U., Khalid, M., & Khader, A. (2014). Artificial immune system for flexible manufacturing system machine loading problem. *ICIC Express Letters, 8*(3), 709-716.

Zhou, L., Ding, L., Qiang, X., & Luo, Y. (2015). An improved discrete firefly algorithm for the traveling salesman problem. *Journal of Computational and Theoretical Nanoscience, 12*(7), 1184-1189.

8

Quality-productivity decision making when turning of Inconel 718 aerospace alloy: A response surface methodology approach

Hamid Tebassi[a]*, Mohamed Athmane Yallese[a], Salim Belhadi[a], Francois Girardin[b] and Tarek Mabrouki[c]

[a]*Mechanics and Structures Research Laboratory (LMS), May 8th 1945 University, P.O. Box 401, Guelma 24000, Algeria*
[b]*Laboratoire Vibrations Acoustique, INSA-Lyon, 25 bis avenue Jean Capelle, F-69621 Villeurbanne Cedex, France*
[c]*Université de Tunis El Manar, Ecole Nationale d'Ingénieurs de Tunis (ENIT), 1002, Tunis, Tunisie*

CHRONICLE	ABSTRACT
Keywords: *Surface roughness* *Productivity* *Response surface methodology* *Box-Cox technique* *Analysis of variance* *Response optimization*	Inconel 718 is among difficult to machine materials because of its abrasiveness and high strength even at high temperature. This alloy is mainly used in aircraft and aerospace industries. Therefore, it is very important to reveal and evaluate cutting tools behavior during machining of this kind of alloy. The experimental study presented in this research work has been carried out in order to elucidate surface roughness and productivity mathematical models during turning of Inconel 718 superalloy (35 HRC) with SiC Whisker ceramic tool at various cutting parameters (depth of cut, feed rate, cutting speed and radius nose). A small central composite design (SCCD) including 16 basics runs replicated three times (48 runs), was adopted and graphically evaluated using Fraction of design space (FDS) graph, completed by a statistical analysis of variance (ANOVA). Mathematical models for surface roughness and productivity were developed and normality was improved using the Box-Cox transformation. Results show that surface roughness criterion *Ra* was mainly influenced by cutting speed, radius nose and feed rate, and that the depth of cut had major effect on productivity. Finally, ranges of optimized cutting conditions were proposed for serial industrial production. Industrial benefit was illustrated in terms of high surface quality accompanied with high productivity. Indeed, results show that the use of optimal cutting condition had an industrial benefit to 46.9 % as an improvement in surface quality *Ra* and 160.54 % in productivity *MRR*.

1. Introduction

Nickel and Cobalt base corrosion, temperature and wear-resistant alloys, such as Inconel 718, are typically used in high temperature applications despite their classification as moderate to difficult when machining. It should be emphasized that these alloys can be relatively machined using conventional production methods at satisfactory rates. The properties that make Inconel 718 an important engineering material are responsible for its poor machinability (Tebassi et al., 2016a; Tebassi et al., 2017; Sharman

* Corresponding author
E-mail: tebassihamid@yahoo.fr (H. Tebassi)

et al., 2006). These properties are commonly the strength maintained during machining, the highly abrasive carbide particles contained in the microstructure and the poor thermal conductivity (Sharman et al., 2001; Li et al., 2002), which leads to elevate the cutting temperature up to 1200 °C at the rake face (Kitagawa et al., 1997). Consequently, the requirements for any cutting tool material used for machining these alloys should include: good wear resistance, high hot hardness, high strength and toughness, good thermal shock properties and adequate chemical stability at high temperature (Ezugwu et al., 1999); such as ceramic tools recommended for machining of these alloys compared with coated carbide at high cutting speed (Darwish, 2000; Nalbant et al., 2007; Gatto & Iuliano, 1997).

Nevertheless, notch wear V_N and flank wear V_B of the SiC whiskers and the Si_3N_4 ceramics, become very large at high speed and/or high feed rate (Narutaki et al., 1993). This large wear at high cutting speed is caused by diffusion between Si in the insert and Inconel 718 (El-Wardany et al., 1996). In addition, tool wear can also be caused by an abrasive rather than by a thermally process, by considering its poor thermal conductivity, which leads to elevate the temperature at the rake face (Kitagawa et al., 1997).The minimum flank wear was observed when using the SNGN tools at low cutting speeds or the RNGN tools at high cutting speeds (Altin et al., 2007). In addition, cutting forces and different types of tool wear were reduced by increasing the feed rate when turning of Inconel 718 with ceramic tool (El-Wardany et al., 1996), this parameter is considered as the most relevant cutting parameter affecting ceramic tool stresses (Kose et al., 2008; Nalbant et al., 2007; Ezugwu & Tang, 1995; Gatto & Iuliano, 1997; Altin et al., 2007).

However, Zhuang et al. (2014) concluded that the tool failure having another form. Certainly, the main failure modes of ceramic cutting tools during machining of Inconel 718 are notch wear and flank wear. This damage may be caused by the hardened layer beneath the workpiece surface. Considering mechanical properties of workpiece, specific shearing energy which is a strong function of feed rate; was the largest at the lowest cutting speed (125 m/min) and reduces subsequently when the cutting speeds increases up to 300 m/min (Pawade et al., 2009). Regarding surface quality and productivity, Yadav et al. (2015) and Tebassi et al. (2016b) obtained that the most influencing factor on **MRR** is depth of cut, whereas spindle speed and depth of cut are the most influencing factors on flank wear. In the same way, at low feed rate; the tendency for built-up edge formation, is also higher than at a higher feed, due to an increase in the size of the plastic deformation area at the interface of the tool and workpiece (Zhou et al., 2012).

Regarding this cited problematic, the main objective of the present work is to investigate the influence of different machining parameters on surface finish and productivity when turning of Inconel 718 super-alloy using the SiC Whisker ceramic tool. Consequently, the current study develops cutting strategies using a combination of the optimal parameters in the goal of part functional requirements while keeping high level of economical and industrial competition. Response surface methodology design approach, has been adopted for the experimental planning during turning of Inconel alloy. The results were analyzed in order to determine the optimal machining parameters settings and achieve optimal surface roughness and productivity. *ANOVA* was performed to investigate the more influencing parameters on the multiple performance characteristics. Mathematical models have been developed based on the full quadratic model which is generally used in RSM problems (Choudhury & El-Baradie, 1999), it can be written as follow:

$$Y = \beta_0 + \sum_{i=1}^{k} \beta_i X_i + \sum_{i=1}^{k} \beta_{ii} X_i^2 + \sum_{i<j}^{k} \beta_{ij} X_i X_j + \varepsilon_{ij} \qquad (1)$$

where, β_0 is constant, β_i, β_{ii} and β_{ij} are the coefficients of linear, quadratic and cross product terms, respectively. (Xi), are the actual variables that correspond to the studied machining parameters. The surface roughness criterion **Ra** and productivity **MRR** are indicated as Y_1 and Y_2 respectively, and analyzed as responses. After modeling, normality is tested and proved using Box-Cox transformation (Osborne, 2010; Sakia, 1992), which defined a family of power transformations and includes any positive or negative power, as well as the *log, power, square*…. etc. The Box-Cox power transformation on observations Yi (i =1, 2… n), is given by Box & Cox (1964) as:

$$Y_i^{(lamda)} = \begin{cases} \dfrac{Y_i^{(lamda)} - 1}{lamda} & \lambda \neq 0 \\ \\ log\, Y_i & \lambda = 0 \end{cases} \qquad (2)$$

where *lamda* is the power transformation parameter and *n* is the sample size. Multi-objective optimization procedure is allowed for minimizing the roughness **Ra** combined with maximal productivity **MRR**, using desirability approach.

2. Experimental procedure

2.1 Material and measurement

The aim of the current experimental work is to investigate the effect of cutting parameters on surface roughness and productivity with developing a correlation between them. In order to reach this objective, cutting speed (*Vc*), feed rate (*f*), depth of cut (*ap*) and nose radius *(r)* are chosen as process parameters. The workpiece material used in this study was Inconel 718 having hardness of 35 HRC and the chemical composition of: 0.08%C; 0.35%Mn; 0.35%Sn; 0.015%P; 0.015%S; 55%Ni+Co; 21%Cr; 12.29%Fe; 3.3%Mo; 1.15%Ti; 0.15%Cu; 0.8%Al; 5.5% (Cb+Ta). The workpiece geometry is a cylindrical bar specimen having 70 mm for diameter , 350 mm for length and cutting length of 20 mm. Straight turning operations have been achieved using a 6.6 kW spindle power   of the lathe (TOS TRENCIN model SN40C) under dry conditions. The experimental setup is shown in Fig.1. Cutting inserts were SiC whisker ceramic with the standard designation (ISO) of SNGN 120408, 120412 and 120416, commercialized by Sandvik under CC670 (Sandvik, 2009). The tool holder used in this experimental study has the standard designation of CSBNR2525M12 with the following angles: $\chi r = 45°$, $\alpha = 6°$, $\gamma = -6°$ and $\lambda = -6°$. Surface roughness measurements have been obtained directly on the machine without disassembling the workpiece; using a roughness meter (Surftest 201 Mitutoyo). Material removal rate **MRR** is calculated using Eq. (3) (Sandvik, 2009; Guo et al., 2012).

$$MRR = 1000 \times Vc \times ap, \qquad (3)$$

where **MRR** is in (mm^3/min), *Vc, ap, f* and *r* are respectively the cutting speed in (m/min), depth of cut in (mm), feed rate in (mm/rev) and nose radius in (mm).

2.2 Experimental design

The experimental approach was carried out in order to investigate the effects of the different factors and their interaction on surface roughness and productivity. Furthermore, three levels are specified for each factor (Table 1). The experimental tests are carried out according to the augmented small central composite design SCCD (16 basics runs with zero center points), replicated three times (48 runs) for reducing its mean standard error, 1 for *alpha* value was carried out. Because, thinking more broadly about what constitutes a good design is important (Anderson-Cook et al., 2009) and it makes imminent sense to compare these designs graphically (Jones, 2009). Two graphical techniques for comparing response surface designs are used (Khuri, 2009; Borkowski, 2009).The technique can be judged as it provides a very nice overview of methods for evaluating and comparing response surface designs (Piepel, 2009). For example, largest design can generate a better (lowest) error for a given design space (Christine et al., 2009). In order to evaluate our current design using graphical tool (FDS plot), Fig. 2 shows the fraction of design space plot for the small replicated design SCCD (48 runs). This graph presents a line graph showing the relationship between the "volume" of the design space (area of interest) and the amount of prediction error (Anderson-Cook et al., 2009; Khuri, 2009; Borkowski, 2009). The curve indicates what fraction (percentage) of the design space has a given prediction error or lower. In general, a lower and flatter FDS curve is better. Lower is more important than flatter. A lower curve translates to a higher Fraction of Design Space - more of the design has useful precision. Indeed, Fig. 2 shows that 78% of design space having a mean standard error inferior or equal than 0.997.

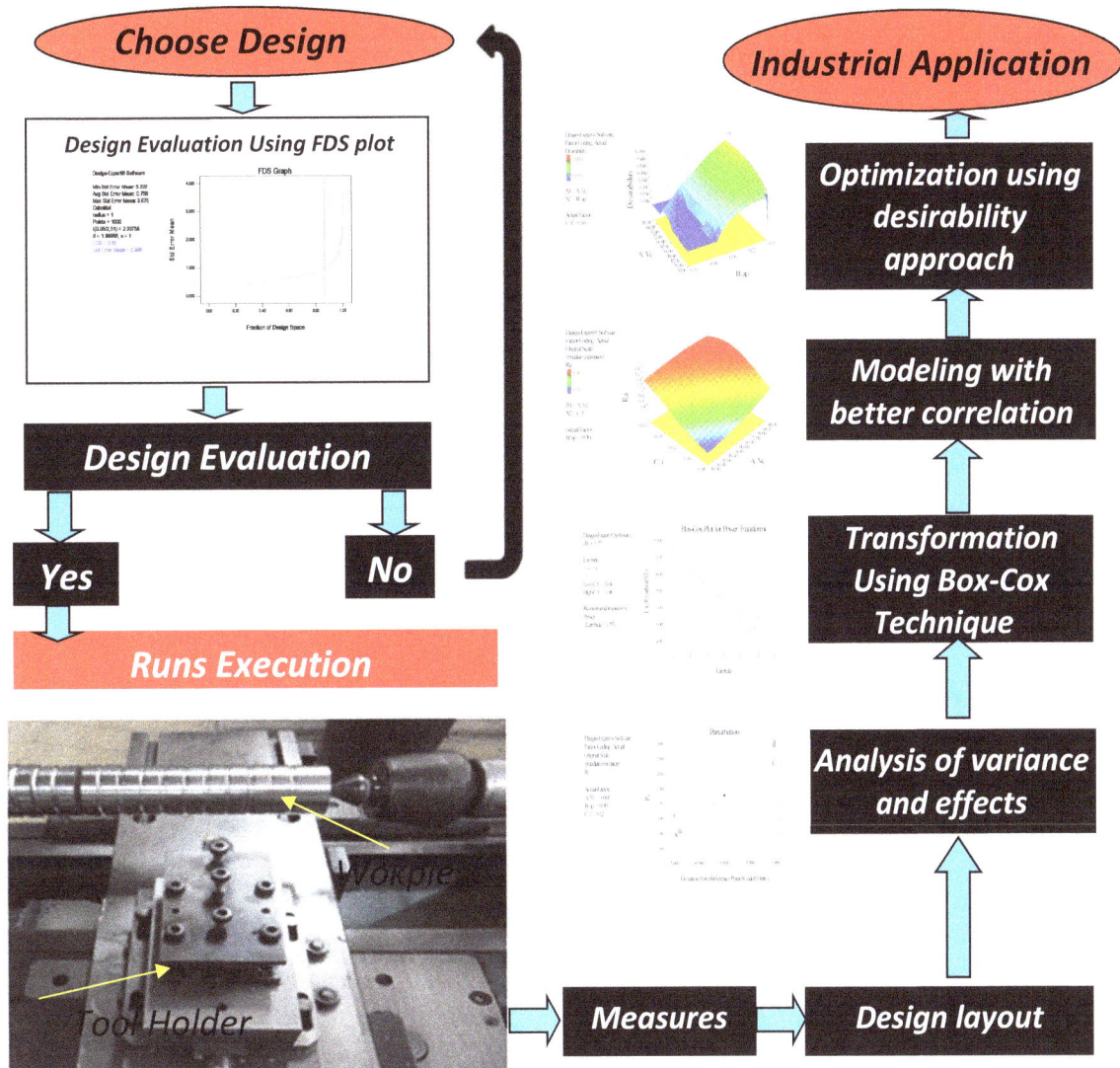

Fig. 1. Set-up and design of experiments

Table 1

Assignment for the levels to the factors

Level	Vc (m/min)	f (mm/rev)	ap (mm)	r (mm)
-1	100	0.08	0.1	0.8
0	150	0.12	0.2	1.2
1	200	0.16	0.3	1.6

The RSM applied in this work is considered as a procedure to identify a relationship between independent input process parameters and output data (process response).This procedure includes commonly six steps (Tebassi et al., 2016b; Gaitonde et al., 2009; Davim et al., 2008):

(1) define the independent input variables and the desired output responses, (2) adopt an experimental design, (3) perform regression analysis with the quadratic model of RSM, (4) perform a statistical analysis of variance (ANOVA) of the independent input variables in order to find parameters which affect the most significantly response, (5) determine the situation of the RSM model and decide whether this model needs screening variables or not and finally (6) optimize, conduct confirmation experiment with verifying the predicted output parameters.

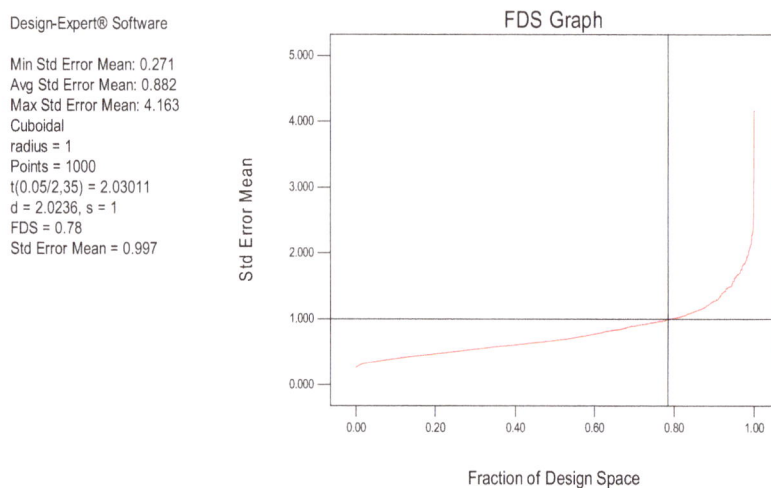

Design-Expert® Software

Min Std Error Mean: 0.271
Avg Std Error Mean: 0.882
Max Std Error Mean: 4.163
Cuboidal
radius = 1
Points = 1000
t(0.05/2,35) = 2.03011
d = 2.0236, s = 1
FDS = 0.78
Std Error Mean = 0.997

FDS Graph

Fig. 2. Fraction of design space plot for Small CCD with 48 runs

3. Results and discussion

The design of experiment was developed for assessing the influence of the cutting speed Vc, feed rate f, depth of cut ap and nose radius r on surface roughness **Ra** and productivity **MRR**. The statistical treatment of the data was made in three phases. The first phase includes the use of ANOVA, with the aim of studying the effect of factors and their interactions. The second phase consists of the choice of the best model transforms law to obtain the highest correlation between the parameters using Box-Cox Plot for Power Transforms (Osborne, 2010; Sakia, 1992). Afterwards, in the final phase, the results have to be optimized.

3.1 Statistical analysis

A variance analysis of the surface roughness and productivity was performed with the objective of analyzing the influence of cutting speed, feed rate, depth of cut and nose radius of cutting tool on the obtained outputs. This analysis was out for a 5% significance level, i.e., for a 95% confidence level.

3.1.1 Surface roughness

It can be shown in Table 2, that the surface roughness **Ra** was obtained in the range of (0.32–1.64) μm and the material removal rate **MRR** was obtained, in the range of (800 –7200) mm^3/min.

According to Table 3, that shows ANOVA for **Ra**, it can be observed that the significant terms on roughness **Ra** were Vc, ap, f and r, the products $Vc*ap$, $Vc*f$, $ap*f$, $f*r$, $ap*r$ and $Vc*r$ and the square f^2. The perturbation plot in Fig. 3 helps to compare the effect of all the factors at a particular point in the design space.

A steep slope for *Vc* and *r* or curvature in a factors *ap* and *f*, shows that the response is sensitive to those factors. Indeed, from this figure, it can be seen that the most significant factors on the parameters **Ra** was the cutting speed *Vc*, feed rate *f* and radius nose *r*.

Table 2
Experimental results for surface roughness and productivity

Run N°	Vc (m/min)	f (mm/rev)	ap (mm)	r (mm)	Ra (µm)	MRR (mm³/min)
1	200	0.12	0.2	1.6	0.41	4800
2	100	0.12	0.2	1.6	0.4	2400
3	100	0.08	0.3	1.6	0.4	2400
4	100	0.16	0.1	1.6	0.68	1600
5	150	0.12	0.3	0.8	0.87	5400
6	100	0.08	0.1	0.8	0.54	800
7	150	0.16	0.3	1.2	0.56	7200
8	150	0.08	0.2	1.2	0.33	2400
9	150	0.12	0.2	1.6	0.35	3600
10	200	0.12	0.1	1.2	0.41	2400
11	150	0.12	0.1	1.2	0.45	1800
12	100	0.08	0.3	1.6	0.4	2400
13	200	0.16	0.1	1.6	0.72	3200
14	200	0.08	0.3	1.6	0.4	4800
15	200	0.08	0.1	0.8	0.38	1600
16	100	0.16	0.2	0.8	1.6	3200
17	200	0.12	0.2	1.6	0.42	4800
18	150	0.16	0.3	1.2	0.57	7200
19	200	0.12	0.1	1.2	0.43	2400
20	100	0.16	0.2	0.8	1.64	3200
21	100	0.08	0.3	1.6	0.35	2400
22	100	0.08	0.3	1.6	0.35	2400
23	150	0.12	0.2	1.6	0.36	3600
24	150	0.08	0.2	1.2	0.32	2400
25	150	0.12	0.3	0.8	0.87	5400
26	200	0.08	0.3	1.6	0.46	4800
27	150	0.12	0.1	1.2	0.46	1800
28	200	0.08	0.1	0.8	0.37	1600
29	100	0.12	0.2	1.6	0.43	2400
30	100	0.16	0.1	1.6	0.69	1600
31	100	0.08	0.1	0.8	0.54	800
32	200	0.16	0.1	1.6	0.73	3200
33	100	0.08	0.3	1.6	0.36	2400
34	150	0.12	0.1	1.2	0.54	1800
35	100	0.08	0.3	1.6	0.36	2400
36	200	0.16	0.1	1.6	0.75	3200
37	200	0.08	0.1	0.8	0.41	1600
38	200	0.12	0.2	1.6	0.45	4800
39	100	0.16	0.1	1.6	0.66	1600
40	150	0.16	0.3	1.2	0.59	7200
41	150	0.08	0.2	1.2	0.35	2400
42	200	0.12	0.1	1.2	0.43	2400
43	100	0.16	0.2	0.8	1.61	3200
44	200	0.08	0.3	1.6	0.48	4800
45	150	0.12	0.3	0.8	0.87	5400
46	100	0.08	0.1	0.8	0.54	800
47	100	0.12	0.2	1.6	0.45	2400
48	150	0.12	0.2	1.6	0.38	3600

Depth of cut has been found with the lowest contribution. In addition, it is clearly observed that the feed rate strongly affects the surface roughness parameter **Ra**. This input parameter has an increasing effect that should be expected. It is well known that the theoretical geometrical surface roughness is primarily a function of the feed rate for a given nose radius and varies with the square of the feed rate value.

This is in good agreement with the established following equation (Davim et al., 2008; Schultheiss et al., 2014).

$$Ra = \frac{f^2}{32.r} \tag{4}$$

where f is the feed rate in (mm/rev) and r is the nose radius of the tool in (mm).

Table 3
Analysis of variance for **Ra**

Source	Sum of Squares	df	Mean Square	FValue	p-value
Model	4.55354735	12	0.37946228	454.016323	<0.0001
A-Vc	0.32731225	1	0.32731225	391.620233	<0.0001
B-f	0.11554926	1	0.11554926	138.251555	<0.0001
C-ap	0.09903889	1	0.09903889	118.497343	<0.0001
D-r	0.28633856	1	0.28633856	342.596325	<0.0001
AB	0.24064576	1	0.24064576	287.926119	<0.0001
AC	0.18507806	1	0.18507806	221.440873	<0.0001
AD	0.34212525	1	0.34212525	409.343581	<0.0001
BC	0.11949016	1	0.11949016	142.966729	<0.0001
BD	0.04523594	1	0.04523594	54.1235752	<0.0001
CD	0.00808913	1	0.00808913	9.67842809	0.00369834
A^2	0.00222189	1	0.00222189	2.65843242	0.11197006
B^2	0.02672432	1	0.02672432	31.9749217	<0.0001
Residual	0.02925265	35	0.00083579		
Lack of Fit	0.01165265	2	0.00582632	10.9243577	<0.0001
Pure Error	0.0176	33	0.00053333		
Cor Total	4.5828	47			

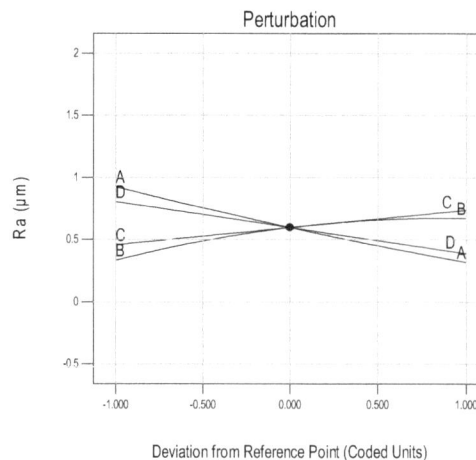

Fig. 3. Perturbation plot for **Ra**

The Fig. 4 shows the Box-Cox plot for **Ra**. This plot provides a guideline for selecting the correct power law transformation (Osborne, 2010; Sakia, 1992).
A recommended transformation is listed, based on the best lambda value, which is found at the minimum point of the curve generated by the natural log of the sum of squares of the residuals. From this figure,

we can see that the current *lambda* value for the chosen law transformation is equal to 1 and its recommended value is equal to 1. For this, the chosen law transformation is shown in Eq. (5).

$$[Ra]^{lamda} = [Ra]^1 = a_0 + \sum_{i=1}^{4} a_i X_i + \sum_{i=1}^{4} a_{ii} X_i^2 \tag{5}$$

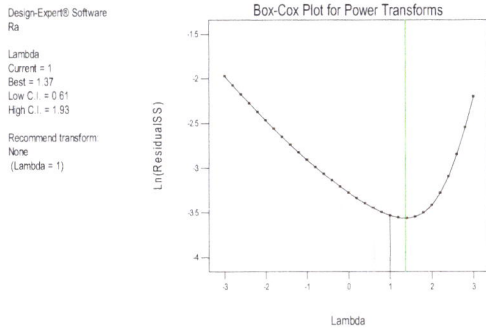

Fig. 4. Box-Cox plot for *Ra*

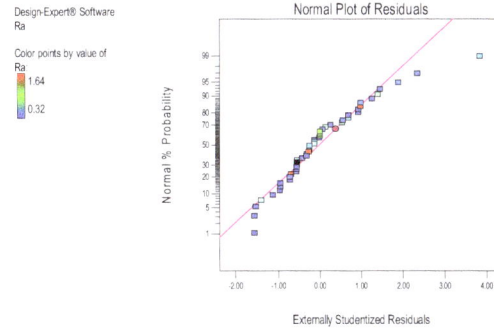

Fig. 5. Normal plot of residuals for *Ra*

The normal plot of residuals for the surface roughness *Ra* was plotted in Fig. 5. The data follows the straight line (Sahoo & Mishra, 2014), closely. This indicates that the transformation of the response provides a good analysis, and the model proposed in Eq. (5) is adequate.

3.1.2 Productivity

The Table 4 shows ANOVA corresponding to the material removal rate *MRR*. This table shows that the effects of cutting speed *Vc*, depth of cut *ap*, feed rate *f* and the products *Vc* ×*ap*, *Vc*×*f* and *ap*×*f* are all significant.

Table 4
Analysis of variance for *MRR*

Source	Sum of Squares	df	Mean Square	FValue	p-value
Model	7.863E+006	6	1.310E+006	3289.54	< 0.0001
A-Vc	2.018E+006	1	2.018E+006	5064.79	< 0.0001
B-f	2.225E+006	1	2.225E+006	5585.19	< 0.0001
C-ap	5.807E+006	1	5.807E+006	14576.89	< 0.0001
AB	69827.27	1	69827.27	175.28	< 0.0001
AC	1.863E+005	1	1.863E+005	467.54	< 0.0001
BC	3.656E+005	1	3.656E+005	917.80	< 0.0001
Residual	16333.58	41	398.38		
Lack of Fit	16333.58	8	2041.70		
Pure Error	0.000	33	0.000		
Cor Total	7.879E+006	47			

The perturbation plot in Fig. 6 shows a steep slope for depth of cut *ap* compared with feed rate *f* and cutting speed *Vc*. Consequently, the response is sensitive to these factors and the highest contribution comes with the depth of cut followed by feed rate and cutting speed. Fig. 7 shows the Box-Cox plot for *MRR*. Consequently, the chosen *lambda* value for the law transformation for *MRR* is to 0.85 as shown in Eq. (6).

$$[MRR]^{lambda} = [MRR]^{0.85} = a_0 + \sum_{i=1}^{4} a_i X_i + \sum_{i=1}^{4} a_{ii} X_i^2 + \sum_{i<j}^{4} a_{ij} X_i X_j \tag{6}$$

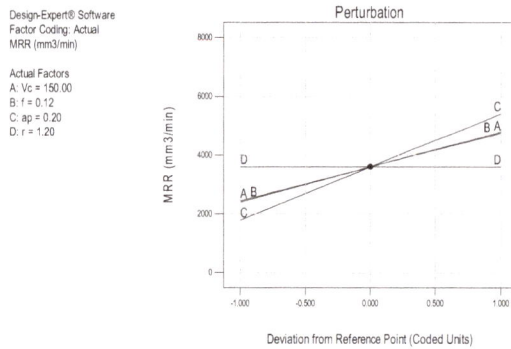

Fig. 6. Perturbation plot for *MRR*   **Fig. 7.** Normal plot of residuals for *MRR*

The normal plot of residuals for the productivity *MRR* was plotted in Fig. 8. The data follows the straight line (Sahoo & Mishra, 2014), closely. This indicates that the response transformation provides a better analysis, and the model proposed in Eq. (6), is adequate.

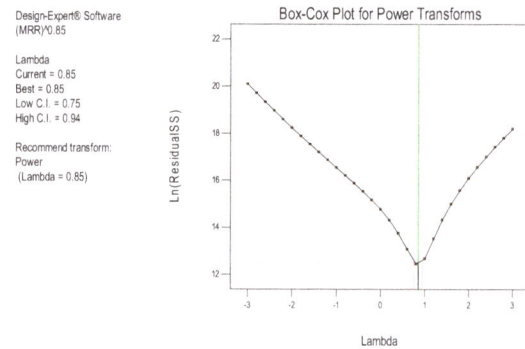

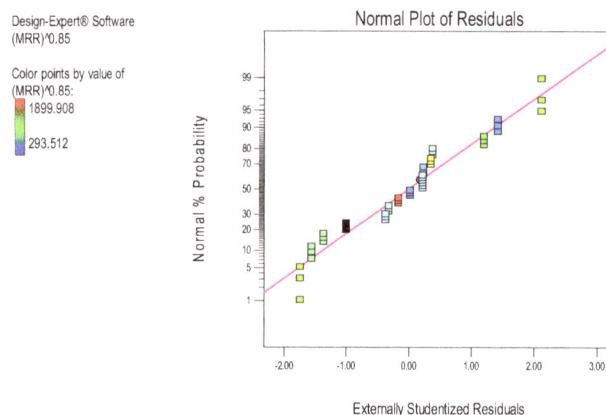

Fig. 8. Normal plot of residuals for *MRR*

3.2 Mathematical models

The initial analysis of the responses obtained from RSM includes all parameters and their interactions. The relationship between the factors and the process response were modeled by full quadratic model (Choudhury & El-Baradie, 1999).

Based on Eq. (5) and Eq. (6), the roughness *Ra* and material removal rate *MRR* models in term of actual factors are given below in Eq. (7) and Eq. (8), respectively.

$$Ra\ (\mu m) = +0.33385 - 1.275\ E\text{-}3 + 36.510 \times f + 14.271 \times ap - 4.802 \times r - 0.138\ Vc \times f - 0.053$$
$$Vc \times ap - 0.016\ Vc \times r - 59.128\ f \times ap + 12.213\ f \times r + 1.885\ ap \times r + 1.030\ E\text{-}5Vc^2 - 59.566\ f^2$$

$$(7)$$

R-Squared = 0.9936;   Adj R-Squared = 0.9914;
Pred R-Squared = 0.9885;   Adeq Precision = 85.298.

According to the Eq. (7), the "Pred R-Squared" of 0.9885 is in reasonable agreement with the "Adj. R-Squared" of 0.9914; i.e. the difference is less than 0.2 "Adeq Precision" measures the signal to noise ratio. A ratio greater than 4 is desirable. The current ratio of 85.298 indicates an adequate signal. Consequently, this model can be used to navigate the design space.

$$MRR = \begin{bmatrix} 405.466 - 2.434\,Vc - 4028.462\,f - 2524.981\,ap + 31.828\,Vc*f + \\ 20.711\,Vc*ap + 32683.751\,f*ap \end{bmatrix}^{0.85}$$

(8)

R-Squared = 0.9979;　　　　　Adj R-Squared = 0.9976;
Pred R-Squared = 0.9970;　　　Adeq Precision = 208.932.

However, from the Eq. (8), the "Pred R-Squared" of 0.9970 is in reasonable agreement with the "Adj R-Squared" of 0.9976; i.e. the difference is less than 0.2. "Adeq Precision" measures the signal to noise ratio. A ratio greater than 4 is desirable. The current value of 208.932 indicates an adequate signal. Consequently, this model can be used to navigate the design space. Based on the model shown in Eq. (7) and from Fig. 9 (a), (b) and (c), that show 3D surface plot for *Ra*, it can be seen that for low feed rate, the surface roughness is highly sensitive to depth of cut (Fig. 9 (a)). An increase in the latter sharply degrades the surface finish. At highest values of feed rate, depth of cut has a contradictory effect, using middle values of cutting speed and radius nose. Nevertheless, this variation becomes smallest with highest values of cutting speed; using middle value of feed rate and radius nose (Fig. 9 (b)). In addition, it revealed that a combination of middle cutting speed along with lower feed rate, high radius nose and middle value of depth of cut is necessary for obtaining better surface finish. The highest value of surface roughness can be shown, when lower values of cutting speed and radius nose are used (Fig. 9 (c)).

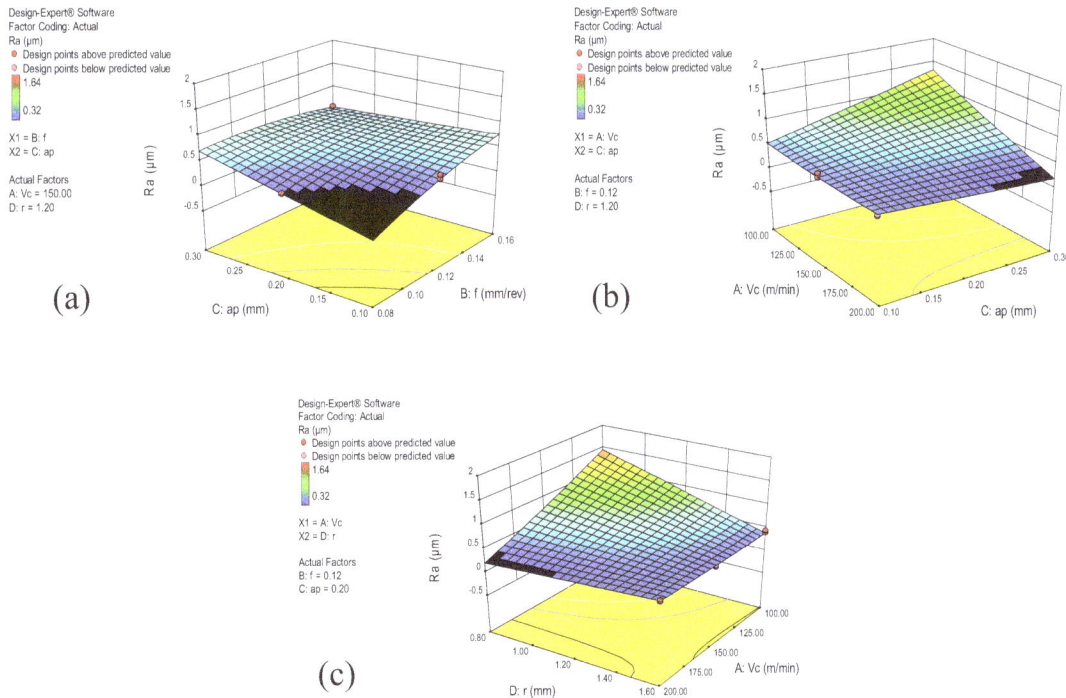

Fig. 9. 3D surface plot for *Ra*

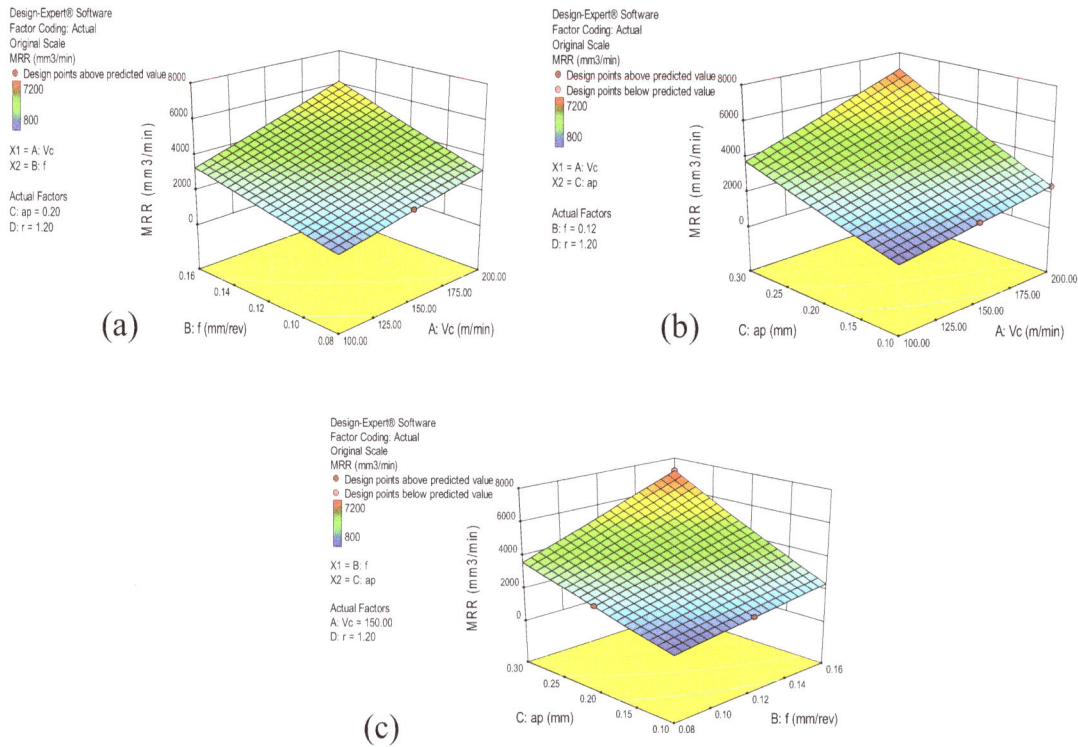

Fig. 10. 3D surface plot for *MRR*

Concerning the material removal rate *MRR* and based on the Eq. (8), it can be observed from Fig. 10 (a), (b) and (c), that the highest *MRR* can be resulted with the combination of the highest *Vc*, higher *ap* and higher *f*. In addition, the lowest *MRR* value can be observed at lower values of *Vc*, *ap* and *f*. Depth of cut maintains the highest effect on productivity. The comparison between actual and predicted response for *Ra* and *MRR* is illustrated in Fig. 11(a) and (b) respectively. The results of comparison were proven to predict the surface roughness and material removal rate close to those readings recorded experimentally with a 95% confidence interval. According to those figures, it can be seen that points split is evenly by the 45 degree line. This reflects the good agreement between experimental values illustrated in Table 2 and predicted values obtained with models shown in Eq. (7) and Eq. (8).

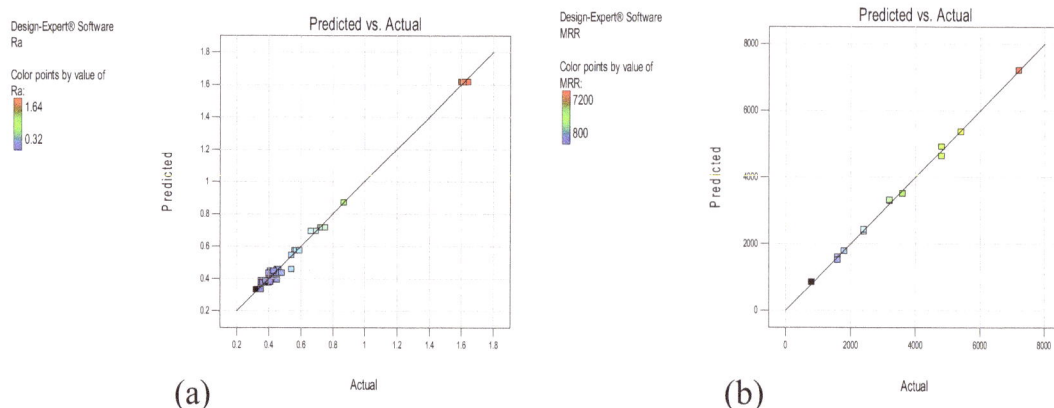

Fig. 11. Comparison between predicted and experimental results **(a)** for *Ra* and **(b)** for *MRR*

One of the main goals for the experiment is to investigate optimal values of cutting parameters. In order to obtain the desired value of the machined surface roughness *Ra* of the industrial product, which presents an agreement with higher productivity; using importance degrees for each output parameter (Table 5), the fixed value of 1.6 mm for tool nose radius *r* is given as an industrial constraint.

Joint optimization must satisfy the requirements for all the responses in the set. Optimization achievement is measured by the composite desirability which is the weighted geometric main of the individual desirability's for the responses on a range from zero to one. Value of 1.0 represents the ideal case and zero indicates that one or more responses are outside acceptable limits (Sahoo & Mishra, 2014; Myers, 2016).

Table 5
Goals and parameter ranges for optimization of cutting conditions

Name	Goal	Lower Limit	Upper Limit	Lower Weight	Upper Weight	Importance
A : Vc	Is in range	100	200	1	1	3
B : f	Is in range	0.08	0.16	1	1	3
C : ap	Is in range	0.1	0.3	1	1	3
D : r	Is equal to 1.6	0.8	1.6	1	1	3
Ra :	Maximise	0.32	1.64	1	1	3
MRR :	Maximise	800	7200	1	1	3

The optimum cutting parameters obtained with the importance degrees of 3 for *Ra* and *MRR* are chosen in term of highest desirability value (Fig. 12) to be cutting speed of 189.51 m/min, feed rate of 0.15 mm/rev and cutting depth of 0.3 mm using tool nose radius of 1.6 mm. The predicted responses are *Ra* = 0.30 μm and 8142.14 mm^3/min for *MRR* with desirability value of 1.00 as shown in Fig. 13; which presents solution ramps of multi-objective optimization (Tebassi et al., 2016b).

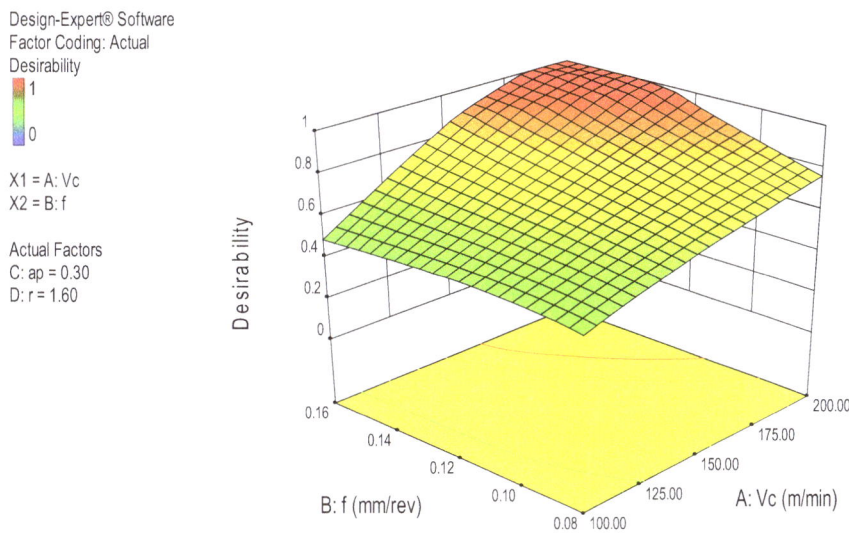

Fig. 12. 3D surface plot for desirability

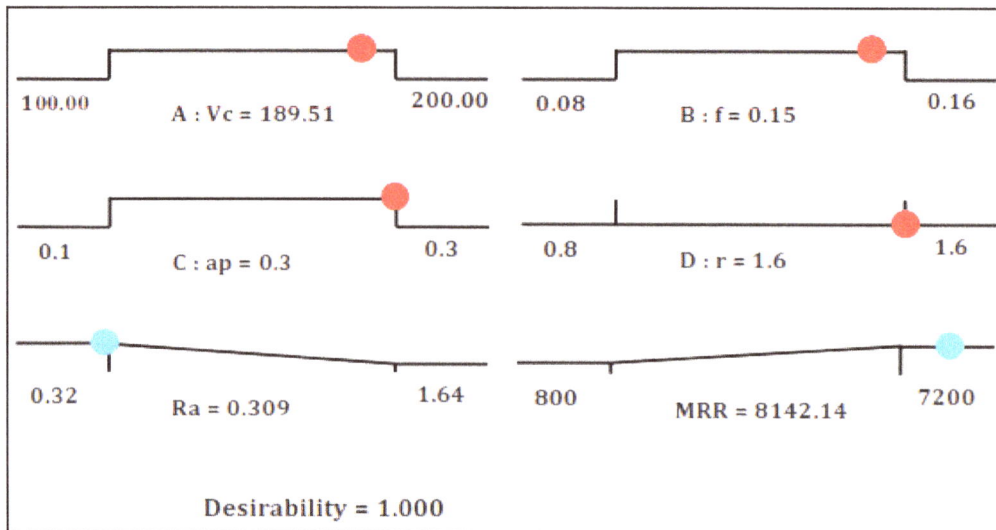

Fig. 13. Solution ramps of multi-objective optimization

3.4 Industrial benefit

In order to summarize the industrial importance of this presented research, the benefits in surface quality and in productivity are illustrated. Indeed, Fig. 14 (a) and (b) show medium value, optimal value and the obtained benefit for *Ra* and *MRR,* respectively. From Fig. 14 (a) it can be seen that the medium value for *Ra* is equal to 0.565 µm and 0.30 µm for its optimal value. Consequently, from this approach, the benefit in surface roughness is equal to 46.9 %. For the productivity, it can be observed in the Fig. 14 (b) that the medium value for *MRR* was 3125 mm³/min and the optimal value was 8142.14 mm³/min. Consequently, the benefit obtained from this approach in productivity was 160.54 %.

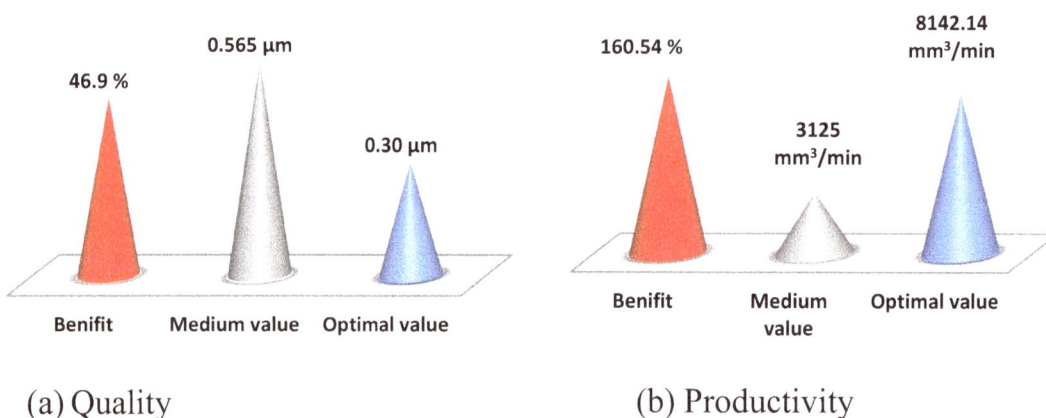

(a) Quality

(b) Productivity

Fig. 14. (a) Improvement in surface quality and **(b)** industrial benefit in productivity

4. Conclusions

The current investigation was based on RSM using fraction of design space (FDS) plot for design evaluation and Box-Cox plots for developing mathematical models and improving their normality. The desirability approach was followed for the optimizing surface roughness and productivity based on cutting parameters (cutting speed, feed rate, depth of cut and nose radius). The important findings can be summarized as follows:

1. Based on the fraction of design space plot, 78 % of the current designs maintained a standard error mean inferior or equal to 0.997 representing the best effectiveness of design;

2. The Box-Cox transformation improved significantly the normality and provided a good correlation for models. Indeed, the R^2 for the surface roughness model was equal to 0.9936 and for the productivity model was equal to 0.9979;

3. The signal to noise ratios (*Adeq Precision*) of about 85.298 for surface roughness model and *208.932* for productivity model have indicated an adequate signal. Consequently, these models can be used to navigate the design space;

4. Cutting speed, feed rate and nose radius have the greatest influences on surface roughness;

5. Depth of cut has the greatest influence on the productivity;

6. It has been found for the tested material that the optimal combination of cutting parameters for the obtained roughness **Ra** (0.30 μm) and the productivity (8142.14 mm^3/min) was 189.51 m/min for **Vc**, 0.3 mm for **ap**, 0.15 mm/rev for **f** and 1.6 mm for **r.**

7. Using optimal cutting conditions, the benefit in productivity registered was about 160.54 % accompanied with an improvement in surface quality of 46.9 %

Regarding the current investigation, the multi-objective optimization methodology proposed can be considered as a powerful approach based on graphical tool for design evaluation (FDS plot) and a design improved modeling step for the best correlation. It can offer to scientific researchers and industrial metalworking a helpful for choice, comparing designs, improvement normality of models used to navigate the design space, and multi-objective optimization procedure for various combinations of input (Workpiece hardness, tool material, tool geometry …) and output (surface finish, productivity, surface integrity …) parameters of machining process.

Acknowledgements

This work was achieved in the laboratories LMS (University of Guelma Algeria) in collaboration with Acoustic Vibration Laboratory (INSA-Lyon, France). The authors would like to thank the Algerian Ministry of Higher Education and Scientific Research (MESRS) and the Delegated Ministry for Scientific Research (MDRS) for granting financial support for CNEPRU Research Project, CODE: A11N01UN240120140013 (University 08 May 1945, Guelma).

References

Altin, A., Nalbant, M., & Taskesen, A. (2007). The effects of cutting speed on tool wear and tool life when machining Inconel 718 with ceramic tools. *Materials & design*, *28*(9), 2518-2522.

Anderson-Cook, C. M., Borror, C. M., & Montgomery, D. C. (2009). Response surface design evaluation and comparison. *Journal of Statistical Planning and Inference*, *139*(2), 629-641.

Borkowski, J. J. (2009). Discussion of "Response surface design evaluation and comparison" by Christine Anderson-Cook, Connie Borror, and Douglas Montgomery. *Journal of Statistical Planning and Inference*, *139*(2), 650-652.

Box, G. E., & Cox, D. R. (1964). An analysis of transformations. *Journal of the Royal Statistical Society. Series B (Methodological)*, *26*(2), 211-252.

Choudhury, I. A., & El-Baradie, M. A. (1999). Machinability assessment of inconel 718 by factorial design of experiment coupled with response surface methodology. *Journal of Materials Processing Technology*, *95*(1), 30-39.

Anderson-Cook, C. M., Borror, C. M., & Montgomery, D. C. (2009). Response surface design evaluation and comparison. *Journal of Statistical Planning and Inference*, *139*(2), 629-641.

Darwish, S. M. (2000). The impact of the tool material and the cutting parameters on surface roughness of supermet 718 nickel superalloy. *Journal of Materials Processing Technology*, *97*(1), 10-18.

Davim, J. P., Gaitonde, V. N., & Karnik, S. R. (2008). Investigations into the effect of cutting conditions on surface roughness in turning of free machining steel by ANN models. *Journal of materials processing technology, 205*(1), 16-23.

El-Wardany, T. I., Mohammed, E., & Elbestawi, M. A. (1996). *Cutting temperature of ceramic tools in high speed machining of difficult-to-cut materials. International Journal of Machine Tools and Manufacture, 36*(5), 611-634.

Ezugwu, E. O., Wang, Z. M., & Machado, A. R. (1999). The machinability of nickel-based alloys: a review. *Journal of Materials Processing Technology,86*(1), 1-16.

Ezugwu, E. O., & Tang, S. H. (1995). Surface abuse when machining cast iron (G-17) and nickel-base superalloy (Inconel 718) with ceramic tools. *Journal of Materials Processing Technology, 55*(2), 63-69.

Gaitonde, V. N., Karnik, S. R., Figueira, L., & Davim, J. P. (2009). Machinability investigations in hard turning of AISI D2 cold work tool steel with conventional and wiper ceramic inserts. *International Journal of Refractory Metals and Hard Materials, 27*(4), 754-763.

Gatto, A., & Iuliano, L. (1997). Advanced coated ceramic tools for machining superalloys. *International Journal of Machine Tools and Manufacture, 37*(5), 591-605.

Guo, Y., Loenders, J., Duflou, J., & Lauwers, B. (2012). Optimization of energy consumption and surface quality in finish turning. *Procedia CIRP, 1*, 512-517.

Jones, B. (2009). Discussion of "Response surface design evaluation and comparison" by Christine Anderson-Cook, Connie Borror and Douglas Montgomery. *Journal of Statistical Planning and Inference, 139*(2), 642-644.

Khuri, A. I. (2009). Discussion of "Response surface design evaluation and comparison" by Christine M. Anderson-Cook, Connie M. Borror, Douglas C. Montgomery. *Journal of Statistical Planning and Inference, 139*(2), 647-649.

Kitagawa, T., Kubo, A., & Maekawa, K. (1997). Temperature and wear of cutting tools in high-speed machining of Inconel 718 and Ti 6Al 6V 2Sn.*Wear, 202*(2), 142-148.

Kose, E., Kurt, A., & Seker, U. (2008). The effects of the feed rate on the cutting tool stresses in machining of Inconel 718. *Journal of Materials Processing Technology, 196*(1), 165-173.

Li, L., He, N., Wang, M., & Wang, Z. G. (2002). High speed cutting of Inconel 718 with coated carbide and ceramic inserts. *Journal of Materials Processing Technology, 129*(1), 127-130.

Myers, R. H., Montgomery, D. C., & Anderson-Cook, C. M. (2016). *Response surface methodology: process and product optimization using designed experiments.* John Wiley & Sons.

Nalbant, M., Altın, A., & Gökkaya, H. (2007). The effect of cutting speed and cutting tool geometry on machinability properties of nickel-base Inconel 718 super alloys. *Materials & Design, 28*(4), 1334-1338.

Narutaki, N., Yamane, Y., Hayashi, K., Kitagawa, T., & Uehara, K. (1993). High-speed machining of Inconel 718 with ceramic tools. *CIRP Annals-Manufacturing Technology, 42*(1), 103-106.

Osborne, J. W. (2010). Improving your data transformations: Applying the Box-Cox transformation. Practical Assessment, *Research & Evaluation, 15*(12), 1-9.

Pawade, R. S., Sonawane, H. A., & Joshi, S. S. (2009). An analytical model to predict specific shear energy in high-speed turning of Inconel 718. *International Journal of Machine Tools and Manufacture, 49*(12), 979-990.

Piepel, G. F. (2009). Discussion of "Response surface design evaluation and comparison" by CM Anderson-Cook, CM Borror, and DC Montgomery. *Journal of Statistical Planning and Inference, 139*(2), 653-656.

Sakia, R. M. (1992). The Box-Cox transformation technique: a review. *The Statistician, 41*(2), 169-178.

Sahoo, A. K., & Mishra, P. C. (2014). A response surface methodology and desirability approach for predictive modeling and optimization of cutting temperature in machining hardened steel. *International Journal of Industrial Engineering Computations, 5*(3), 407- 416.

Sandvik., C. (2009). *Catalogue General, Outils de coupe Sandvik Coromant.* Tournage – Fraisage – perçage – Alésage – Attachements.

Schultheiss, F., Hägglund, S., Bushlya, V., Zhou, J., & Ståhl, J. E. (2014). Influence of the minimum chip thickness on the obtained surface roughness during turning operations. *Procedia CIRP, 13*, 67-71.

Sharman, A., Dewes, R. C., & Aspinwall, D. K. (2001). Tool life when high speed ball nose end milling Inconel 718™. *Journal of Materials Processing Technology, 118*(1), 29-35.

Sharman, A. R. C., Hughes, J. I., & Ridgway, K. (2006). An analysis of the residual stresses generated in Inconel 718™ when turning. *Journal of Materials Processing Technology, 173*(3), 359-367.

Tebassi, H., Yallese, M., & Meddour, I. (2016a). A new method for evaluation nominal coefficient of friction and frictional forces in turning and inserts characterization using cutting forces profiles. *Engineering Solid Mechanics, 4*(1), 1-10.

Tebassi, H., Yallese, M., Khettabi, R., Belhadi, S., Meddour, I., & Girardin, F. (2016b). Multi-objective optimization of surface roughness, cutting forces, productivity and Power consumption when turning of Inconel 718. *International Journal of Industrial Engineering Computations, 7*(1), 111-134.

Tebassi, H., Yallese, M., Meddour, I., Girardin, F. & Mabrouki, T. (2017). On the modeling of surface roughness and cutting force when turning of inconel 718 using artificial neural network and response rurface methodology: accuracy and benefit. *Periodica Polytechnica Mechanical Engineering*, 61(1), 1-11.

Yadav, R. K., Abhishek, K., & Mahapatra, S. S. (2015). A simulation approach for estimating flank wear and material removal rate in turning of Inconel 718.*Simulation Modelling Practice and Theory, 52*, 1-14.

Zhou, J. M., Bushlya, V., & Stahl, J. E. (2012). An investigation of surface damage in the high speed turning of Inconel 718 with use of whisker reinforced ceramic tools. *Journal of Materials Processing Technology, 212*(2), 372-384.

Zhuang, K., Zhu, D., Zhang, X., & Ding, H. (2014). Notch wear prediction model in turning of Inconel 718 with ceramic tools considering the influence of work hardened layer. *Wear, 313*(1), 63-74.

9

Hybridized genetic-immune based strategy to obtain optimal feasible assembly sequences

Bala Murali Gunji[a], B. B. V. L. Deepak[a*], M V A Raju Bahubalendruni[b] and Bibhuti Bhusan Biswal[b]

[a]Department of Industrial Design, National Institute of Technology- Rourkela,769008, India
[b]Department of mechanical engineering, GMR Institute of Technology –Andrapradesh, 532127, India

CHRONICLE	ABSTRACT
Keywords: *Assembly sequence planning* *Artificial immune system* *Genetic algorithm* *Assembly automation* *Feasible assembly sequence* *Assembly automation*	An appropriate sequence of assembly operations increases the productivity and enhances product quality there by decrease the overall cost and manufacturing lead time. Achieving such assembly sequence is a complex combinatorial optimization problem with huge search space and multiple assembly qualifying criteria. The purpose of the current research work is to develop an intelligent strategy to obtain an optimal assembly sequence subjected to the assembly predicates. This paper presents a novel hybrid artificial intelligent technique, which executes Artificial Immune System (AIS) in combination with the Genetic Algorithm (GA) to find out an optimal feasible assembly sequence from the possible assembly sequence. Two immune models are introduced in the current research work: (1) Bone marrow model for generating possible assembly sequence and reduce the system redundancy and (2) Negative selection model for obtaining feasible assembly sequence. Later, these two models are integrated with GA in order to obtain an optimal assembly sequence. The proposed AIS-GA algorithm aims at enhancing the performance of AIS by incorporating GA as a local search strategy to achieve global optimum solution for assemblies with large number of parts. The proposed algorithm is implemented on a mechanical assembly composed of eleven parts joined by several connectors. The method is found to be successful in achieving global optimum solution with less computational time compared to traditional artificial intelligent techniques.

1. Introduction

Industrial robots are widely used in manufacturing applications for material handling and assembly operations. Typical assembly process consumes major stake of overall manufacturing cost, an appropriate assembly sequence can reduce the assembly time and effort and thus offers high productivity. Achieving an optimal assembly sequence is a major problem faced by the manufacturers as there exists millions of possible combinations. Many traditional methods have been proposed by the researchers for optimization of assembly sequence with large search space and high computational time (Linn & Liu, 1999). Computer

* Corresponding author
E-mail: bbv@nitrkl.ac.in (B.B.V.L. Deepak)

aided design (CAD) integrated knowledge based methods have been implemented to extract the predicate criteria such as liaison data (Bahubalendruni & Biswal, 2014a, 2014b; Bahubalendruni & Biswal, 2015a) mechanical feasibility (Bahubalendruni et al., 2015b, 2015c), stability (Bahubalendruni & Biswal, 2015d), and geometrical feasibility (Bahubalendruni et al., 2016 & Bahubalendruni et al., 2016) to solve optimum assembly sequence planning.

Knowledge based methods solve Assembly Sequence Planning (ASP) accurately but these methods take more computational time due to their huge search space utilization (Dong et al., 2007). To overcome this problem researchers implemented several Artificial Intelligent (AI) techniques to achieve optimum assembly sequence. In most AI techniques, the input supplied to the process is generated manually which is again time consuming and needs accuracy without error. Due to human intervention while supplying inputs, these techniques have the drawback of high redundancy and premature convergence during the optimization of assembly sequence. To validate the quality of an assembly sequence, it must be tested for assembly predicates. Besides liaison and geometrical feasibility predicates, stability and mechanical feasibility are two more essential assembly predicates to yield the appropriate results (Zha et al., 1998). Limitations of AI based optimal ASP methods in terms of assembly predicate consideration, premature convergence and high computational time are listed in Table 1.

Table 1
Assembly predicate considerations with AI in the cited research literature

Reference	Algorithm	Objective function (minimization of)	Assembly predicate criteria	Premature convergence/ Local Optima	Computational time
(Hsu et al., 2011)	KBM	Computational time	C	C	NC
(Kashkoush & ElMaraghy, 2015)		Computational time	C	C	NC
(Chen & Liu, 2001)	GA	Computational time	C	C	NC
(Chen et al., 2002)		Computational time	C	C	NC
(Sinanoglu & Riza Börklü, 2005)	NN	Computational time	C	C	NC
(Bahubalendruni et al., 2015)	PSO	Computational time	C	C	NC
(Wang & Liu, 2010)		Computational time	C	C	C
(Nayak et al., 2015)	SA	Computational time	C	C	NC
(Lee & Gemmill, 2001)		Assembly cost	C	C	NC
(Bahubalendruni et al., 2016)	AIS	Computational time	C	C	NC
(Biswal et al., 2103)		Assembly cost	C	C	NC
(Chang et al., 2009)		Computational time	C	C	C
(AkpıNar et al., 2013)	HA	Computational time	C	C	NC
(Xing & Wang, 2012)		Computational time	C	C	C
(Shan et al., 2009)		Computational time	C	C	C

PSO-Particle Swarm Optimization, NN-Neural Network, SA-Simulating Annealing, KBM-Knowledge Based Method, GA-Genetic Algorithm, HA- Hybrid Algorithm.
Note: C- considered, NC- Not considered.

Even though AI with Computer Aided Design (CAD) integrated algorithms have been developed, but these have failed to achieve the global optimum solution for ASP. To solve many of the engineering problems, GA and AIS have been implemented successfully. The current research work is focused on the hybridization of these algorithms to achieve the global optimization solution of ASP. In this paper, a hybrid AIS-GA based algorithm along with CAD integration for extraction of liaison data, stability feasibility, mechanical feasibility and geometrical feasibility is explained to obtain an Optimal Feasible Assembly Sequence (OFAS) with less computational time.

2. Assembly Information Extraction

Generally, performing the required input data for generating the algorithm to optimize the assembly manually is a time consuming process. This section details about the methods to extract liaison matrix, interference matrices, stability matrix and mechanical feasibility matrix from CAD environment. Computer Aided Three-dimensional Interactive Application Version-5 (CATIA V5R17) is used as CAD tool and programming is done in Visual Basic (VB scripting) to extract the outcomes. A transmission assembly consists of eleven parts is considered in this investigation is as shown in the Fig.1. The six possible directions of part movements during assembly process can be represented as: x+ → 1, x- → 2, y+ → 3, y- → 4, z+ → 5 and z- → 6.

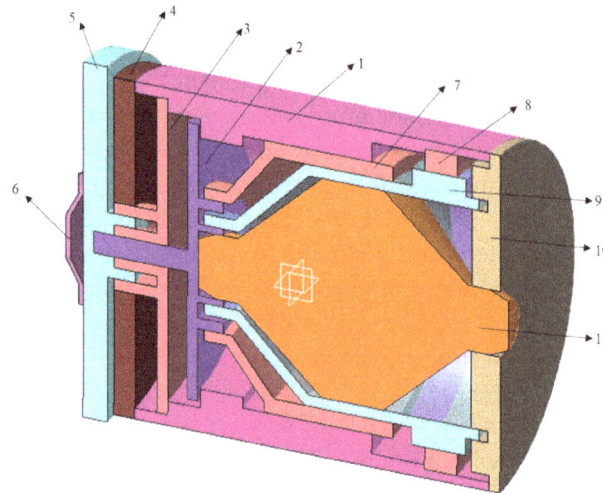

Fig. 1. Cut section view of transmission assembly (De Fazio and Whitney, 1987)

2.1. Liaison Matrix Extraction

Liaison data gives the information about all possible mating surfaces of an assembly. Two weights 1 &0 are assigned for mating and not mating parts respectively. Since in the current investigation an eleven-part assembly structure has been considered, a matrix with 11-by-11 is extracted from the CAD tool: CATIA-V5R17. The obtained liaison matrix for the considered eleven-part mechanical assembly structure is as follows:

$$
\text{Liaisoan matrix} =
\begin{array}{c}
\\ 1 \\ 2 \\ 3 \\ 4 \\ 5 \\ 6 \\ 7 \\ 8 \\ 9 \\ 10 \\ 11
\end{array}
\begin{array}{c}
\begin{array}{ccccccccccc}
1 & 2 & 3 & 4 & 5 & 6 & 7 & 8 & 9 & 10 & 11
\end{array} \\
\left(
\begin{array}{ccccccccccc}
0 & 0 & 1 & 1 & 0 & 0 & 1 & 0 & 0 & 1 & 1 \\
0 & 0 & 1 & 0 & 1 & 0 & 1 & 1 & 1 & 0 & 0 \\
1 & 1 & 0 & 1 & 1 & 0 & 0 & 0 & 0 & 0 & 0 \\
1 & 0 & 1 & 0 & 0 & 0 & 0 & 0 & 0 & 0 & 0 \\
0 & 1 & 1 & 0 & 0 & 1 & 0 & 0 & 0 & 0 & 0 \\
0 & 0 & 0 & 0 & 1 & 0 & 0 & 0 & 0 & 0 & 0 \\
1 & 1 & 0 & 0 & 0 & 0 & 0 & 1 & 0 & 0 & 0 \\
0 & 1 & 0 & 0 & 0 & 0 & 1 & 0 & 1 & 1 & 1 \\
0 & 1 & 0 & 0 & 0 & 0 & 0 & 1 & 0 & 0 & 1 \\
1 & 0 & 0 & 0 & 0 & 0 & 0 & 1 & 0 & 0 & 0 \\
1 & 0 & 0 & 0 & 0 & 0 & 0 & 1 & 1 & 0 & 0
\end{array}
\right)
\end{array}
$$

2.2. Interference Matrix Extraction

The interference matrices give the information about all possible interferences of the components during the assembly. Two weights 0 &1 are assigned for the interference and not interference respectively. Part movements during the assembly process are of six directions, six interference matrices corresponding to each direction will be obtained. Since in the current investigation an eleven-part assembly structure has been considered, a matrix with 11-by-11 is extracted from the CAD tool.

The obtained interference matrices for the considered eleven-part mechanical assembly structure in X+ and X- directions are as follows:

X+

	1	2	3	4	5	6	7	8	9	10	11
1	0	0	0	1	0	1	0	0	0	0	0
2	0	0	0	0	0	1	0	0	0	1	1
3	0	0	0	0	0	1	1	1	1	1	1
4	1	0	0	0	0	1	1	1	1	1	1
5	0	0	0	0	0	1	1	1	1	1	1
6	1	1	1	1	1	0	1	1	1	1	1
7	0	0	1	1	1	1	0	0	0	1	1
8	0	0	1	1	1	1	0	0	0	0	0
9	0	0	1	1	1	1	0	0	0	0	0
10	0	1	1	1	1	1	1	0	0	0	1
11	0	1	1	1	1	1	1	0	0	1	0

X-

	1	2	3	4	5	6	7	8	9	10	11
1	0	0	0	1	0	1	0	0	0	0	0
2	0	0	0	0	0	1	0	0	0	1	1
3	0	0	0	0	0	1	1	1	1	1	1
4	1	0	0	0	0	1	1	1	1	1	1
5	0	0	0	0	0	1	1	1	1	1	1
6	1	1	1	1	1	0	1	1	1	1	1
7	0	0	1	1	1	1	0	0	0	1	1
8	0	0	1	1	1	1	0	0	0	0	0
9	0	0	1	1	1	1	0	0	0	0	0
10	0	1	1	1	1	1	1	0	0	0	1
11	0	1	1	1	1	1	1	0	0	1	0

The obtained interference matrices for the considered eleven-part mechanical assembly structure in Y+ and Y- directions are as follows:

Y+

	1	2	3	4	5	6	7	8	9	10	11
1	0	1	1	1	1	1	0	0	1	0	0
2	0	0	1	1	1	1	0	0	0	0	0
3	0	0	0	1	1	1	0	0	0	0	0
4	0	0	0	0	1	1	0	0	0	0	0
5	0	0	0	0	0	1	0	0	0	0	0
6	1	0	0	0	0	0	0	0	0	1	0
7	1	1	1	1	1	1	0	0	0	1	0
8	1	1	1	1	1	1	1	0	0	1	0
9	1	1	1	1	1	1	1	1	0	1	0
10	1	1	1	1	1	1	1	1	1	0	0
11	1	1	1	1	1	1	1	1	1	1	0

Y-

	1	2	3	4	5	6	7	8	9	10	11
1	0	0	0	0	0	1	1	1	1	1	1
2	1	0	0	0	0	0	1	1	1	1	1
3	1	1	0	0	0	0	1	1	1	1	1
4	1	1	1	0	0	0	1	1	1	1	1
5	1	1	1	1	0	0	1	1	1	1	1
6	1	1	1	1	1	0	1	1	1	1	1
7	0	0	0	0	0	0	0	1	1	1	1
8	0	0	0	0	0	0	0	0	1	1	1
9	1	0	0	0	0	0	0	0	0	1	1
10	0	0	0	0	0	1	1	1	1	0	1
11	0	0	0	0	0	0	0	0	0	0	0

The obtained interference matrices for the considered eleven-part mechanical assembly structure in Z+ and Z- directions are as follows:

Z+

	1	2	3	4	5	6	7	8	9	10	11
1	0	0	0	1	0	1	0	0	0	0	0
2	0	0	0	0	0	1	0	0	0	1	1
3	0	0	0	0	0	1	1	1	1	1	1
4	1	0	0	0	0	1	1	1	1	1	1
5	0	0	0	0	0	1	1	1	1	1	1
6	1	1	1	1	1	0	1	1	1	1	1
7	0	0	1	1	1	1	0	0	0	1	1
8	0	0	1	1	1	1	0	0	0	0	0
9	0	0	1	1	1	1	0	0	0	0	0
10	0	1	1	1	1	1	1	0	0	0	1
11	0	1	1	1	1	1	1	0	0	1	0

Z-

	1	2	3	4	5	6	7	8	9	10	11
1	0	0	0	1	0	1	0	0	0	0	0
2	0	0	0	0	0	1	0	0	0	1	1
3	0	0	0	0	0	1	1	1	1	1	1
4	1	0	0	0	0	1	1	1	1	1	1
5	0	0	0	0	0	1	1	1	1	1	1
6	1	1	1	1	1	0	1	1	1	1	1
7	0	0	1	1	1	1	0	0	0	1	1
8	0	0	1	1	1	1	0	0	0	0	0
9	0	0	1	1	1	1	0	0	0	0	0
10	0	1	1	1	1	1	1	0	0	0	1
11	0	1	1	1	1	1	1	0	0	1	0

2.3 Stability Matrix Extraction

Stability matrix gives the information about the stability of the parts during assembly whether the parts are incomplete stable, partial stable or complete stable. In this study three weights, 0, 1 & 2 have been allotted for incomplete stable, partial stable and complete stable respectively. Since in the current investigation an eleven-part assembly structure has been considered, a matrix with 11-by-11 is extracted from the CAD tool-CATIA (V5R17).

Combined stability matrix:

$$
\begin{array}{c}
 \\
1 \\ 2 \\ 3 \\ 4 \\ 5 \\ 6 \\ 7 \\ 8 \\ 9 \\ 10 \\ 11
\end{array}
\begin{array}{c}
\begin{array}{ccccccccccc}
1 & 2 & 3 & 4 & 5 & 6 & 7 & 8 & 9 & 10 & 11
\end{array} \\
\left(\begin{array}{ccccccccccc}
0 & 0 & 2 & 2 & 0 & 0 & 1 & 0 & 0 & 1 & 2 \\
0 & 0 & 1 & 0 & 2 & 0 & 1 & 1 & 1 & 0 & 0 \\
2 & 1 & 0 & 1 & 1 & 0 & 0 & 0 & 0 & 0 & 0 \\
2 & 0 & 1 & 0 & 0 & 0 & 0 & 0 & 0 & 0 & 0 \\
0 & 2 & 1 & 0 & 0 & 2 & 0 & 0 & 0 & 0 & 0 \\
0 & 0 & 0 & 0 & 2 & 0 & 0 & 0 & 0 & 0 & 0 \\
1 & 1 & 0 & 0 & 0 & 0 & 0 & 1 & 0 & 0 & 0 \\
0 & 1 & 0 & 0 & 0 & 0 & 1 & 0 & 1 & 1 & 1 \\
0 & 1 & 0 & 0 & 0 & 0 & 0 & 1 & 0 & 0 & 1 \\
1 & 0 & 0 & 0 & 0 & 0 & 0 & 1 & 0 & 0 & 0 \\
2 & 0 & 0 & 0 & 0 & 0 & 0 & 1 & 1 & 0 & 0
\end{array}\right)
\end{array}
$$

2.4. Mechanical Feasibility Extraction

Mechanical feasibility matrix gives the information about, whether two parts can be joined by a physical connector in the presence of already existed part without any interference. The order of the matrix is represented by n-by-n-by-n. The third dimension represents, whether the part represented in it offers any interference to place hard connections between parts represented in first two dimensions. The mechanical feasibility matrix for 11-part assembly is shown in the Fig. 2.

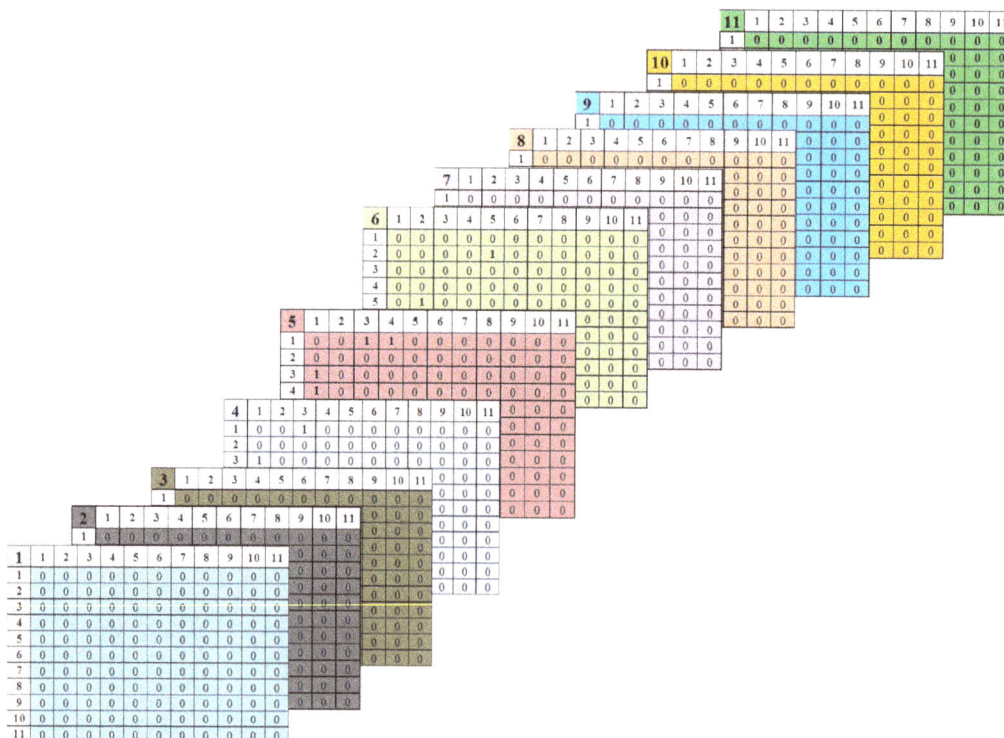

Fig. 2. Mechanical feasibility matrix for 11-part assembly

In the Fig. 2. it is clearly shown that parts (1&3) cannot be joint with a physical connector in the presence of part 4. Similarly, parts (1&3) and (1&4) can't be joint with physical connectors in the presence of part 5 and also parts (2&5) can't be joint with physical connector in the presence of part 6.

3. Fitness Function Evaluation

Once an assembly sequence qualifies liaison, stability and geometrical feasibility, the sequence is treated to be a valid sequence which will be further sent for fitness value calculation. Therefore a fitness function (m) is to be developed which should satisfy the following three criteria in the assembly tasks.

i). Assembly operations should associate with less number of end-effectors/tool changes (n_{tc}). An end effector is used to hold an object to perform assembly task. Since an assembled product consists of various mechanical parts there is a requirement of using several end effectors to hold the specific object. It is required to minimize the number of changes of end effector during assembly process because changing an end effector requires lot of time.

$$\Rightarrow m_1 = w_{tc} * n_{tc} \qquad (1)$$

Where (w_t) is the weighing parameter is corresponds to end-effectors/tool change

ii). Assembly operations should associate with less number of direction changes (n_{dc}).It is prefers to perform the assembly operation for all parts in a single direction. Changing the direction frequently leads to the time consuming.

$$\Rightarrow m_2 = w_{dc} * n_{dc} \qquad (2)$$

Where$(w_{dc})$is the weighing parameter is corresponds to directional changes

iii).Assembly operations should associates with less number of part movements (n_{pm}). During assembly operation parts to be aligned in correct position for quicker assembly because parts are hold by the gripper, which leads to time and energy consumption.

$$\Rightarrow m_3 = w_{pm} * n_{pm} \qquad (3)$$

Where$(w_{pm})$is the weighing parameter is corresponds to part movements

By combining the Eq. (1), Eq. (2) and Eq. (3) the final form of the fitness will be as follows

$$m = m_{1+}m_2 + m_3$$

$$m = w_{tc} * n_{tc} + w_{dc} * n_{dc} + w_{pm} * n_{pm} \qquad (4)$$

By observing the equation (4), the fitness value of an optimal assembly sequence must be low. In the meanwhile, the assembly sequence should satisfy the basic predicate tests. Therefore, it is required to apply an AI based optimization technique in order to minimize the total manufacturing time and cost

The proposed method is implemented for various combinations of weights aiming minimum gripper change, minimum assembly direction and minimum part movements. However, the selection of weights is significantly dependent on the industrial engineer's decision based on the assembly tool facilities and time to change grippers.

4. Artificial Intelligent Techniques in Assembly Sequence

Several AI techniques are presented to achieve the optimized assembly sequence. But these techniques which are using are not sufficient to get the optimum feasible sequence because of huge search time and manual data extraction. Instead of manual data extraction, CAD platform has been used here while subjected to four basic criteria. This research work has been carried out by combining the AIS-GA algorithms with integration of CAD to achieve global optimal solution while reducing the system redundancy and computational time.

4.4. Artificial Immune System

The interest in studying the immune system is increasing over the last few years. Computer scientists, engineers, mathematicians, philosophers and other researchers are particularly interested in the capabilities of this system, whose complexity is comparable to that of the brain (Deepak & Parhi 2016). Many properties of the immune system are of great interest for computer scientists and engineers:

- *Uniqueness*: each individual possesses its own immune system.
- *Recognition of Foreigners*: the antigens are recognized and eliminated from body.
- *Anomaly Detection:* the immune system can detect and react to pathogens that the body has never encountered before.
- *Distributed Detection:* the cells are distributed all over the body and are not to subject any centralized control.
- *Noise Tolerance:* the system is flexible since the recognition of the antigens is not required.
- *Reinforcement Learning and Memory:* future responses to the same pathogens are faster and stronger since the immune system can "learn" the structures of pathogens.

Because of its special features, several models have been introduced to solve various engineering problems (Deepak & Parhi 2013). The following immune inspired algorithms are used in this investigation:

- *Bone Marrow Models:* For generating possible assembly sequences.
- *Negative Selection Model:* To reduce the system redundancy and predicate check.

4.4.1. Bone marrow model for Assembly Sequence Generation

In this model two gene libraries are considered with five-bit memory in order to generate an antibody (assembly sequence). First gene library i.e. library-1 corresponds to the part allocation and second gene library i.e. library-2 corresponds to the memory allocation of the part. The population selection is done randomly by the AIS algorithm shown in the Fig. 3. Once an allocation is filled in the either of the library, that specific part/memory allocation will not be repeated during the current assembly sequence generation. For example, if part A is selected from library-1 and memory allocation-1 from library-2, then part A will be in allocation-1 and these two allocations will not be repeated.

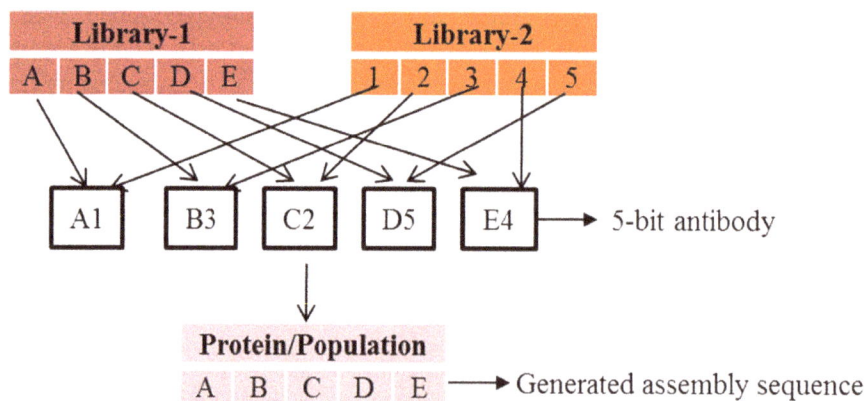

Fig. 3. Selection of population using AIS method

In order to avoid the system redundancy / generation of same antibody (assembly sequence) string current anti body string is subjected to reinforcement learning.

4.4.2. Negative selection algorithm

The negative selection algorithm is applied for the stored anti body string (assembly sequence) generated from the bone marrow model is tested with the predicated criteria. If this criterion is satisfying, then this model stores the current assembly sequence as a feasible assembly sequence which shown in the Fig.4. Later the stored feasible assembly sequence will be evaluated for its optimal criteria.

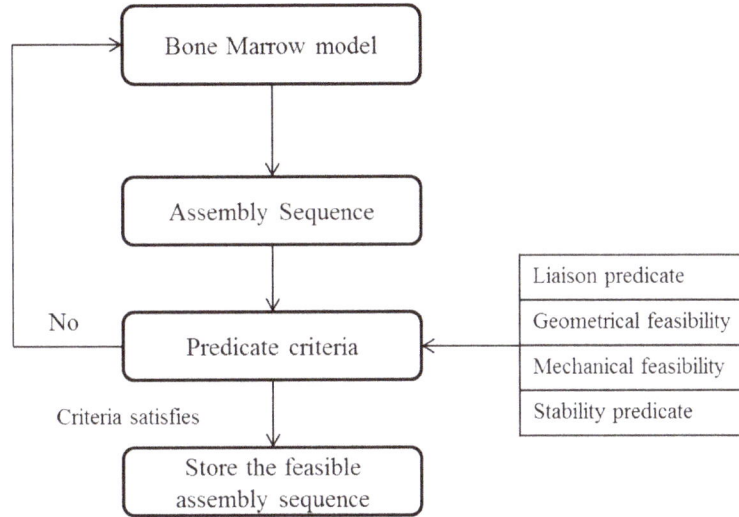

Fig. 4. Negative selection algorithm

4.5. Genetic Algorithm

Genetic algorithm is one of the artificial intelligent technique which uses the heretic search to minimize the process of natural selection. This is generally used for the search problems to generate the useful solution for optimization which is inspired by the inspired by natural evolution techniques, such as inheritance, mutation, selection, and crossover. In this algorithm population initialization is from negative selection algorithm. For the generated population fitness value is to be calculated as discussed section-3. For the evaluated population apply mutation and cross over genetic operators to generate the off-springs as shown in the Fig. 5 & 6.

Mutation: Mutation is swapping of two randomly chosen parts. A typical example is shown in the Fig. 4 for the mutation where C3 and D4 are chosen randomly in the parent, after swapping it generates the off spring.

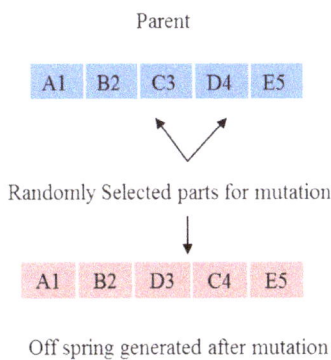

Fig. 5. Mutation for off-spring generation

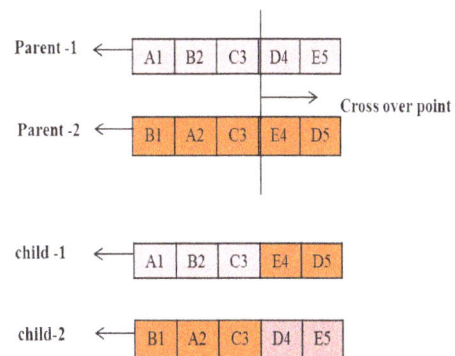

Fig. 6. Cross over for generating off-spring

Cross over: Cross over is selecting a common point of two parents so that two new off -springs will be generated. A typical example for crossover implementation is shown in Fig. 5, where a common point is selected for the parent-1& parent-2. After cross over two off-springs are generated having characteristics of both parents. Flow chart for the implementation of GA is explained in detail with the flow chart as shown in the Fig. 7.

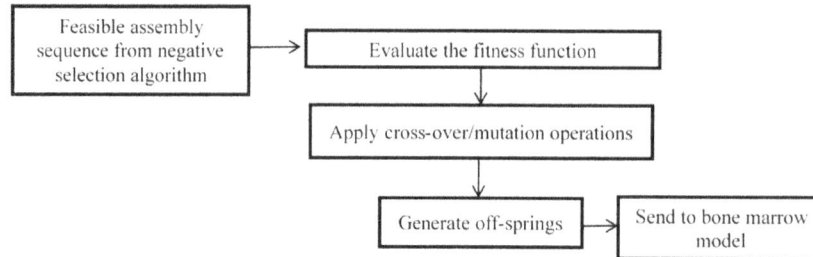

Fig. 7. Flow chart of GA

4.6. Hybrid AIS-GA Architecture

The Hybrid AIS-GA based algorithm for generating the optimal feasible assembly sequence is explained in detail with the steps as below:

Step 1. Generate the population (assembly sequence) using bone marrow model.
Step 2. Check the predicate criteria (Liaison data, Stability, Geometric feasibility and Mechanical feasibility) for the generated assembly sequence through negative selection model.
Step 3. Send the feasible assembly sequence to GA model to evaluate its fitness value.
Step 4. Apply mutation and cross over genetic operators for producing the off-springs (new assembly sequences).
Step 5. Send the generated off-springs to the bone marrow model for its storage.
Step 6. Continue step 2 to step 5 until it reaches optimal criteria.

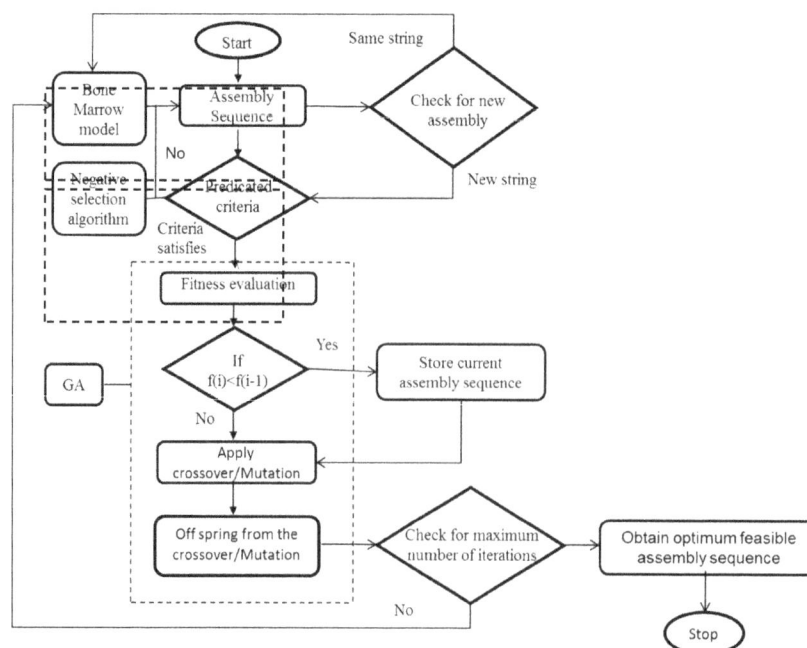

Fig. 8. Flow chart of the AIS-GA algorithm

5. Results & Discussion

The developed methodology for obtaining an optimum feasible assembly sequence has been implemented for a transmission assembly. This product consists of 11 parts (parts labelled with 1 to 11) as shown in Fig. 1 and the possible sequences are 11-factorial. Hybrid AIS-GA model which is discussed in section 4.3 produces the feasible assembly sequence planning. From the 11! Number of possible sequences, AIS-GA produced 1808 feasible sequences as listed in the supplementary material. For these feasible sequences optimization criteria to be applied with respect to end effector changes, directional changes and part movements.

The developed algorithm has been implemented to a standard transmission assembly as considered by Wang et al. (2005) and Smith (2004). Classical genetic algorithm has been implemented by Smith (2004) and ant colony optimization has been implemented by Wang et al. (2005) to produce OFAS while considering the directional changes as the objective function. From the past experience, it is found that the obtained OFAS is 11-9-8-7-10-1-2-3-4-5-6. However, the past strategies are dealt with only considering the directional changes during the assembly process. Whereas in the current research work, all criteria such as directional changes, gripper changes and energy of the parts are considered for obtaining the optimal feasible assembly sequence. More over the results are compared with the GA and AIS, which take less time for obtaining the optimal feasible assembly sequence as shown in the table 4 &table 5 respectively. Three weighing parameters corresponding to the end effectors (w_{tc}), directional changes (w_{dc}) and part movements (w_{pm}) are to be decided. In this investigation following two cases are considered for investing the weighing parameters

Case 1: Performing assembly operation by considering equal priority to tool changes, directional changes and part movements (i.e. $w_{tc} = w_{dc} = w_{pm} = 0.33$)

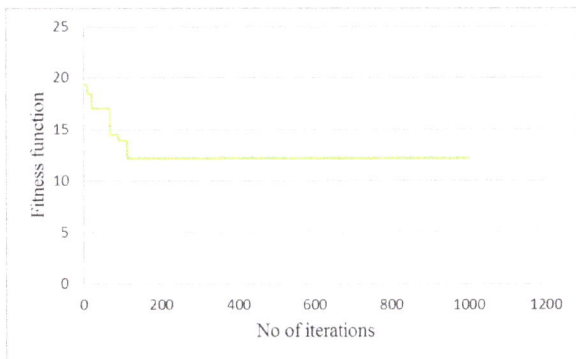

Case 2: Performing assembly operation by considering full priority to end-effectors/tool changes (i.e. $w_{tc} = 1, w_{dc} = 0, w_{pm} = 0$)

According to the nature of the assembly process the weights for each parameter can be decided by the user. In this paper, analysis has been performed for the above two cases of industrial settings.

Case 1: Performing assembly operation while considering equal priority to tool changes, directional changes and part movements (i.e. $w_{tc} = w_{dc} = w_{pm} = 0.33$)

Fig. 9. Convergence of fitness over number of iterations

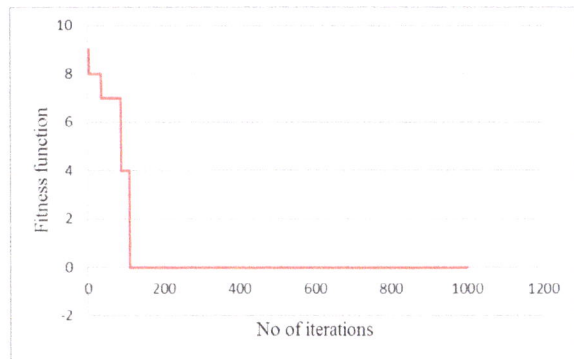

Fig. 10. Convergence of fitness over number of iterations

From the Fig. 9, it is noticed that the model is converged at 112th iteration with the fitness value 7.676 and the optimized assembly sequence is listed in the Table 2.

Table 2
Optimal feasible assembly sequence for Case-1

Iteration number	Assembly sequence	Assembly direction	Gripper changes	Fitness value
112	1-7-8-10-9-11-2-3-4-5-6	3-3-3-3-3-3-4-4-4-4-4	1-7-8-10-9-11-2-3-4-5-6	7.676

Case 2: Performing assembly operation while considering full priority to end-effectors/tool changes (i.e. $w_{tc} = 1, w_{dc} = 0, w_{pm} = 0$)

It is observed from Fig. 10 that the model is converged at 110th iteration having the fitness value '0' and the optimized assembly sequence is listed in the table 3.

Table 3
Optimal feasible assembly sequences for Case-2

Iteration number	Assembly sequence	Assembly direction	Gripper changes	Fitness value
110	11-9-8-7-10-1-2-3-4-5-6	4-4-4-4-4-4-4-4-4-4-4	11-9-8-7-10-1-2-3-4-5-6	0

5.4. Comparison Among GA, AIS and AIS-GA

For the comparison purpose this analysis is focused only on assembly directional changes. Therefore the weights of fitness function are $w_{dc} = 1$, $w_{tc} = 0, w_{pm} = 0$. Here the analysis is carried out for GA, AIS and AIS-GA up to 1000 iterations. The fitness value of '1' achieved by GA, AIS and AIS-GA at 414th, 479th and 101st iteration respectively. Individual GA and AIS algorithms couldn't generate the optimal sequence with fitness value '0' for 1000 iterations. Whereas, the hybrid AIS-GA model achieved the optimal feasible assembly sequence with '0' fitness value at 110th iteration. A detail analysis results are illustrated in table 4 and Fig.11.

Table 4
Optimum assembly sequences for the assembly directional changes

Algorithm	No. of iterations	Time taken to generate OFAS (Sec)	Generated OFAS	Assembly direction of OFAS	No. of directional changes
GA	Up to 500	12.4	10-8-9-11-7-1-2-3-4-5-6	3-3-3-3-4-4-4-4-4-4-4	1
	Up to 1000	18.6	10-8-9-11-7-1-2-3-4-5-6	3-3-3-3-4-4-4-4-4-4-4	1
AIS	Up to 500	13.5	10-8-9-11-7-1-2-3-4-5-6	3-3-3-3-4-4-4-4-4-4-4	1
	Up to 1000	19.5	10-8-9-11-7-1-2-3-4-5-6	3-3-3-3-4-4-4-4-4-4-4	1
Hybrid (AIS-GA)	Up to 500	8.6	11-9-8-7-10-1-2-3-4-5-6	4-4-4-4-4-4-4-4-4-4-4	0
	Up to 1000	10.6	11-9-8-10-7-1-2-3-4-5-6	4-4-4-4-4-4-4-4-4-4-4	0

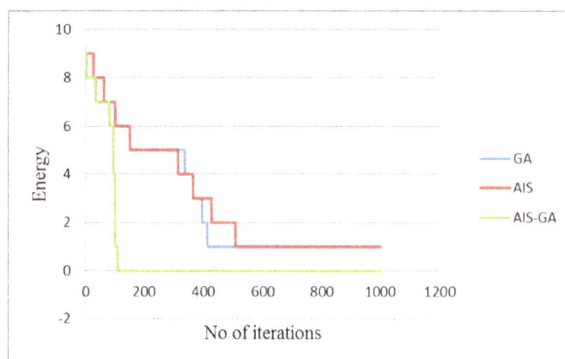

Fig. 11. A graph for developed AI techniques with directional changes with respect to number of iterations

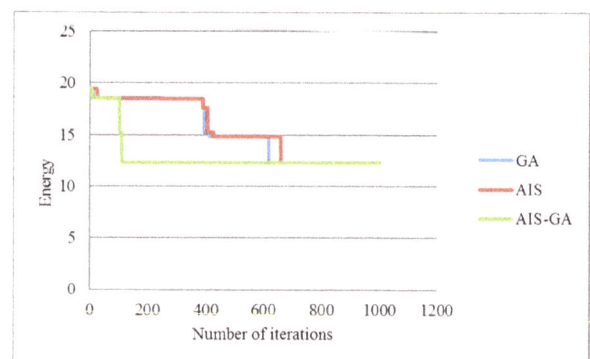

Fig. 12. A graph for different AI techniques with energy and Number of iterations

For the comparison purpose, this analysis is focused only on energy i.e. part movements. Therefore the weights of fitness function are $w_{dc} = 0$, $w_{tc} = 0, w_{pm} = 1$. Here the analysis is carried out for GA, AIS and AIS-GA up to 1000 iterations. The minimum fitness value of '12.262' achieved by GA, AIS and AIS-GA at 616th, 658th and 110th iteration respectively. A detailed analysis results are illustrated in table 5 and Fig. 12.

Table 5

Optimum assembly sequences for the energy

Algorithm	No. of iterations	Time taken to generate OFAS (Sec)	Generated OFAS	Assembly direction of OFAS	Fitness value
GA	Up to 500	13.1	10-8-9-7-1-11-2-3-4-5-6	3-3-3-4-4-3-4-4-4-4-4	14.810
	Up to 1000	19.2	10-8-9-11-7-1-2-3-4-5-6	3-3-3-3-4-4-4-4-4-4-4	12.262
AIS	Up to 500	14.4	10-8-9-7-11-1-2-3-4-5-6	3-3-3-4-3-4-4-4-4-4-4	12.302
	Up to 1000	20.1	10-8-9-11-7-1-2-3-4-5-6	3-3-3-3-4-4-4-4-4-4-4	12.262
Hybrid(AIS-GA)	Up to 500	8.8	11-9-8-7-10-1-2-3-4-5-6	4-4-4-4-4-4-4-4-4-4-4	12.062
	Up to 1000	11.1	11-9-8-10-7-1-2-3-4-5-6	4-4-4-4-4-4-4-4-4-4-4	12.062

In the current study, two cases: $w_{tc} = w_{dc} = w_{pm} = 0.33$; $w_{tc} = 1, w_{dc} = 0, w_{pm} = 0$ are considered for understanding purpose. But the selection of these weights depends on the user/industrialist requirement.

6. Conclusion

Hybrid algorithm with CAD assistance has been introduced in this research work to obtain optimal feasible assembly sequence for a given product. In the present investigation, two separate algorithms namely GA and AIS are combined to generate OFAS. The developed methodology is found successful in succeeding all sets of valid assembly sequences for different configurations of assemblies in less computational time. Unlike the traditional optimization algorithms as mentioned in introduction section, the hybrid model does not generate redundant assembly sequence and thus saves lots of computational time and drives toward global optima. Moreover, the proposed hybrid algorithm has the flexibility for user defined weights: selection for gripper changes, directional changes and part movements unlike the existed artificial intelligent techniques. As a future work the developed hybrid algorithm can be applied effectively for the identification of sub-assembly to make more flexible for parallel assembly systems.

References

Akpınar, S., Bayhan, G. M., & Baykasoglu, A. (2013). Hybridizing ant colony optimization via genetic algorithm for mixed-model assembly line balancing problem with sequence dependent setup times between tasks. *Applied Soft Computing, 13*(1), 574-589.

Bahubalendruni, M. V. A., & Biswal, B. B. (2014a). An algorithm to test feasibility predicate for robotic assemblies. *Trends in Mechanical Engineering & Technology, 4*(2), 11-16.

Bahubalendruni, M. R., & Biswal, B. B. (2014b). Computer aid for automatic liaisons extraction from cad based robotic assembly. *In Intelligent Systems and Control (ISCO), 2014 IEEE 8th International Conference on (pp. 42-45)*. IEEE.

Bahubalendruni, M. R., & Biswal, B. B. (2015a). A novel concatenation method for generating optimal robotic assembly sequences. *Proceedings of the Institution of Mechanical Engineers, Part C: Journal of Mechanical Engineering Science*, 0954406215623813.

Bahubalendruni, M. R., Biswal, B. B., Kumar, M., & Nayak, R. (2015b). Influence of assembly predicate consideration on optimal assembly sequence generation. *Assembly Automation, 35*(4), 309-316.

Bahubalendruni, M. R., & Biswal, B. B. (2015c). A review on assembly sequence generation and its automation. *Proceedings of the Institution of Mechanical Engineers, Part C: Journal of Mechanical Engineering Science*, 0954406215584633.

Bahubalendruni, M. R., & Biswal, B. B. (2015d). An intelligent method to test feasibility predicate for robotic assembly sequence generation. *In Intelligent Computing, Communication and Devices* (pp. 277-283). Springer India.

Bahubalendruni, M. R., & Biswal, B. B. (2016). Liaison concatenation–A method to obtain feasible assembly sequences from 3D-CAD product. *Sadhana, 41*(1), 67-74.

Bahubalendruni, M. R., Biswal, B. B., Kumar, M., & Deepak, B. B. V. L. (2016). A Note on Mechanical Feasibility Predicate for Robotic Assembly Sequence Generation. In *CAD/CAM, Robotics and Factories of the Future*(pp. 397-404). Springer India.

Bahubalendruni, M. R., Deepak, B. B. V. L., & Biswal, B. B. (2016). An advanced immune based strategy to obtain an optimal feasible assembly sequence. *Assembly Automation, 36*(2), 127-137.

Bahubalendruni, M. V. A., Biswal, B. B., & BB, V. (2015). Optimal Robotic Assembly Sequence generation using Particle Swarm Optimization. *Journal of Automation and Control Engineering, 4*(2), 89-95

Biswal, B. B., Deepak, B. B., & Rao, Y. (2013). Optimization of robotic assembly sequences using immune based technique. *Journal of Manufacturing Technology Management, 24*(3), 384-396.

Chang, C. C., Tseng, H. E., & Meng, L. P. (2009). Artificial immune systems for assembly sequence planning exploration. *Engineering Applications of Artificial Intelligence, 22*(8), 1218-1232.

Chen, S. F., & Liu, Y. J. (2001). An adaptive genetic assembly-sequence planner. *International Journal of Computer Integrated Manufacturing, 14*(5), 489-500.

Chen, R. S., Lu, K. Y., & Yu, S. C. (2002). A hybrid genetic algorithm approach on multi-objective of assembly planning problem. *Engineering Applications of Artificial Intelligence, 15*(5), 447-457.

De Fazio, T., & Whitney, D. (1987). Simplified generation of all mechanical assembly sequences. *IEEE Journal on Robotics and Automation, 3*(6), 640-658.

Deepak, B. B. V. L., & Parhi, D. R. (2016). Control of an automated mobile manipulator using artificial immune system. *Journal of Experimental & Theoretical Artificial Intelligence, 28*(1-2), 417-439.

Deepak, B. B. V. L., & Parhi, D. (2013). Intelligent adaptive immune-based motion planner of a mobile robot in cluttered environment. *Intelligent Service Robotics, 6*(3), 155-162.

Dong, T., Tong, R., Zhang, L., & Dong, J. (2007). A knowledge-based approach to assembly sequence planning. *The International Journal of Advanced Manufacturing Technology, 32*(11-12), 1232-1244.

Hsu, Y. Y., Tai, P. H., Wang, M. W., & Chen, W. C. (2011). A knowledge-based engineering system for assembly sequence planning. *The International Journal of Advanced Manufacturing Technology, 55*(5-8), 763-782.

Kashkoush, M., & ElMaraghy, H. (2015). Knowledge-based model for constructing master assembly sequence. *Journal of Manufacturing Systems, 34*, 43-52.

Lee, H. R., & Gemmill, D. D. (2001). Improved methods of assembly sequence determination for automatic assembly systems. *European Journal of Operational Research, 131*(3), 611-621.

Linn, R. J., & Liu, H. (1999). An automatic assembly liaison extraction method and assembly liaison model. *IIE transactions, 31*(4), 353-363.

Nayak, R., Bahubalendruni, M. R., Biswal, B. B., & Kumar, M. (2015, September). Comparison of liaison concatenation method with simulated annealing for assembly sequence generation problems. *In Next Generation Computing Technologies (NGCT), 2015 1st International Conference on (pp. 531-535). IEEE.*

Shan, H., Zhou, S., & Sun, Z. (2009). Research on assembly sequence planning based on genetic simulated annealing algorithm and ant colony optimization algorithm. *Assembly Automation, 29*(3), 249-256.

Sinanoglu, C., & Riza Börklü, H. (2005). An assembly sequence-planning system for mechanical parts using neural network. *Assembly Automation, 25*(1), 38-52.

Smith, S. S. F. (2004). Using multiple genetic operators to reduce premature convergence in genetic assembly planning. *Computers in Industry, 54*(1), 35-49.

Wang, Y., & Liu, J. H. (2010). Chaotic particle swarm optimization for assembly sequence planning. *Robotics and Computer-Integrated Manufacturing, 26*(2), 212-222.

Wang, J. F., Liu, J. H., & Zhong, Y. F. (2005). A novel ant colony algorithm for assembly sequence planning. *The International Journal of Advanced Manufacturing Technology, 25*(11-12), 1137-1143.

Xing, Y., & Wang, Y. (2012). Assembly sequence planning based on a hybrid particle swarm optimisation and genetic algorithm. *International Journal of Production Research, 50*(24), 7303-7312.

Zha, X. F., Lim, S. Y., & Fok, S. C. (1998). Integrated knowledge-based assembly sequence planning. *The International Journal of Advanced Manufacturing Technology, 14*(1), 50-64.

10

A two-layer genetic algorithm for the design of reliable cellular manufacturing systems

Hassan Rezazadeh[a]* and Amin Khiali-Miab[a]

[a]*Department of Industrial Engineering, Faculty of Mechanical Engineering, University of Tabriz, Iran*

CHRONICLE	ABSTRACT
	This study presents a new mathematical model for the design of reliable cellular manufacturing systems, which leads to reduced manufacturing costs, improved product quality and improved total reliability of the manufacturing system. This model is expected to provide a more noticeable improvement in time and solution quality in comparison with other existing models. Each part to be manufactured may select each of the predefined manufacturing routes, such that the total reliability of the system is increased. On the other hand, the model adopts to categorize the machines to determine the manufacturing cells (cell formation) and reduce the transportation costs. Thereby, both criteria of system reliability and manufacturing costs will be simultaneously improved. Due to the complexity of cell formation problems, a two-layer genetic algorithm is applied on the problem in order to achieve near optimal solutions. Furthermore, the performance of the proposed algorithm is shown for solving some computational experiments. Finally, the results of a practical study for designing a cellular manufacturing system as a case study in Iranian Diesel Engine Manufacturing Co., Tabriz, Iran are present.
Keywords: *Cell Formation* *Reliability* *Mathematical Model* *Two-Layer Genetic Algorithm*	

1. Introduction

System definition is an approach to develop sciences for involvement of interactions in the problem and realization of the decisions and results more than before. Reliability index can be defined for whatever known as the system. It is for many year, all effective components in manufacturing a product are known as the manufacturing system. In other words, having a same goal (producing a qualified product) the components have an interactive cooperation and influence. Taking into account the above mentioned definitions, a reliable manufacturing system can also be discussed. In the literature, the previous studies about reliability theory are mainly limited to its definition in the field of physical product, although a reliable product will be manufactured in a reliable system. There are some papers on this context, but we could say that some characteristics of the suggested mathematical models have made them rather impractical or of limited industrial application. The evolution of these papers will be addressed in the next section.

* Corresponding author
E-mail: hrezazadeh@tabrizu.ac.ir (H. Rezazadeh)

The remainder of this paper is organized as follows: In Section 2, we review relevant literature on the cellular manufacturing system (CMS). Section 3 presents the reliability considerations in designing CMS. Section 4 and 5 presents the Problem description and mathematical formulation for the CMS. In Section 6, we introduce a brief review of Two-Layer Genetic Algorithm. Computational results are reported in Section 7 and the conclusion is given in Section 8.

2. Literature Review

Machinery is the main element of a manufacturing system. They play a key role in quality and price of the products. Maintenance and repair conditions are also of major concerns on these two factors of the products (Das et al., 2007a). Frequent breakdowns and repeated repairing and resetting the machinery lead to non-uniformity and reduced quality of the products in addition to incur greater maintenance and labor costs. Therefore, organization of the machinery in the form of an effective manufacturing system may basically improve quality and costs of the products. Conventional manufacturing paradigms (Job shop and Flow shop) have found it difficult to satisfy the competitive advantages as well as control and improve the manufacturing efficiency (Wemmerlov & Hyer, 1989; Wemmerlov & Johnson, 1997; Askin, 1999; Das et al., 2007b). Although advantages of the cellular manufacturing systems are properly addressed by either users or researchers, some others have criticized about drawbacks of these systems (Flynn & Jacobs, 1986; Morris & Tersine, 1990; Boughton & Arokiam, 2000; Agarwal & Sarkis, 1998; Das et al., 2007b). Their results can be summarized as below:

1- Cellular Manufacturing System (CMS) decreases flexibility of the system (in comparison to Job shop manufacturing),
2- CMS reduces utilization rate of machine due to allocation of them to cells,
3- Breakdowns of machines significantly influence due date of operations,
4- Increased in-process inventories due to allocation of machines to cells (Das et al., 2007b),
5- Among parameters which influence the CMS, allocation of machines to cells, combination of manufactured parts and reliability of machinery seem to play the most significant role in performance of a CMS (Seifoddini and Djassemi, 2001; Das et al., 2007b).

A considerable amount of researchers on the field of CMS previously have presumed the machinery 100% reliable, while more recent works have embedded the reliability in a rather complicated and undesirable way in the CMS problems (Das et al., 2007a; Das et al., 2007b; Jabal Ameli et al., 2007; Jabal Ameli & Arkat, 2008; Saxena & Jain, 2011). Das et al. have treated reliability of the manufacturing system as being serial so they have deemed breakdown of a machine equal to that of the whole manufacturing system (Das et al., 2007a; Das et al., 2007b). Approximately 75% of the manufacturing systems produce their products as a large set of parts via batch processing (Akturk and Turkcan, 2000 ; Saxena & Jain, 2011). In this case, the whole system does not necessarily stop with breakdown of a machine, because the system will never stop if repairing the failed machine lasts for some time shorter than the current batch operation. Thereby, the system will be incurred heavy costs in order to retain a reliability level which is not needed in practice. Complexity of the manufacturing systems in comparison with other common engineering systems will make it impossible to be dealt with them like electronic systems. This will practically limit application of the paper of K. Das et al.However, other authors have just considered the costs of system breakdown in a way that can be improved (Jabal Ameli et al., 2007; Arkat, 2008; Saxena & Jain, 2011).

A complete set of reliability-related costs have been addressed in this paper by adding the relevant term of maintenance labor force to the objective function. These costs are: (1) calculation of hardware costs for each machine, (2) calculation of maintenance labor force in each breakdown, (3) breakdown costs of production line. This three mentioned reliability-related costs in addition to transportation costs will form the objective function.

Other characteristic of the objective function presented in this paper is proper linearization which reduces the number of integer variables of this model in comparison with other models suggested in (Jabal Ameli et al., 2007; Jabal Ameli and Arkat, 2008). This is expected to reduce the computational effort for solving by exact algorithms. In average, 20-50% of costs in the manufacturing systems are associated with intercellular and intracellular transportations as well as proper innovative planning in allocation of machinery and transportation. This can decrease 10-30% of the production costs (Tompkins et al., 2003; Saxena and Jain, 2011). Another issue which must be discussed in relation with the objective function of this model is that the costs of intercellular transportation are considered in it, which plays a significant role in efficiency of the model. A great number of studies have developed their own model by assuming only one processing route (Lokesh & Jain, 2008, 2009, 2010a, 2010b). Some authors have discussed on the flexibility obtained from alternative process routing problems for the parts (Gupta et al., 1996; Sofianopoulou, 1999; Zhao and Wu, 2000; Kim et al., 2004). This assumption has several features in design phase, as summarized below: less needed machinery, greater utilization rate of the machinery, less intercellular transportation (less relation between the cells), and improved output of the system (Kusiak, 1987). Moreover, multiple processing routes were mentioned in some studies with the effect and importance of this assumption being shown on the results obtained from this model (Defersha and Chen, 2006a; 2006b, 2007, 2008; Ahkioon et al., 2009, Bulgak & Bektas, 2009; Nsakanda, 2006).

3. Problem description and formulation

3.1. Notations

3.1.1. Indexes

i	Index of part type where $i=1, 2, ..., n$
j	Index of process routing where $j=1, 2, ..., q_i$
k	Index of machine type where $k=1, 2, ..., m$
l	Index of cell number where $l=1, 2, ..., C$
r	Index of process in a rout $r=1,2, ...,k_{ij}$

3.1.2. Input parameters

k_{ij}	Number of processes in jth rout of part i
A	Intercellular movement unit cost
$T_{i\left(u_{ij}^{(r)}\right)}$	The process time of part i on r^{th} machine from u_{ij} processing route
$TBF(m)$	Stochastic variable for time between failures for machine m
$MTBF_{\left(u_{ij}^{(r)}\right)}$	The mean time between failures for r^{th} machine from u_{ij} processing route
$TTR(m)$	Stochastic variable for time to repair machine m
MTTR(m)	The mean time to repair machine m
$t_p(m)$	Total production time on machine m
$T_{i\left(u_{ij}^{(r)}\right)}$	Process time for part i on machine $u_{ij}^{(r)}$
$B_{\left(u_{ij}^{(r)}\right)}$	The hardware cost of rth machine from u_{ij} processing route;
H	The man-hour cost of the maintenance labor force;
P_i	Production volume for part i
$B(m)$	Hardware breakdown Costs for machine m
V	Maximum number of machines in each cell
$r = 1,2, ...,k_{ij}$	the counter of machines in u_{ij} processing route;

NOTE1: $\left\{u_{ij}^{(1)}, u_{ij}^{(2)}, ...,u_{ij}^{(r)}, ..., u_{ij}^{(k_{ij})}\right\}$ is the index matrix of the machines required in u_{ij} processing route.

NOTE2: The machine associated with operation r^{th} through u_{ij} processing route might be k^{th} machine in the system, while the index k (machine number) acts as name of the machine.

3.1.3. Decision variables

$$x_{ijrl} = \begin{cases} 1 & if & processr of route j of part i is allocated to cell l \\ \\ 0 & & otherwise \end{cases}$$

$$Y_{kl} = \begin{cases} 1 & if & if\ machine\ k\ is\ allocated\ to\ cell\ l \\ \\ 0 & & otherwise \end{cases}$$

$$Z_{ij} = \begin{cases} 1 & if & processr of route j of part i is allocated to cell l \\ \\ 0 & & otherwise \end{cases}$$

f_{ijrl}, g_{ijrl} : Auxiliary variables for linearization of intercellular transportation cost of parts

3.2. Reliability Considerations in Designing CMS

One of the most important features of CMS, which can improve this model, is the effect of breakdowns of the machine on the system. Conventional CMS models the design of systems which are vulnerable against breakdown of the machinery. The conventional models used to work with 100% reliability of the machinery in designing the cells and allocation of their tasks and components. However, the machinery may experience breakdown in practice. Breakdowns of the machines have the most significant effect on delivery times, even though the system represents a great flexibility. Taking into consideration the reliability of machinery would contribute to select more realistic processing routes in designing models of the manufacturing cells with multiple process routes. Very few studies have considered the effect of machine breakdown in cell formation problems (Saxena & Jain, 2011). Reliability of the manufacturing system is generally defined as the probability of desired performance of the system in a given time and under certain conditions (Saxena & Jain, 2011). The reliability costs may include hardware costs (spare parts) and labor costs.

3.3. Hardware Costs

The hardware costs of every breakdown including the shut downs as well as spare parts, resetting and restarting costs belonged to the machine m can be calculated as below.

$$BC(m)=\{(Production\ demand \times Operation\ time \times unit\ breakdown\ cost\ of\ machine\ per\ breakdown)\ /\ MTBF(m)\}$$

Where, TBF follows an exponential distribution with λ parameter:

$$TBF(m) \sim \exp(\lambda(m)) \tag{1}$$

A common technique to calculate MTBF(m), which can be used during design phase of CMS can be extracted as follows:

$$MTBF(m) = 1/\lambda(m) \tag{2}$$

So, the number of breakdowns for the machine m during its operation time (t_p) can be calculated as below:

$$N(m, t_p) = t_p/MTBF(m) \tag{3}$$

While the breakdown costs of the machine m could be derived from this formula:

$$BC(m, t_p) = (t_p * B(m)/MTBF(m))$$ (4)

3.4. Labor Costs

The following equation can be used to find the labor costs of every breakdown for the machine m:

HC(m)={(Production demand × Operation time × unit human cost of machine per hour) × MTTR(m) / MTBF(m)}

Assuming that $TTR(m)$ follows an exponential distribution, that is:

$$TTR(m) \sim \exp(\mu(m))$$ (5)

A straightforward method to calculate $MTTR(m)$ is given below, where $\mu(m)$ is repairing rate of the machine m:

$$MTTR(m) = 1/\mu(m)$$ (6)

Having multiplied the number of failures $N(m, t_p)$ by the mean time to repair $MTTR(m)$, the total time allocated to repair during the manufacturing process would be obtained as below:

$$t_{rep}(m, t_p) = t_p * MTTR(m)/MTBF(m)$$ (7)

The total software (labor) cost associated with the system reliability will be obtained if the product above is multiplied in the per hour labor cost.

Reliability costs of the system (RC) are entered the model as below:

$$RC = \left[\sum_{i=1}^{n} \sum_{j=1}^{q_i} \sum_{r=1}^{k_{ij}-1} Z_{ij} \left[B_{\left(u_{ij}^{(r)}\right)} \frac{P_i T_{i\left(u_{ij}^{(r)}\right)}}{MTBF_{\left(u_{ij}^{(r)}\right)}} + H \frac{P_i T_{i\left(u_{ij}^{(r)}\right)} MTTR_{\left(u_{ij}^{(r)}\right)}}{MTBF_{\left(u_{ij}^{(r)}\right)}} \right] \right]$$ (8)

3.5. Calculation of Intercellular Transportation Costs in Objective Function

Consider the following equations:

$$IC = A \left[\sum_{i=1}^{n} \sum_{j=1}^{q_i} \sum_{r=1}^{k_{ij}-1} \sum_{l=1}^{c} P_i \quad Z_{ij} \quad Y_{u_{ij}^{(r)}l} \left(1 - Y_{u_{ij}^{(r+1)}l} \right) \right]$$ (9)

IC represents a nonlinear objective function for minimizing the cost of intercellular flows, such that considering each pair of the existing consecutive machines in a given processing route, a number is counted once allocated in different cells and is then added to the total value which is obtained by multiplying total number of "one"s in the cost parameter A after counting the "one"s.

It was difficult to solve large scale problems with this nonlinear objective function, in spite of the few variables introduced to the model. Limitation in using nonlinear objective function for large problems is still deemed as an important concern. Thus, some studies have been directed in the past for linearization of this function.

3.5.1. Linearization of the proposed Intercellular Transportation Costs in Objective Function

An objective function very similar to the one in section 4-1 is reported in the papers of Jabal ameli et al. (2007; 2008) It is obvious that y and z are binary variables. They have suggested other binary variables in combination with a set of limitations for linearization. It should be noted that as mentioned in the results section, this kind of linearization needs a rather long time to be solved by optimization algorithms after programming due to the significantly great number of variables and limitations. It will additionally make programming of exact solution (Simplex) and meta-heuristic methods rather difficult. As a result, the problems with large dimensions would become almost unsolvable.

This paper proposes the following objective function for this problem by defining the decision variable based on allocation of the operations to the cells (and not the machines to the cells).

$$IC = \frac{A}{2} \left(\sum_{i=1}^{n} \sum_{j=1}^{q_i} \sum_{r=1}^{k_{ij}-1} \sum_{l=1}^{c} |x_{ij,r+1,l} - x_{ijrl}| \right) \tag{10}$$

Eq. (10) counts the number of operations done in separate cells, which will be entered the main model as shown below using the variables g_{ijrl} and f_{ijrl} :

$$IC = \frac{A}{2} \left(\sum_{i=1}^{n} \sum_{j=1}^{q_i} \sum_{r=1}^{k_{ij}-1} \sum_{l=1}^{c} (f_{ijrl} + g_{ijrl}) \right). \tag{11}$$

The limitations for linearization are represented as below:

$$x_{ijrl} \leq Y_{u_{ij}^{(r)}l} \tag{12}$$

This limitation is for showing the relation between two decision variables $y_{u_{ij}^r,l}$ and x_{ijrl} , once a

process is allocated to that cell. The required machine must be allocated to that cell for processing the operation. Furthermore,

$$x_{ij,r+1,l} - x_{ijrl} = f_{ijrl} - g_{ijrl} \tag{13}$$

is used for linearization of the term containing absolute value in calculating the total costs of intercellular transportation.

3.6. proposed mixed integer linear model

In this section a mixed linear integer programming model is introduced for the problem by definition of the decision variable based on allocation of processes to cells:

$$\min \quad C = IC + RC \tag{14}$$

subject to

$$IC = \frac{A}{2} \left(\sum_{i=1}^{n} \sum_{j=1}^{q_i} \sum_{r=1}^{k_{ij}-1} \sum_{l=1}^{c} (f_{ijrl} + g_{ijrl}) \right) \tag{15}$$

$$MC = \left[\sum_{i=1}^{n} \sum_{j=1}^{q_i} \sum_{r=1}^{k_{ij}-1} Z_{ij} \left[B_{\left(u_{ij}^{(r)}\right)} \frac{P_i T_{i\left(u_{ij}^{(r)}\right)}}{MTBF_{\left(u_{ij}^{(r)}\right)}} + H \frac{P_i T_{i\left(u_{ij}^{(r)}\right)} MTTR_{\left(u_{ij}^{(r)}\right)}}{MTBF_{\left(u_{ij}^{(r)}\right)}} \right] \right] \tag{16}$$

$$\sum_{l=1}^{c} x_{ijrl} = z_{ij} \quad , \quad i = 1,2,\dots,n \ , \quad j = 1,2,\dots,q_i \ , \quad r = 1,2,\dots,k_{ij} \tag{17}$$

$$x_{ijrl} \leq Y_{u_{ij}^{(r)}l} \quad , \quad i = 1,2,\dots,n \ , \quad j = 1,2,\dots,q_i \ , \quad r = 1,2,\dots,k_{ij} \ , l = 1,2,\dots,c \tag{18}$$

$$x_{ij,r+1,l} - x_{ijrl} = f_{ijrl} - g_{ijrl} \ , \quad i = 1,2,\dots,n \ , j = 1,2,\dots,q_i \ , r = 1,2,\dots,k_{ij} - 1, l = 1,2,\dots,c \tag{19}$$

$$\tag{20}$$

$$\sum_{k=1}^{p} Y_{kl} \leq V \quad , \quad l = 1,2,\dots,c$$

$$\sum_{l=1}^{c} Y_{kl} = 1 \quad , \quad k = 1,2,\dots,p \tag{21}$$

$$\sum_{j=1}^{q_i} Z_{ij} = 1 \quad , \quad i = 1,2,\dots,n \tag{22}$$

$$z_{ij}, y_{kl} \in \{0,1\} \ , \quad i = 1,2,\dots,n, \quad j = 1,2,\dots,q_i; \quad k = 1,2,\dots,p; \quad l = 1,2,\dots,c \tag{23}$$

$$x_{ijrl} \geq 0 \qquad , \qquad i = 1,2,\dots,n \ , \quad j = 1,2,\dots,q_i \ , r = 1,2,\dots,k_{ij} \ ; \ l = 1,2,\dots,c \tag{24}$$

As previously discussed, in this model the objective function is the sum of intercellular transportation cost of the materials, and repair and maintenance cost of the machinery. Eq. (15) calculates the total intercellular transportation cost of the material, which is in fact a linearized version of the nonlinear Eq. (25) below.

$$IC = \frac{A}{2} \left(\sum_{i=1}^{n} \sum_{j=1}^{q_i} \sum_{r=1}^{k_{ij}-1} \sum_{l=1}^{c} p_i |x_{ij,r+1,l} - x_{ijrl}| \right) \tag{25}$$

Eq. (16) calculates the total repair and maintenance costs of the machinery, in which the first term gives the hardware cost of repair and maintenance whereas the second term provides the labor force cost required for it. Constraint (17) represents the relation between decision variables Z_{ij} and X_{ijrl}. It further ensures that: (a) the rth process of manufacturing route j from part i is only allowed when the manufacturing route j for the part i is selected, and (b) the rth process of manufacturing route j from part i must be allocated only to one cell if the manufacturing route j is selected for the part i. Constraint (18) demonstrates the relation between decision variables $y_{u_{ij}^r,l}$ and X_{ijrl}. It additionally states that the

machine needed for processing that operation must be also allocated to that cell if an operation is allocated to a cell. Constraint (19) is employed to linearize the term containing absolute value in calculation of the total intercellular transportation costs of the materials (Eq. (25)). Constraint (20) controls the number of machinery for each cell and sets upper limit of V for this number. Constraint (21) indicates that each machine must be allocated to one cell, while constraint (22) guarantees that only one manufacturing route is selected for each part. Constraints (23) and (24) are determinant for the type of decision variables. As previously discussed in section 4, there are other linear models in the literature for this problem or other similar problems. A pure 0-1 linear integer model is proposed in these models using the nonlinear model introduced in section 4-1 by Jabal ameli et al. (2007). This model influences the great number of variables and its performance constraints. Table 1 illustrates a comparison between the nonlinear model developed in Section 4-1, the absolute 0-1 linear integer model proposed in Section 4-2, and the mixed linear integer model suggested by this paper in Section 4-3, where Q denotes the total number of existing manufacturing routes, R denotes the total number of existing processes in these routes where $Q = \sum_{i=1}^{n} q_i$ are $R = \sum_{i=1}^{n} \sum_{j=1}^{q_i} k_{ij}$, respectively. It is observed from table 1 that the numbers of continuous variables and constraints of the proposed model are considerably smaller than that of the Jabal Ameli & Arkat's model (2008) (except for the case with 3 or less machines, which is not practically applicable). Therefore, solving their release model (linear, Lagrangian or etc.) becomes more extensive in large scale problems. As a result, the proposed model will be both faster and more efficient:

Table 1
Comparison between proposed model and other existing models

Model	Number of variables			Number of constraints
	0-1	continuous	total	
NILP	Q+MC	0	Q+MC	N+M+C
pure ILP (literature)	Q+MC	RMC	Q+MC+RMC	N+M+C+4RMC
proposed MILP	Q+MC	3RC	Q+MC+3RC	N+M+C+R+RC+(R-Q)C

$$R = \sum_{i=1}^{n}\sum_{j=1}^{q_i} k_{ij} \ , \ Q = \sum_{i=1}^{n} q_i \ ,$$ M: number of machines, C: number of cells, N: number of parts

4. Two-Layer Genetic Algorithm

Genetic Algorithm (GA), introduced by Holland (1975) and developed by Goldberg (1989) is a widespread, parallel, stochastic search and optimization method that has been employed for solving numerical optimization problems in a wide variety of application fields including engineering, biology, economics, agriculture, business, telecommunications, and manufacturing (Goldberg , 1989; Gen and Cheng, 1997; Man et al., 1999; Onwubolu & Mutingi, 2001). It is a model of machine learning which derives its behaviour from a metaphor of the processes of evolution in nature. GA starts with encoding the solutions in a chromosome structure containing several genes and then, an initial population of chromosomes will be formed as the initial solutions for the problem. Afterwards, the superior chromosomes are selected by the selection operator for the new mating pool. The evolution is simulated by using reproduction operators such as crossover operator, imitating propagation, and mutation operator, imitating random changes occurring on chromosomes in nature. Reproduction operators are employed on the mating pool to generate new solutions, called off springs. Then the replacement operator chooses some chromosomes among former generation and off springs to form new generation. This procedure will be continued until the termination criterion is met.

Because of the efficient and multi-direction searching capability of GA using a population of solutions instead of single solution, its simple and understandable nature, its ability in preventing from getting trapped in local optimum and its successful applications in designing manufacturing systems, it has been used and a bi-level hybrid genetic algorithm (HGA) has been developed here in order to solve the cell formation and routing selection problems simultaneously considering the reliability issues. Information like operation sequence of routes nominated for manufacturing the products, processing time of operations, production volumes, alternative routes of parts, maximum size of cells, number of cells, unit cost of intercellular material handling, unit cost of repair and maintenance and its required labor force are used in the proposed algorithm.

The first level of this algorithm seeks to find the best manufacturing route for the parts, while the second level tries to find the best configuration of the cells based on the manufacturing routes that are selected at the first level. There is a relationship between these two levels which implies that when the first level's search is implemented to get a better combination of the manufacturing routes, after meeting the termination criterion of this level, the search process is transferred to the second level in order to address the best configuration of the cells using the solution obtained from the first level. Meanwhile, the search process is transferred to the first level once the termination criterion of the second level is met. This process will be continued until the termination criterion of the algorithm is met. The proposed algorithm explores the search space adequately using mutation operator, elimination of repetitive chromosomes and acceptance of worse solutions. Besides, it exploits the best solutions using roulette wheel selection, crossover operator and elitism technique. As a result of improved exploration and exploitation, the algorithm can hopefully achieve the optimum or near-optimum solutions. The components of the proposed GA and its operators are explained in the following subsections.

4.1. Notation used in the proposed HGA

$popSize$	Population size
f_i	The objective function of chromosome i
F_i	The fitness of chromosome i
$best\_fitt$	The fitness of best chromosome in the current population
$best\_partial\_fitt$	The best partial fitness found so far
$best\_overal\_fitt$	The best fitness found so far
$MaxNoIMP1$	Number of consecutive iterations with no fitness improvement for terminating first level of algorithm
$MaxNoIMP2$	Number of consecutive iterations with no fitness improvement for terminating second level of algorithm
$NoIMP$	Number of consecutive iterations with no fitness improvement so far
$MaxTime$	The runtime of algorithm
μ_c	Crossover rate
μ_m	Mutation rate

4.2. Encoding Scheme

Each chromosome is a string of size $N + M$ and indicates a feasible solution of the search space. It is composed of two parts. The first one represents a solution for routing selection problem and contains N genes in which, the allele (value of a characteristic) located in the ith locus (position of a characteristic in the chromosome) demonstrates the selected manufacturing route for part type i. The second part of the chromosome shows a solution for cell formation problem and is composed of M genes, where the allele located in the kth locus indicates the cell selected for machine k. Fig. 1 depicts a chromosome for a problem with 3 parts, 13 machines, 3 cells, 2 process routes for each part and maximum cell size of 5.

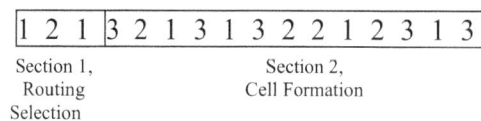

1 2 1	3 2 1 3 1 3 2 2 1 2 3 1 3
Section 1, Routing Selection	Section 2, Cell Formation

Fig. 1. Configuration of a chromosome

4.3. Fitness Function Evaluation

The fitness value is a measure to determine the quality level of a chromosome. The fitness values are used to select parent chromosomes to create the next generation. A certain transformation is required to use the objective function values as the fitness values. Since the objective function of the considered problem is the minimization of total cost, transformation can be carried out by Eq. (26).

$$F_i = BIGM - f_i \tag{26}$$

where BIGM is a big positive number. Therefore, the better a chromosome's fitness is, the better its quality will be.

4.4. Selection Operator

The roulette wheel selection procedure has been used in the proposed HGA. The purpose of this selection operator is to allow the fittest individuals to be considered more often to reproduce children for the next generation. Each individual is assigned a probability of being selected based on its fitness value. Although better individuals will have a higher selection probability, all individuals in the population will have a chance to be selected.

4.5. Crossover Operator

In the GAs, each pair of selected parents can generate two children using crossover operator. After two parent chromosomes have been selected for recombination, with a probability of μ_c, the parents are recombined and two new chromosomes are generated as off springs and with a probability of $1-\mu_c$, no crossover occurs and both parents pass unchanged as off springs. In the proposed algorithm, uniform crossover operator is used for the first and the second parts of the chromosomes. For the first part, a binary mask of size N is randomly generated and thereby, the first (second) child will receive its *ith* gene from the first (second) parent if the *ith* element of the binary mask is equal to 1 and from the second (first) parent if it is equal to 0. For the second part of chromosomes, a binary mask of size M is randomly generated, and the same procedure is applied. However, applying this crossover to the second part may lead no feasible offsprings because of violating the constraint (7) of the mathematical model. In this situation, modification operator must be applied. Fig. 2 illustrates how the crossover operator works.

Parent 1	1	2	1	3	2	1	3	1	3	2	2	1	2	3	1	3
Parent 2	2	2	2	1	3	2	1	2	3	1	2	3	1	2	1	2
Binary Mask	0	0	1	0	1	1	1	0	1	0	0	1	1	0	1	0
Offspring 1	2	2	1	1	2	1	3	2	3	1	2	1	2	2	1	2
Offspring 2	1	2	2	3	3	2	1	1	3	2	2	3	1	3	1	3

Fig. 2. Crossover operator

4.6. Mutation Operator

For better exploration of the search space and prevention of getting trapped in local optimum, at last one gene of the first part and at last two genes of the second part of the generated off springs are mutated with a probability of μ_m. The mutation operator for the first part of chromosome is such that if a gene is chosen for being mutated, that gene will take a value different from its current value (the manufacturing route is actually changed for the part corresponding to that gene). For the second part of the chromosome assume that the kth gene is selected for mutation. In this case, a gene with different value is randomly selected and then these two genes swap their values (the cell of two machines are actually swapped). It should be noted that, applying the mutation operator never leads to infeasible solutions.

4.7. Modification Operator

As mentioned before, crossover of the second part of chromosomes may lead no feasible offsprings. In this situation, the infeasible offsprings must be modified. In the modification schema, for each cell of each offspring, the number of machines allocated to that cell is counted and if this number exceeds the maximum cell size, the additional machines will be randomly allocated to the cells which have unfilled places. Fig. 3 illustrates how the modification operator works (the maximum cell size is equal to 5 here):

After crossover	Offspring 1	2	2	1	1	2	1	3	2	3	1	2	1	2	2	1	2
	Offspring 2	1	2	2	3	3	2	1	1	3	2	2	3	1	3	1	3
After modification	Offspring 1	2	2	1	1	2	1	3	2	3	1	2	1	3	2	1	2
	Offspring 2	1	2	2	3	3	2	1	1	3	2	2	3	1	3	1	1

Fig. 3. Modification operator

4.8. Replacement Operator and Elitism

In the proposed HGA, the population at a given generation $k+1$ is obtained by replacement operator. This operator selects the best individuals among generation k and its generated offsprings and forms the generation $k+1$. Since all previous and current best population members are included in generation $k+1$, elitism is ensured.

4.9. Acceptance of Worse Solutions

Genetic algorithms often use mutation operator for prevention of getting trapped in local optimum. To ensure further exploration of the search space, a mechanism similar to simulated annealing (SA) algorithm has been used in the proposed algorithm. The only difference is that, unlike SA, in the developed algorithm, after a quarter of algorithm's runtime, the worse solutions are accepted with constant probability of 5%. In other words, the proposed algorithm lists all the chromosomes with worse fitness than *best_fitt* and then accepts one of them as *best_partial_fitt*. Afterwards, all individuals with better fitness than *best_partial_fitt* are replaced with random chromosomes.

4.10. Elimination of Repetitive Solutions

Since genetic algorithm is based on population and selection of the chromosomes as the parents is done here based on their fitness values, after some iterations, several repetitive solutions may exist in the current population. This will significantly reinforce the probability of selecting two similar chromosomes as the parents by increasing the number of iterations. Moreover, off springs similar to the previous population will be created and probably, the algorithm will get trapped in local optimum. To avoid this, completely similar chromosomes are eliminated from the population, after each iteration, and they are replaced with random chromosomes.

4.11. Proposed Hybrid Genetic Algorithm

Proper determination of the parameters for each algorithm significantly influences its convergence rate and the quality of the obtained solutions. To determine the value of these parameters, one must always evaluate and compare the quality of the obtained solutions as well as the computational efficiency of the algorithm. In the proposed HGA, some parameters like size of the population, termination criteria, crossover rate and mutation rate have been addressed by implementing several tests and comparing quality of the obtained solutions and convergence rate of the algorithm. The candidate values and the selected value of the parameters in the proposed algorithm are as follows:

popSize	20 (candidate options: 20, 30, 40 and 50)
MaxNoIMP1	Q (candidate options: $\frac{Q}{2}, Q, \frac{3Q}{2}, 2Q$)
MaxNoIMP2	$\frac{3M}{2}$ (candidate options: $\frac{M}{2}, M, \frac{3M}{2}, 2M$)
MaxTime	The algorithm converges within 10-1000 seconds according to the size of the problem
μ_c	0.99 (candidate options: 0.90, 0.95, 0.97, 0.99)
μ_m	0.05 (candidate options: 0.01, 0.02, 0.05, 0.10)

Fig. 4 and 5 shows the flowcharts of the proposed bi-level HGA.

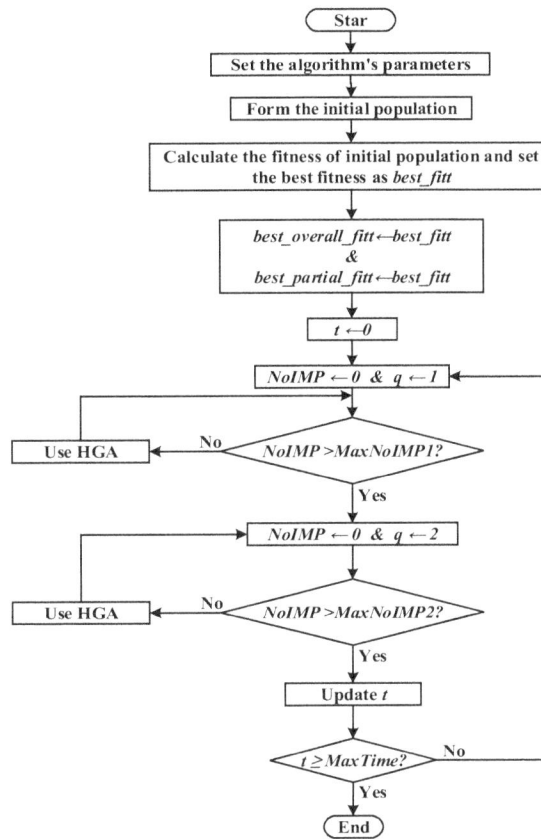

Fig. 4. Flowchart of the bi-level HGA

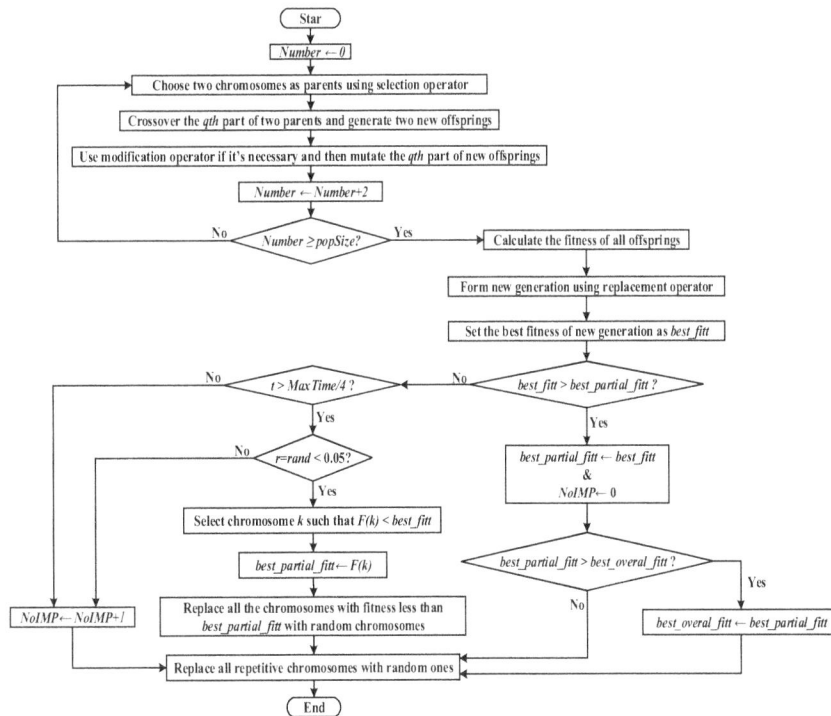

Fig. 5. Flowchart of HGA for the first and second parts of chromosomes

5. Results of Calculations

5.1. Performance Results and Evaluation of Proposed Linear Model

Some 13 numerical examples are used to assess and demonstrate efficiency of the linear model and mixed genetic algorithm developed. All the problems except for problems 2 and 3 are problems with random numbers. Problems 2 and 3 are case studies and their parameters are obtained from space of a mean manufacturing system (parameters and solution for this problem can be found in appendix (A)). Table 2 summarizes the results obtained from solving the numerical examples by the proposed model and also the existing models in the literature through exact method of branch & bound using LINGO 8 software.

Table 2

Solution of numerical examples with the proposed model and existing models

Problem No.	Parameters				INLP		pure ILP (literature)		proposed ILP	
	N	Q	M	C	OFV	time	OFV	time	OFV	time
1	3	5	9	3	22463 *	00:00:05	22463 *	00:00:32	22463.81 *	00:00:02
2	3	9	22	3	22573.13 *	00:02:14	25073.13 *	00:32:27	22573.13 *	00:00:30
3	3	12	22	3	22330.97 *	00:24:55	22330.97 *	01:43:08	22330.97 *	00:01:59
4	3	6	13	4	46867.32 *	00:02:29	46867.32 *	00:51:00	46867.32 *	00:00:42
5	5	10	8	2	83800.6 *	00:00:03	83800.6 *	00:00:11	83800.57 *	00:00:02
6	5	10	15	3	68893.7 *	00:01:53	68893.7 *	01:23:00	68893.7 *	00:00:41
7	5	15	16	3	47465.6 #	>7:00:00	47465.6 #	>7:00:00	39606.7 *	0.002095
8	8	16	15	3	169481.19 #	>7:00:00	170241 #	>7:00:00	166545.47 *	0.004086
9	8	18	20	4	315718.45 #	>7:00:00	114843 $	>7:00:00	241462.62 *	0.080625
10	8	22	22	4	231471.34 #	>7:00:00	66772.4 $	>7:00:00	168032.63 *	0.225914
11	10	20	20	4	293536.26 #	>7:00:00	144491 $	>7:00:00	278148 *	0.090787
12	10	25	27	5	97603.7 $	>7:00:00	97603.7 $	>7:00:00	250004 #	>7:00:00
13	10	29	43	6	68023.14 $	>7:00:00	out of memory		158806 $	>7:00:00

* optimal solution was found

best objective(not optimal) was found in predetermined time

$ only best lower bound was found

The obtained results were global optimum answers of the problems, which show that the proposed MILP model is practically more efficient that other existing models. These answers are being used in the next section for evaluation of the developed genetic algorithm.

5.2. Performance Results and Evaluation of Proposed Genetic Algorithm

Table 3 shows the obtained results from solving the numerical examples with the proposed genetic algorithm. The obtained results demonstrate that the proposed algorithm is able to give good answers for the problems with various sizes in a reasonable time. Taking into consideration these results, it can be seen that the proposed algorithm has attained to an optimal answer in 93.64% of the cases. Meanwhile, mean error percentage obtained from optimal value in the worst case was 0.87%, with error content of the worst answer from optimal value being equal to 2.77% in the worst case. Computational cost of the proposed algorithm is also much smaller than that of MILP model suggested here. Therefore, the algorithm would take just a very little time to reach a good answer (the algorithm reaches a good answer at every quarter of its running time). However, most of the time of the algorithm is assigned to implement techniques to escape from local optimum, do more improved search within the search space and try to reach a better possible answer.

Table 3

Solution of numerical examples with the suggested genetic algorithm

	Parameters					GA (number of runs:10)						
Problem number	N	Q	M	C	Optimum objective	best obtained objective	number of optimum solution	mean obtained objective	worst obtained objective	time	Error percentage of mean objective from optimum (%)	Error percentage of worst objective from optimum (%)
1	3	5	9	3	22463.8	22463.8	10	22463.8	22463.8	00:00:10	0.00	0.00
2	3	6	22	3	22573.1	22573.1	9	22736.3	24402.3	00:00:30	0.72	8.10
3	3	12	22	3	22331.0	22331.0	10	22331.0	22331.0	00:00:30	0.00	0.00
4	3	6	13	4	46867.3	46867.3	10	46867.3	46867.3	00:00:20	0.00	0.00
5	5	10	8	2	83800.6	83800.6	10	83800.6	83800.6	00:00:10	0.00	0.00
6	5	10	15	3	68893.7	68893.7	10	68893.7	68893.7	00:00:30	0.00	0.00
7	5	15	16	3	39606.7	39606.7	7	39953.0	40704.1	00:00:30	0.87	2.77
8	8	16	15	3	166545.5	166545.5	10	166545.5	166545.5	00:00:30	0.00	0.00
9	8	18	20	4	241462.6	241462.6	9	241756.6	244402.2	00:00:40	0.12	1.22
10	8	22	22	4	168032.6	168032.6	9	168106.4	168769.7	00:00:40	0.04	0.44
11	10	20	20	4	278148.0	278148.0	9	278195.5	278623.5	00:00:45	0.02	0.17
12	10	25	27	5		247803.7	3	249969.9	253317.6	00:01:00		
13	10	29	43	6		412025.3	7	416035.3	437025.3	00:01:20		

Fig. 6 depicts search process of the algorithm for the problem number 2 with 3 parts, 6 manufacturing routes, 22 machines and 3 cells within 30 seconds. It is observed that the algorithm has reached an optimal answer at 1115th repetition. From 1116th to 3823rd repetition, it is tried to search the answer space more and further ensure about prevention of getting trapped in the local optimum.

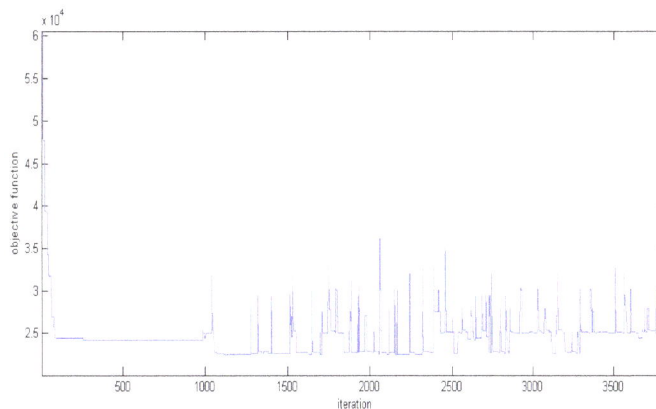

Fig. 6. Search process of the algorithm for the problem number

6. Conclusion

We proposed a mixed linear integer numerical model for designing the manufacturing cells in order to reduce the manufacturing costs, increase quality of the products and enhance reliability of the system. Since this model is mainly developed to deal with a natural issue, it has some characteristics which make them more practical in comparison with the other existing models. For instance, some of the models which consider the reliability in a CF problem have suggested multi-objective models. Thereby, the main purposes are to improve the reliability and reduce the manufacturing costs. Calculating the reliability value is difficult for the system and rather useless for the system, so solving this model will be very difficult. A logical relation is used in this paper such that a cost approach can also enhance the reliability value. However, it is not calculated in this numerical method, since the reliability of this system is improved by decreasing the number of breakdowns during system operation, while the repair and maintenance costs are reduced as well. The little number of limitations and integer variables, and also linear behavior of the model have all added to its efficiency for both small and the large scale problems. This model has been solved for some ranges of small to large scale problems using LINGO software with its improved efficiency being shown in Table 2. A multi-layer genetic algorithm is recommended to be developed to facilitate solving all kinds of problems and especially the large scale one. Efficiency of this algorithm is illustrated in Table 2 for the both small and large scale problems.

References

Agarwal, A., & Sarkis, J.A. (1998). Review and analysis of comparative performance studies of functional and cellular manufacturing layouts. *Computers and Industrial Engineering, 34*(1) 77–89.

Ahkioon, S., Bulgak, A. A., & Bektas, T. (2009). Cellular manufacturing systems design with routing flexibility, machine procurement, production planning and dynamic system reconfiguration. *International Journal of Production Research, 47*(6), 1573-1600.

Akturk, M. S., & Turkcan, A. (2000). Cellular manufacturing system design using a holonistic approach. *International Journal of Production Research, 38*(10), 2327-2347.

Ameli, M. S. J., Arkat, J., & Barzinpour, F. (2008). Modelling the effects of machine breakdowns in the generalized cell formation problem. *The International Journal of Advanced Manufacturing Technology, 39*(7-8), 838-850.

Ameli, M. S. J., & Arkat, J. (2008). Cell formation with alternative process routings and machine reliability consideration. *The International Journal of Advanced Manufacturing Technology, 35*(7-8), 761-768.

Askin, R.G., & Estrada, S. (1999). Investigation of cellular manufacturing practices, S.A. (Ed.) *Handbook of Cellular Manufacturing Systems (Chapter1)* John Wiley, New York, 25-34.

Boughton, N.J., & Arokiam, I.C. (2000). The application of the cellular manufacturing: a regional small to medium enterprise perspective. *Proceedings of the Institution of the Mechanical Engineers, 214*(Part B), 751–754.

Bulgak, A. A., & Bektas, T. (2009). Integrated cellular manufacturing systems design with production planning and dynamic system reconfiguration. *European Journal of Operational Research, 192*(2), 414-428.

Das, K., Lashkari, R. S., & Sengupta, S. (2007). Machine reliability and preventive maintenance planning for cellular manufacturing systems. *European Journal of Operational Research, 183*(1), 162-180.

Das, K., Lashkari, R. S., & Sengupta, S. (2007). Reliability consideration in the design and analysis of cellular manufacturing systems. *International Journal of Production Economics, 105*(1), 243-262.

Defersha, F. M., & Chen, M. (2006). A comprehensive mathematical model for the design of cellular manufacturing systems. *International Journal of Production Economics, 103*(2), 767-783.

Defersha, F. M., & Chen, M. (2006). Machine cell formation using a mathematical model and a genetic-algorithm-based heuristic. *International Journal of Production Research, 44*(12), 2421-2444.

Defersha, F. M., & Chen, M. (2008). A parallel genetic algorithm for dynamic cell formation in cellular manufacturing systems. *International Journal of Production Research, 46*(22), 6389-6413.

Defersha, F. M., & Chen, M. (2008). A parallel multiple Markov chain simulated annealing for multi-period manufacturing cell formation problems. *The International Journal of Advanced Manufacturing Technology, 37*(1-2), 140-156.

Flynn, B. B., & Robert Jacobs, F. (1986). A simulation comparison of group technology with traditional job shop manufacturing. *International Journal of Production Research, 24*(5), 1171-1192.

Gen, M., & Cheng, R. (1997). *Genetic algorithms and engineering design.* Wiley, MA.

Goldberg, D.E. (1989). *Genetic Algorithms in Search Optimization & Machine Learning.* Addison Wesley.

Gupta, Y., Gupta, M., Kumar, A., & Sundaram, C. (1996). A genetic algorithm-based approach to cell composition and layout design problems. *International Journal of Production Research, 34*(2), 447-482.

Holland, J. H. (1975). *Adaptation in natural and artificial systems*. Ann Arbor: The University of Michigan Press.

Kusiak, A. (1987). The generalized group technology concept. *International Journal of Production Research*, *25*(4), 561-569.

Lokesh, K., & Jain, P.K. (2008). Part-machine group formation with operation sequence, time and production volume. *International Journal of Simulation Modelling, 7*(4), 198–209.

Lokesh, K., & Jain, P.K. (2009). Part-Machine group formation with ordinal-ratio level data and production volume. *International Journal of Simulation Modelling, 8*(2), 90–101.

Lokesh, K., & Jain, P.K. (2010). Concurrently part-machine group formation with important production data. *International Journal of Simulation Modelling, 9*(1), 5–16.

Lokesh, K., & Jain, P.K. (2010). Dynamic cellular manufacturing systems design-a comprehensive model & HHGA. *Advances in Production Engineering & Management Journal, 5*(3), 151–162.

Man, K.F., Tang, K.S., & Kwong, S. (1999). *Genetic algorithms: concepts and design*. Springer, London.

Morris, S.J., & Tersine, R.J. (1990). A simulation analysis of factors influencing the attractiveness of group technology cellular layouts. *Management Science*, *36*(12), 1567–1578.

Nsakanda, A. L., Diaby, M., & Price, W. L. (2006). Hybrid genetic approach for solving large-scale capacitated cell formation problems with multiple routings. *European Journal of Operational Research*, *171*(3), 1051-1070.

Onwubolu, G. C., & Mutingi, M. (2001). A genetic algorithm approach to cellular manufacturing systems. *Computers & industrial engineering*, *39*(1), 125-144.

Ouk Kim, C., Baek, J. G., & Baek, J. K. (2004). A two-phase heuristic algorithm for cell formation problems considering alternative part routes and machine sequences. *International Journal of Production Research*, *42*(18), 3911-3927.

Saxena, L. K., & Jain, P. K. (2011). Dynamic cellular manufacturing systems design—a comprehensive model. *The International Journal of Advanced Manufacturing Technology*, *53*(1-4), 11-34.

Seifoddini, H., & Djassemi, M. (2001). The effect of reliability consideration on the application of quality index. *Computers & Industrial Engineering*, *40*(1), 65-77.

Sofianopoulou, S. (1999). Manufacturing cells design with alternative process plans and/or replicate machines. *International Journal of Production Research*, *37*(3), 707-720.

Tompkins, J.A., White, J.A., Bozer, Y.A., & Tanchoco J.M.A. (2003). *Facility planning*. Wiley, New York.

Wemmerlov, U., Hyer, N.L. (1989). Cellular manufacturing in the U S industry: a survey of users, *International Journal of Production Research*, *27*(9), 1511–1530.

Wemmerlov, U., & Johnson, D. J. (1997). Cellular manufacturing at 46 user plants: implementation experiences and performance improvements. *International journal of production research*, *35*(1), 29-49.

Zhao, C., & Wu, Z. (2000). A genetic algorithm for manufacturing cell formation with multiple routes and multiple objectives. *International Journal of Production Research, 38*(2), 385–395.

Appendix (A)

Data for example 3

Machine	Encoding(K parameter)	Machine	Encoding(K parameter)
TC-201	1	MU-202	12
TC-202	2	MU-203	13
TC-203	3	MU-204	14
TC-204	4	MU-205	15
MC-206	5	MU-206	16
MC-208	6	DR-210	17
MC-201	7	DR-201	18
DR-206	8	MU-207	19
GC-202	9	DR-205	20
TU-201	10	MU-101	21
MU-201	11	MU-102	22

P.N	P.V	R.N	P.S	P.T
1	2500	1	5 10 10 2 20 15 14 12 4 7 8 8 9 8 17	24 1.83 1.5 5.75 0.92 2.5 3 1.83 3 5.5 1.5 1.5 2 1.5 3
		2	6 10 10 2 20 19 14 12 4 7 8 8 9 8 17	26 1.83 1.5 5.75 0.92 2.4 3 1.83 3 5.5 1.5 1.5 2 1.5 3
2	2500	1	11 16 16 17 17 20 11 18 18 22 21 21 20	2.5 2 2 2.5 2 1 2 1.5 1.5 2 1.5 3 2.5
		2	11 15 15 17 17 20 11 18 18 22 21 21 20	2.5 2.1 2.1 2.5 2 1 2 1.5 1.5 2 1.5 3 2.5
3	6000	1	6 10 10 1 14 3 13 12 9 8	24 1.2 1.1 5 4.3 3 3.5 1.7 2 1.5
		2	5 10 10 1 13 3 13 12 9 8	26 1.2 1.1 5 4.5 3 3.5 1.7 2 1.5

P.N: part no.

P.V: production volume

R.N: routing no.

P.S: production sequence

P.T: production time

Scheduled production plan

P.N	R.N	P.S	P.T
1	1	5 10 10 2 20 15 14 12 4 7 8 8 9 8 17	24 1.83 1.5 5.75 0.92 2.5 3 1.83 3 5.5 1.5 1.5 2 1.5 3
2	1	11 16 16 17 17 20 11 18 18 22 21 21 20	2.5 2 2 2.5 2 1 2 1.5 1.5 2 1.5 3 2.5
3	2	5 10 10 1 13 3 13 12 9 8	26 1.2 1.1 5 4.5 3 3.5 1.7 2 1.5

Machines in each cell (Design of cells)

Cell 1	Cell 2	Cell 3
TC-202	TC-204	TC-201
MU-201	MC-208	TC-203
MU-206	MC-201	MC-206
DR-210	MU-204	DR-206
DR-201	MU-205	GC-202
DR-205	MU-207	TU-201
MU-101		MU-202
MU-102		MU-203

11

An adaptive large neighborhood search heuristic for solving the reliable multiple allocation hub location problem under hub disruptions

S. K. Chaharsooghi[a]*, Farid Momayezi[a] and Nader Ghaffarinasab[b]

[a]*Department of Industrial & Systems Engineering, Tarbiat Modares University, Tehran, Iran*
[b]*Department of Industrial Engineering, University of Tabriz, Tabriz, Iran*

CHRONICLE	ABSTRACT
Keywords: *Hub location problem* *Reliability* *Stochastic programming* *Adaptive large neighborhood search*	The hub location problem (HLP) is one of the strategic planning problems encountered in different contexts such as supply chain management, passenger and cargo transportation industries, and telecommunications. In this paper, we consider a reliable uncapacitated multiple allocation hub location problem under hub disruptions. It is assumed that every open hub facility can fail during its use and in such a case, the customers originally assigned to that hub, are either reassigned to other operational hubs or they do not receive service in which case a penalty must be paid. The problem is modeled as two-stage stochastic program and a metaheuristic algorithm based on the adaptive large neighborhood search (ALNS) is proposed. Extensive computational experiments based on the CAB and TR data sets are conducted. Results show the high efficiency of the proposed solution method.

1. Introduction

Hubs are intermediate facilities that perform a set of tasks such as consolidation, break-bulk, sorting, etc. in transportation and telecommunication networks. In other words, the traffic flows (cargo, passengers, or data) in the network rather than being sent directly from their origins to their destinations, are routed via these intermediate facilities. Therefore, smaller number of connections with large flow volumes are used in the network which, in turn, makes it possible to exploit economies of scale in transportation costs, especially on the inter-hub connections.

Hub location problem (HLP) deals with locating the hub facilities in the network and determine the pattern based on the non-hub nodes assignment to each hub so that a specific objective function is optimized. Regarding the non-hub nodes assignment to hubs in the HLP, we have two types of allocations. First type, called single allocation, each non-hub node can be allocated to exactly one hub in the network, whereas in the second type, called multiple allocation, each non-hub node can simultaneously be allocated to more than one hub in the network. In both mentioned schemes, the hub

* Corresponding author
E-mail: SKCH@modares.ac.ir (S. K. Chaharsooghi)

nodes as well as the network links can have limited or no capacity. Therefore, HLPs are divided into four main categories in the literature:

I.   Capacitated single allocation hub location problem
II.   Uncapacitated single allocation hub location problem
III. Capacitated multiple allocation hub location problem
IV.   Uncapacitated multiple allocation hub location problem

HLPs are frequently used in some industries such as transportation, communication and computer network design. Most of the HLP studies, assume that all of the established hubs in network function perfectly well throughout the planning horizon and hence they will be accessible to all the customers. However, the infrastructures of supply chains are always under risk of disruption due to environmental, technological and international damages. Natural disasters like flood, hurricane and earthquake can affect extensive geographical areas and make transportation and other network elements un-operational (Matisziw et al., 2010).

As the functions such as consolidation, break-bulk, and sorting are executed at hub nodes, disruption at hubs can result in high costs for the customers as well as the network operators. Therefore, reassignment of the customers from disrupted hubs to the non-affected hubs is of crucial importance for reducing network costs. In this paper, we have considered the reliable uncapacitated multiple hub location problem under hub disruption. It is assumed that when a disruption happens at each hub and makes it un-operational, all the assigned non-hub nodes to the disrupted hub should be reassigned to other operational hubs or if the costs for serving these nodes via the operational hubs are too high, then serving these nodes can be cancelled and a penalty is paid for each unit not served. We have modeled different possibilities of hub disruption using a group of scenarios in which a random subset of hubs is un-operational due to disruptions. The problem has been modeled as a two-stage stochastic program in which the decisions on hub locations are made in the first phase. In second phase when disruption scenario has occurred, the allocation of non-hub nodes to hubs takes place in second phase with regard to the operational hubs. To solve the proposed model, a metaheuristic algorithm which is based on the Adaptive Large Neighborhood Search (ALNS) is presented and the effectiveness of this algorithm is tested by solving large set of instances from the CAB and TR data sets.

The remainder of this paper is organized as follows. The literature review is presented in the next section. The mathematical models are presented in third section. Section four describes the proposed solution algorithm in detail. Numerical results are presented in section five and finally, conclusions and some directions for future research are presented in last section.

2. Literature

O'Kelly (1986) presented first mathematical formulation for the single allocation p-hub median problem as quadratic model. Mixed integer linear models for different versions of the single and multiple allocation hub location problem such as p-hub median, p-hub center and hub covering problems were proposed for the first time by Campbell (1994). Later, other mathematical formulation for hub location problem were proposed by Ernst and Krishnamoorthy (1994), Skorin-Kapov et al. (1996) and Ebery (2001).

Many studies on HLP assumed that the established hubs would always be operational. Nevertheless, these facilities may fail due to different reasons in practice. As an example, unexpected weather conditions can adversely affect the availability of an airport serving as a hub in air transportation industry. The same problem can occur in supply chain and logistics systems, where facilities, same as hubs, play the central role and their locations are derived using facility location models (An et al., 2014). Therefore, considering reliability in HLP is of utmost importance. Snyder and Daskin (2005) study facility location in which some of cases with definite probability become unusable and assume that customers would be served by facilities which are not affected by disruption. Berman et al. (2009) and Shen et al. (2011) who are inspired by this model developed new location problem models with disruption consideration. They supposed that facilities are not completely reliable and customers do not have any information about a facility being operational or not and it is supposed that every facility may be non-operation by a definite probability. Shen et al. (2011) present reliability subject in this area where some facilities are disrupted temporarily. If a facility becomes defected, other allocated customers shall be reassigned to other operational facility. Authors develop 2 step stochastic program in a non-linear integer model. Wang and Ouyang (2013) present continuous probability approach in order to identify competitive facility location at risk of disruption condition. They use models related to games theory in order to optimize location of facility services in condition of facilities competition and facility disruption risk. They believe that customer demand share in market depends on server facilities performance and competitor presence in closed place because customers are usually following up nearest way. Author's model which are based on game's theory, merge these complicated factors in an integrated framework. They use experimental and hypothesis data in order to evaluate their suggestive models and monitored impact of competition, disruption risk in facilities and transportation cost on optimized plan.

Medal et al. (2014) present a multi-objective model for the facility problem. They use two methods for decreasing the risk of disruption: a) identifying facility location strategically, and b) using the rigid and reliable facilities. Authors merge these two theories in their suggestive model decline farthest distance from demand points to nearest available facility after disruption in facilities. It is supposed that decider is reluctant to the risk and eager to decrease disruption of facilities with maximum output therefore a multi objective mixed integer model is suggested. Matisziw et al. (2010) present a multi objective optimized model for the first time in order to restoration network after disruption and time scheduling of possible restoration scenario when network nodes and arcs are lost and disruption takes place.

In order to enhance the reliability of hub location problems, Kim (2008) proposes a single allocation p-hub protection with primary and secondary routes. Kim and O'Kelly (2009) propose single and multiple allocation models to derive an optimal network structure that maximizes the expected network flow given that each arc or hub has a given reliability. Their work does not consider backup hubs and alternative routes and a tabu search heuristic is utilized to solve the real instances with up 20 nodes. Zeng et al. (2010) address the reliable single and multiple allocation hub location models by considering hub unavailability where alternative routes have been developed and a heuristic algorithm has been proposed. In another study, Taghipourian et al. (2012) studied hub services to non-hub nodes in forecasted disruption. In this research they considered some non-hub nodes as a virtual hub, so as main disrupted hubs should be closed and virtual ones open and serves other hubs in forecastable inappropriate weather condition and other forecastable disruptions. Next, they proceed to present a nonlinear fuzzy mathematical model in order to decrease costs. Parvaresh et al. (2013) study imperative disruption of hubs. They model their study as Steklberg game that includes a leader and follower as two steps with consideration of bi-objective model. First objective function is used in order to minimize means of transportation cost and second objective function is used to maximize cost of disruption imposed on network. Parvaresh et al. (2014) develop model presented in 2013 and present it in two-level with three objective functions and add a decision variable with two other constraints to modify solution algorithm. In a newer research, Azizi et al. (2014) propose a new formulation with conservation of different theories. In this research one backup hub is selected from the available hub of network for disrupted hub and all of allocated flows of defected hub are allocated to backup hub. An et al. (2015) propose a set of reliable hub-and-spoke network design models where the selection of backup hubs and alternative routes are taken into consideration to proactively handle hub disruptions. They develop Lagrangian relaxation and Branch-and-Bound methods to solve these nonlinear mixed integer formulation. More recently, Mohammadi et al. (2016) introduce a different perspective to reliable hub-and-spoke network design. The authors categorize the disruption into two classes: a) complete disruption (accessibility disruption), and b) partial disruption (capacity disruption). They also assume that hub network is incomplete and the connections between the hubs is tree.

As hub location problem is an NP-hard problem, exact solutions for the large and real-sized instances are very time consuming and even sometimes impossible to reach. O'Kelly (1985) propose two solution methods for solving the p-hub median problems. Both of these algorithms consider all of possible scenarios for choosing p-hub locations. In first algorithm, demand nodes are allocated to nearest hub and in second method, allocations are determined based on the value of the objective function between the first nearest and the second hub. He used CAB data set to test his solution method. Metaheuristic methods have successfully been implemented for solving hub location problems by many researches in this filed. Skorin-Kapov (1994) uses taboo search (TS) algorithm for solving the single allocation p-hub median problem. In another research, Abdinnour-Helm (2001) propose a simulated annealing (SA) solution method for the single allocation p-hub median problem. Later, Perez et al. (2007) present heuristic algorithm as hybrid of two metaheuristics: variable neighborhood search (VNS) and path relinking (PR) for the uncapacitated single allocation HLP. Their algorithm shows a better performance in comparison to the simulated annealing and taboo search described above. Lin et al. (2012) uses genetic algorithm (GA) method for solving the p-hub median problem with integral constraints. Marti et al. (2014) use scatter search in order to solve the uncapacitated p-hub median problem. They also strengthened their algorithm by hybridization with path relinking.

Although every metaheuristic algorithm has its own characteristics but recently adaptive large neighborhood search (ALNS) has been used extensively in routing and allocation problems and in many cases higher quality solutions are obtained in comparison to other metaheuristics on the same problems. The ALNS heuristic which is generalization of the Large Neighborhood Search (LNS) algorithm has been presented for the first time by Ropke and Pisinger (2006) for solving pickup and delivery vehicle routing problem (VRP). Hemmelmayr et al. (2012) use the ALNS for solving two-echelon VRP and the location routing problem (LRP). They show that the solution obtained by the ALNS are better than other solution methods for the two-echelon VRP and excellent results have been obtained in case of the LRP. In another study, Demir et al. (2012) use the ALNS for solving the pollution-routing problem (PRP) and effectiveness of their applied algorithm is demonstrated on a large set of test instances. Mauri (2012) uses the ALNS algorithm for solving the berth allocation problem (BAP). Results show that the ALNS improve the best known solutions in so many cases in comparison with other algorithms which are used for solving the same problem. More recently, Grangier et al. (2016) use this algorithm for solving the two-echelon multiple-trip vehicle routing problem and obtained superior solutions.

3. Mathematical formulations

Let $G=(N,E)$ be a graph, where N is the set of nodes and E is the set of edges such that $E \subseteq N \times N$. A subset $J \subseteq N$ of nodes would be selected as the hubs with remaining $|N|-|J|$ spokes being allocated to these hubs. The following parameters are used in our model:

f_k: fixed cost of establishing a hub at node $k \in N$

w_{ij}: amount of flow originated at node $i \in N$ and destined to node $j \in N$

c_{ij}: transportation cost per unit of traffic between nodes $i \in N$ and $j \in N$

α: discount factor ($0 \le \alpha \le 1$) representing the scale economies on the inter-hub connections

c_{ijkm}: unit transportation cost between nodes $i \in N$ and $j \in N$ that is routed via hubs $k \in N$ and $m \in N$ calculated as:

$$C_{ijkm} = c_{ik} + \alpha c_{km} + c_{mj}$$

θ: unit penalty cost for the traffic that is not routed because of hub disruptions.

$I_{k(\xi)}$: binary parameter representing the operational status of hubs equal to 1 if hub k is operational and equal to 0 if the hub is disrupted.

We define the following sets of decision variables:

- $Z_k \in \{0,1\}$ is 1 if a hub is opened at node $k \in N$ and 0, otherwise;
- $Y_{km} \in \{0,1\}$ is 1 if node $i \in N$ is assigned to the hub node $k \in N$ and 0, otherwise
- $X_{ijkm}(\xi) \geq 0$ is the fraction of flow originated from origin i and destination to node j $(i, j \in N)$ that is routed through hubs located at nodes k and m $(k, m \in N)$ in that order.
- $V_{ij}(\xi) \geq 0$ is the fraction of flow from origin i to destination j $(i, j \in N)$ that is not routed (for which penalty cost is incurred).

Based on the parameters and the variables define above, the two-stage stochastic programming model for our problem can be written as follows:

$$\min \sum_k f_k Z_k + E[Q(Z, \xi)] \tag{1}$$

subject to:
$$Z_k \in \{0,1\} \qquad\qquad \forall\, k \tag{2}$$

where
$$Q(Z, \xi) = \min \sum_i \sum_j \sum_k \sum_m w_{ij} c_{ijkm}. X_{ijkm}(\xi) + \sum_i \sum_j w_{ij} V_{ij}(\xi) \tag{3}$$

$$\sum_k \sum_m X_{ijkm}(\xi) + V_{ij}(\xi) = 1 \qquad\qquad \forall\, i, j \tag{4}$$

$$\sum_m X_{ijkm}(\xi) + \sum_{m|m \neq k} X_{ijmk}(\xi) \leq Z_k \qquad\qquad \forall\, i, j, k \tag{5}$$

$$\sum_m X_{ijkm}(\xi) + \sum_{m|m \neq k} X_{ijmk}(\xi) \leq I_k(\xi) \qquad\qquad \forall\, i, j, k \tag{6}$$

$$X_{ijkm}(\xi) \geq 0 \qquad\qquad \forall\, i, j, k, m \tag{7}$$

$$V_{ij}(\xi) \geq 0 \qquad\qquad \forall\, i, j \tag{8}$$

In above the formulation E_ξ denotes the mathematical expectation with respect to ξ and Ξ is the support of ξ. The objective function (1) minimizes the sum of the first-stage cost of opening hub facilities and the transportation cost and penalty (as calculated in equation (3)). Constraint (4) state that each origin-destination flow must either be routed via some pair of hubs or a penalty must be occurred if it is not routed. Constraints (5) and (6) prohibit commodities from being routed via an unopened hub or a disrupted hub, respectively. Finally Eq. (2), Eq. (7) and Eq. (8) are domain constraint for associated decision variables.

Let the uncertainty associated with operational status of the hub facilities be described by a finite set of scenarios $(s \in S)$ each of which having a probability (p_s) that is assumed to be known. Under each scenario s, denote the realized value of the random variable $I_k(\xi)$ as $I_{k,s}$. We can now write the so-called extensive from or the deterministic equivalents of the above two-stage stochastic problem as follows.

$$\min \sum_k f_k Z_k + \sum_s p_s \left(\sum_i \sum_j \sum_k \sum_m w_{ij}. C_{ijkm}. X_{ijkm}^s + \sum_i \sum_j \theta. w_{ij}. V_{ij}^s \right) \tag{9}$$

$$\sum_{k}\sum_{m}X_{ijkm}^{s} + V_{ij}^{s} = 1 \qquad\qquad \forall\, i,j,s \qquad\qquad (10)$$

$$\sum_{m}X_{ijkm}^{s} + \sum_{m|m\neq k}X_{ijkm}^{s} \leq Z_{k}I_{k,s} \qquad\qquad \forall i,j,k,s \qquad\qquad (11)$$

$$X_{ijkm}^{s}, V_{ij}^{s} \geq 0 \qquad\qquad \forall i,j,k,m,s \qquad\qquad (12)$$

$$Z_{k} \in \{0,1\} \qquad\qquad \forall\, k \qquad\qquad (13)$$

4. Solution Method

As mentioned earlier, the ALNS metaheuristic is a generalized version of the LNS algorithm and was first presented by Ropke and Pisinger (2006) for solving the pickup and delivery VRP. The LNS method which has been presented first time by Shaw (1997) for solving the VRP, try to improve initial solutions of a combinatorial optimization problem by changing the solutions locally one at a time. Since the selection of neighborhood directly affects the process of generating new solutions within the search space, it should be handled carefully and in a smart manner. Let x be a feasible solution to our reliable hub location problem and X be the set of all feasible solutions to this problem. For each solution $x \in X$ we define a neighborhood $N(x) \subseteq X$ as a function $N : X \rightarrow P(X)$. In neighborhood search method of function N is created by combination of destroy and repair operators. The basic idea behind this method is that some part of solution is destroyed and then is repaired in the following steps. The main purpose of the destroy operator is to remove a part of a given solution so that the repair operator could rebuild that part resulting in a new solution (Lutz, 2015). Unlike the LNS algorithm in which only one destroy and one repair operator is used, the ALNS is able to use several operators for the repair and several operators for the destroy functions, simultaneously. Then algorithm will allocate a weight for each operator that reflects success level of related function in the previous steps. Operator selection is random in each stage and will be according to related weights. If $D = \{d_i | i = 1,...,k\}$ is a group of "k" destroy operators and $R = \{r_i | i = 1,...,l\}$ is a group of "l" repaired operators and primary weights of operators are defined as $w(d_i)$ and $w(r_i)$, so operator selecting probability is as below:

$$P(d_i) = \frac{w(d_i)}{\sum_{j=1}^{k} w(d_j)} \qquad\qquad P(r_i) = \frac{w(r_i)}{\sum_{j=1}^{l} w(r_j)}$$

Adjusting the operators weights plays essential role for increasing the probability of using more successful functions in comparison with less successful ones. Success of a given operator varies for different problems. Other factors such as instance size can also affect the usefulness of an operator in the same problem.

In order to solve the reliable hub location problem using the ALNS algorithm, we developed and used four destroy and three repair operators. For the proposed algorithm, destroy 1, 2 and 3 operators are able to be used in combination with any of the repair 1 and 2 operators. However, the destroy 4 and repair 3 operators are used together.

4.1 Destroy 1

Destroy 1 operator randomly selects 40% of the opened hubs in the solution and changes them to non-hub nodes, the goal of this operator is to destroy the hub location part of the solutions.

4.2 Destroy 2

In destroy 2 operator, 40% of the non-hub nodes are randomly selected and then are changed to hub nodes. The goal of this operator is to destroy the part of solutions which determines the non-hub nodes as well as their allocation to the hub nodes.

4.3 Destroy 3

Destroy 3 operator randomly selects 60% of entire nodes in the netwok and changes their status randomly. In other words, the selected hub nodes are changed to non-hubs or stay as hub with equal probabilities (0.5). Also the status of a selected non-hub node is changed to or still stay as a non-hub node with equal probability.

4.4 Repair 1

In repair 1 operator, the average failure probability of each node in the network is estimated based on the realized scenario matrix. Then, 10% of hubs in the solution obtained from the destroy heuristics are randomly selected. For every selected hub if the failure probability of that hub is more than the corresponding probability for the nearest node to that hub, then the hub becomes non-hub and the nearest node becomes hub.

4.5 Repair 2

In repair 2 operator, the nodes in the network are sorted based on the total distance from other nodes in non-decreasing order. Then 10% of the hub nodes in the solution (obtained from destroy heuristic) that have the largest total distances are turned to non-hub nodes. On the other hand, the same number of nodes which have the least total distances are turned to hubs.

4.6 Destroy 4 and repair 3

In destroy 4 operator, 20% of hub nodes are randomly selected and for each of these hub nodes, the repair 3 operator calculates the network cost in two cases: a) network cost assuming the considered hub stays as hub, b) network cost assuming that the considered hub becomes a non-hub node. If the cost of case (b) is smaller than case (a), then the node stays as hub, otherwise it is turned into a non-hub node. Results show that the use of destroy 4 operator in combination with the repair 3 operator makes more qualified and successful solutions compared with combination of other destroy and repair heuristics. However, this combination needs more time to be performed compared with the other operator combinations. The pseudo-code for the proposed ALNS algorithm for the reliable hub location problem under hub disruptions is shown in Fig. 1.

Algorithm: Adaptive Large Neighborhood Search (ALNS)

Input: Initial solution $x_0 \in X$, Maximum iterations *MaxIt*,
Current solution $x = x_0$, best solution $x_{best} = x_0$;
while stopping criteria not met **do**
    for *Iter* = 1, . . . , *MaxIt* **do**
        select $r \in R$, $d \in D$ according to probabilities p
        $x = r(d(x))$
        if accept (x, x') **then**
            $x = x'$
            if $f(x) < f(x_{best})$ **then**
                $x_{best} = x$
        adjust the weights w and probabilities p for the heuristics
return x_{best}

Fig. 1. Pseudo-code for the ALNS algorithm for the reliable hub location problem under hub disruptions

The stopping criterion set for the proposed algorithm is the number of iterations performed. Based on a set of preliminary experiments, it was shown that in most cases setting the maximum of 500 iterations provide a good trade-off between the solution quality and the run time of the algorithm.

5. Numerical experiment

To test the efficiency of the proposed solution algorithm as well as the validity of the proposed mathematical formulation, we have conducted a set of computational experiments. For this purpose, we have employed two famous data sets from the literature of HLP, namely the CAB and TR data sets which have extensively been used in the literature of HLP. The CAB data set is based on the airline passenger interactions between 25 US cities in 1970 evaluated by the Civil Aeronautics Board (CAB). Since the CAB data set does not contain fixed hub establishment, we have used four different values for this parameter as: 100, 150, 200, and 250 like most of the works in the literature. The second data set that is used in our computational experiments is the TR data set which is based on the cargo flows between 81 cities of Turkey. The fixed hub establishment costs in the TR data set are scaled by three different scaling factors (CF) as 0.1, 0.3, and 0.5. For both the data sets, the parameter α is considered at five levels: 0.2, 0.4, 0.6, 0.8, and 1. Mathematical models are solved using ILOG CPLEX 12.6 optimization software and the ALNS algorithm is coded in MATLAB R2013a. All the experiments are conducted on a computer with 3.3-GHz Intel Core i3 CPU and 4-GB of RAM under Windows 7 operating system.

The results obtained by solving the problem on CAB data set using the proposed ALNS heuristics as well as CPLEX for penalty coefficient $\theta=2000$ are presented in Table 1. The first column in this table includes applied discount factor for transportation cost on inter-hub connections. The second column shows fixed cost of hub establishment in the network. Next three columns present the solution results which are obtained by CPLEX. These three columns include the optimum value of objective function, the opened hubs in the optimum solutions, and the CPU time (in seconds) it took to reach the optimum solution. Result for solving the problem by ALNS are shown in next three columns and the last column indicates the optimality gap between the objective value obtained by the ALNS and the corresponding optimal value obtained by CPLEX.

Table 1
Results for the CAB data set with $\theta = 2000$

α	F	CPLEX			ALNS			%GAP
		Opt	Hubs	CPU (s)	Opt	Hubs	CPU (s)	
0.2	100	1251.896	4,12,16,18,22	28.985	1251.896	4,12,16,18,22	5.389	0.00
	150	1434.326	4,18,22	31.394	1434.326	4,18,22	3.262	0.00
	200	1493.734	18	25.021	1493.734	18	2.427	0.00
	250	1543.734	18	25.437	1543.73	18	2.404	0.00
0.4	100	1347.617	4,12,18	26.274	1347.617	4,12,18	3.642	0.00
	150	1443.734	18	23.389	1443.734	18	2.673	0.00
	200	1493.734	18	23.699	1493.734	18	2.283	0.00
	250	1543.734	18	23.10	1543.734	18	2.449	0.00
0.6	100	1371.645	4,18	22.324	1371.645	4,18	2.916	0.00
	150	1443.734	18	22.138	1443.734	18	2.350	0.00
	200	1493.734	18	22.399	1493.734	18	2.187	0.00
	250	1543.734	18	22.621	1543.734	18	2.257	0.00
0.8	100	1380.987	4,18	22.280	1380.987	4,18	2.869	0.00
	150	1443.734	18	21.853	1443.734	18	2.466	0.00
	200	1493.734	18	22.168	1493.734	18	2.497	0.00
	250	1543.734	18	22.00	1543.734	18	2.288	0.00
1	100	1385.640	4,18	22.191	1385.640	4,18	2.718	0.00
	150	1443.734	18	22.432	1443.734	18	2.557	0.00
	200	1493.734	18	23.270	1493.734	18	2.270	0.00
	250	1543.734	18	23.169	1543.734	18	2.270	0.00

We observe from Table 1 that the gap percentage between objective function of two methods is equal to zero for all the instances which shows that the proposed ALNS algorithm is capable of obtaining the optimum solutions. From solution times perspective, it is also seen that all the instances are solved within a time span of maximum five seconds which is an indication of the efficiency of the proposed solution algorithm. Results also indicate that if fixed cost of opening hubs increases, the number of opened hubs in optimum solution will decrease. In the meantime, lower values of discount factor results in increased number of opened hubs in the optimum solution. Numerical results for the CAB dataset based on penalty coefficient values as $\theta=3000$ and $\theta=4000$ are shown respectively in Tables 2 and Table 3.

Table 2
Results for the CAB data set with $\theta = 3000$

α	F	CPLEX			ALNS			%GAP
		Opt	Hubs	CPU (s)	Opt	Hubs	CPU (s)	
0.2	100	1252.452	4,12,16,18,22	28.938	1252.452	4,12,16,18,22	5.590	0.00
	150	1479.197	4,12,18	58.478	1479.197	4,12,18	3.340	0.00
	200	1629.197	4,12,18	79.813	1629.197	4,12,18	3.432	0.00
	250	1753.000	12,18	57.447	1753.000	12,18	2.854	0.00
0.4	100	1374.248	4,12,16,18,22	46.887	1374.248	4,12,16,18,22	5.037	0.00
	150	1548.254	4,12,18	26.951	1548.254	4,12,18	3.476	0.00
	200	1698.254	4,12,18	45.095	1698.254	4,12,18	3.305	0.00
	250	1788.190	18	30.934	1788.190	18	2.216	0.00
0.6	100	1452.102	4,12,18	31.697	1452.102	4,12,18	3.844	0.00
	150	1602.102	4,12,18	24.386	1602.102	4,12,18	3.371	0.00
	200	1724.165	11,18	23.123	1724.165	11,18	2.771	0.00
	250	1788.190	18	22.715	1788.190	18	2.200	0.00
0.8	100	1490.242	4,8,18	36.496	1490.242	4,8,18	3.560	0.00
	150	1638.979	11,18	33.271	1638.979	11,18	3.152	0.00
	200	1738.190	18	22.812	1738.190	18	2.554	0.00
	250	1788.190	18	22.303	1788.190	18	2.154	0.00
1	100	1512.400	4,8,18	24.543	1512.400	4,8,18	3.184	0.00
	150	1647.549	11,18	22.764	1647.549	11,18	2.854	0.00
	200	1738.190	18	22.286	1738.190	18	2.108	0.00
	250	1788.190	18	21.955	1788.190	18	2.273	0.00

Table 3
Results for the CAB data set with $\theta = 4000$

α	F	CPLEX			ALNS			%GAP
		Opt	Hubs	CPU (s)	Opt	Hubs	CPU (s)	
0.2	100	1252.452	4,12,16,18,22	26.168	1252.452	4,12,16,18,22	5.572	0.00
	150	1479.499	4,12,18,22	48.115	1479.499	4,12,18,22	4.351	0.00
	200	1638.692	4,12,18	73.779	1638.692	4,12,18	3.363	0.00
	250	1788.692	4,12,18	91.573	1788.692	4,12,18	2.919	0.00
0.4	100	1374.248	4,12,16,18,22	37.323	1374.248	4,12,16,18,22	5.00	0.00
	150	1557.772	4,12,18	41.788	1557.772	4,12,18	3.173	0.00
	200	1707.772	4,12,18	50.570	1707.772	4,12,18	3.387	0.00
	250	1840.891	12,18	51.327	1840.891	12,18	2.636	0.00
0.6	100	1454.257	4,8,18,22	38.631	1454.257	4,8,18,22	4.442	0.00
	150	1611.688	4,12,18	30.354	1611.688	4,12,18	3.439	0.00
	200	1761.688	4,12,18	30.708	1761.688	4,12,18	3.379	0.00
	250	1862.172	11,18	23.920	1862.172	11,18	2.66	0.00
0.8	100	1499.759	4,8,18	37.602	1499.759	4,8,18	3.587	0.00
	150	1649.759	4,8,18	24.856	1649.759	4,8,18	3.257	0.00
	200	1776.986	11,18	24.956	1776.986	11,18	2.759	0.00
	250	1876.986	11,18	22.717	1876.986	11,18	2.642	0.00
1	100	1521.918	4,8,18	33.902	1521.918	4,8,18	3.332	0.00
	150	1671.918	4,8,18	33.144	1671.918	4,8,18	3.119	0.00
	200	1785.556	11,18	22.776	1785.556	11,18	2.706	0.00
	250	1883.082	18	22.531	1883.082	18	2.075	0.00

As observed in Table 2 and Table 3, the proposed algorithm could get optimal solutions for all the instances in a very short computational times. Numerical experiment results based on the TR data set with different penalty coefficient values (1000, 1500 and 2000) are shown in Table 4. Since it is not possible to solve the large-sized instances of the TR data set by CLPEX in our computer, we only solved these instances using the proposed ALNS algorithm. The first column in Table 4 indicates applied coefficient discount for transportation cost on the inter-hub connections. Second column shows scaling factor for the fixed hub establishment costs. Next columns show the results of solving the problem based on different penalty coefficients. For each value of penalty coefficient (θ), three columns include respectively the optimum value of objective function, the optimum set of opened hubs, and the corresponding CPU times required for solving each instance.

Table 4
Results for the TR data set

α	CF	$\theta = 1000$			$\theta = 1500$			$\theta = 2000$		
		Opt	Hubs	CPU (s)	Opt	Hubs	CPU (s)	Opt	Hubs	PU (s)
0.2	0.1	748.275	6,25,33,34,35	85.70	748.2750	6,25,33,34,35	171.71	748.275	6,25,33,34,35	179.746
	0.3	966.6485	6,34	95.20	1060.564	6,33,34,35	138.64	1089.487	6,33,34,35	138.64
	0.5	1000	-	372.03	1218.558	6,34	372.03	1260.419	6,34	87.14
0.4	0.1	820.55	5,33,34,35	126.69	842.905	6,25,33,34,35	175.04	842.912	6,25,33,34,35	182.74
	0.3	979.297	34	87.31	1120.419	34,35,38	89.78	1130.039	6,33,34,35	98.13
	0.5	1000	-	63.82	1234.561	6,34	89.73	1276.659	6,34	94.21
0.6	0.1	855.031	6,33,34,35	147.61	916.338	6,25,33,34,35	154.78	927.523	6,33,34,35	165.31
	0.3	979.297	34	87.98	1138.133	6,34	77.50	1168.968	6,33,34	101.48
	0.5	1000	-	84.32	1246.166	6,34	83.93	1288.489	6,34	79.06
0.8	0.1	872.19	6,34,35	130.46	966.42	6,33,34,35,44	150.77	969.15	6,25,33,34,35	152.72
	0.3	979.29	34	109.16	1147.745	6,34	118.87	1206.580	38,41	108.80
	0.5	1000	-	74.05	1255.778	6,34	88.88	1298.611	6,34	93.61
1	0.1	880.32	6,34,35	118.17	998.71	33,35,38,41	152.74	997.12	6,25,33,34,35	161.76
	0.3	979.29	34	99.41	1193.884	34,35,38	103.07	1215.246	38,41	90.63
	0.5	1000	-	68.62	1256..859	6	73.68	1305.907	6,34	92.10

We can observe from Table 4 that the proposed algorithm solves all the problem instances for the large-sized TR data set in less than 3 minutes. Based on the obtained results we can conclude that the proposed ALNS algorithm has a high efficiency for solving large size problems. Significant and interesting point is based on the value of penalty coefficient which can be seen from these tables. No hub is established in network in optimum solution when fixed cost of hub establishment is high therefore none of available flows in network has been routed instead penalty of non-transporting has been paid. Also as observed from these tables, the values of the expected transportation cost have been increased by increasing of penalty coefficient and fixed cost of hub establishment.

6. Conclusion

In this paper, we have proposed formulation for the reliable uncapacitated multiple allocation hub location problem under hub disruptions. It was assumed that every open hub facility can fail after installation. If a hub fails, customers originally assigned to that hub, are either reassigned to other hubs that are still operational or they do not receive service in which case a penalty should be paid because of high expenses of reallocation. The problem was modeled as two-stage stochastic program and then transformed into its deterministic equivalents (extended forms) by defining a set of scenarios and associating with each scenario, the corresponding probability of occurrence. As our proposed problem is a NP-hard problem, a metaheuristic algorithm based on the was developed for solving it. Computational experiments are conducted to show the efficiency of our solution method. It was shown that the proposed algorithm obtains optimal solutions for all instances of the CAB data set in short computational times. Our results show that the structure of the solution changes when uncertainty is considered. In general, when the uncertainty in the operational status of hubs is considered, the number of hubs in optimal solution is greater than the classical counterpart in which it is assumed that the hubs are not subject to failure. Also results indicate that if fixed cost value for opening hubs increases, the number of opened hub in optimum solution will decrease and also if value of discount factor between hub connections increases, the number of opened hubs will drop.

References

Abdinnour-Helm, S. (2001). Using simulated annealing to solve the p-hub median problem. *International Journal of Physical Distribution & Logistics Management, 31*(3), 203-220.

An, Y., Zhang, Y., & Zeng, B. (2015). The reliable hub-and-spoke design problem: Models and algorithms. *Transportation Research Part B: Methodological, 77*, 103-122.

An, Y., Zeng, B., Zhang, Y., Zhao, L. (2014). Reliable p-median facility location problem: the two-stage robust models and algorithms. *Transportation Research Part B, 64*, 54-72.

Azizi, N., Chauhan, S., Salhi, S., & Vidyarthi, N. (2016). The impact of hub failure in hub-and-spoke networks: Mathematical formulations and solution techniques. *Computers & Operations Research, 65*, 174-188.

Berman, O., Krass, D., & Menezes, M. B. (2009). Locating facilities in the presence of disruptions and incomplete information. *Decision Sciences, 40*(4), 845-868.

Campbell, J. F. (1994). Integer programming formulations of discrete hub location problems. *European Journal of Operational Research, 72*(2), 387-405.

Demir, E., Bektaş, T., & Laporte, G. (2012). An adaptive large neighborhood search heuristic for the pollution-routing problem. *European Journal of Operational Research, 223*(2), 346-359.

Ebery, J. (2001). Solving large single allocation p-hub problems with two or three hubs. *European Journal of Operational Research, 128*(2), 447-458.

Ernst, A. T., & Krishnamoorthy, M. (1996). Efficient algorithms for the uncapacitated single allocation p-hub median problem. *Location Science, 4*(3), 139-154.

Grangier, P., Gendreau, M., Lehuédé, F., & Rousseau, L. M. (2016). An adaptive large neighborhood search for the two-echelon multiple-trip vehicle routing problem with satellite synchronization. *European Journal of Operational Research, 254*(1), 80-91.

Hemmelmayr, V. C., Cordeau, J. F., & Crainic, T. G. (2012). An adaptive large neighborhood search heuristic for two-echelon vehicle routing problems arising in city logistics. *Computers & Operations Research, 39*(12), 3215-3228.

Kim, H. (2008). Reliable p-hub location problems and protection models for hub network design (*Doctoral dissertation, The Ohio State University*).

Kim, H., & O'Kelly, M. E. (2009). Reliable p-hub location problems in telecommunication networks. *Geographical Analysis, 41*(3), 283-306.

Lin, C. C., Lin, J. Y., & Chen, Y. C. (2012). The capacitated p-hub median problem with integral constraints: An application to a Chinese air cargo network. *Applied Mathematical Modelling, 36*(6), 2777-2787.

Lutz, R. (2015). Adaptive Large Neighborhood Search. *Bachelor thesis at Ulm University*.

Martí, R., Corberán, Á., & Peiró, J. (2015). Scatter search for an uncapacitated p-hub median problem. *Computers & Operations Research, 58*, 53-66.

Matisziw, T. C., Murray, A. T., & Grubesic, T. H. (2010). Strategic network restoration. *Networks and Spatial Economics, 10*(3), 345-361.

Mauri, G. R., Ribeiro, G. M., Lorena, L. A. N., & Laporte, G. (2016). An adaptive large neighborhood search for the discrete and continuous Berth allocation problem. *Computers & Operations Research, 70*, 140-154.

Medal, H. R., Pohl, E. A., & Rossetti, M. D. (2014). A multi-objective integrated facility location-hardening model: Analyzing the pre-and post-disruption tradeoff. *European Journal of Operational Research, 237*(1), 257-270.

Mohammadi, M., Tavakkoli-Moghaddam, R., Siadat, A., & Dantan, J. Y. (2016). Design of a reliable logistics network with hub disruption under uncertainty. *Applied Mathematical Modelling, 40*(9), 5621-5642.

O'kelly, M. E. (1987). A quadratic integer program for the location of interacting hub facilities. *European Journal of Operational Research, 32*(3), 393-404.

Parvaresh, F., Golpayegany, S. H., Husseini, S. M., & Karimi, B. (2013). Solving the p-hub median problem under intentional disruptions using simulated annealing. *Networks and Spatial Economics, 13*(4), 445-470.

Parvaresh, F., Husseini, S. M., Golpayegany, S. H., & Karimi, B. (2014). Hub network design problem in the presence of disruptions. *Journal of Intelligent Manufacturing, 25*(4), 755-774.

Pérez, M. P., Rodríguez, F. A., & Moreno-Vega, J. M. (2007). A hybrid VNS–path relinking for the p-hub median problem. *IMA Journal of Management Mathematics, 18*(2), 157-171.

Ropke, S., & Pisinger, D. (2006). An adaptive large neighborhood search heuristic for the pickup and delivery problem with time windows. *Transportation Science, 40*(4), 455-472.

Shaw, P. (1998, October). Using constraint programming and local search methods to solve vehicle routing problems. *In International Conference on Principles and Practice of Constraint Programming (pp. 417-431). Springer Berlin Heidelberg.*

Shen, Z. J. M., Zhan, R. L., & Zhang, J. (2011). The reliable facility location problem: Formulations, heuristics, and approximation algorithms. *INFORMS Journal on Computing, 23*(3), 470-482.

Skorin-Kapov, D., & Skorin-Kapov, J. (1994). On tabu search for the location of interacting hub facilities. *European Journal of Operational Research, 73*(3), 502-509.

Skorin-Kapov, D., Skorin-Kapov, J., & O'Kelly, M. (1997). Tight linear programming relaxation of uncapacitated p-hub median problems. *Location Science, 5*(1), 68-69.

Snyder, L. V., & Daskin, M. S. (2005). Reliability models for facility location: the expected failure cost case. *Transportation Science, 39*(3), 400-416.

Taghipourian, F., Mahdavi, I., Mahdavi-Amiri, N., & Makui, A. (2012). A fuzzy programming approach for dynamic virtual hub location problem. *Applied Mathematical Modelling, 36*(7), 3257-3270.

Wang, X., & Ouyang, Y. (2013). A continuum approximation approach to competitive facility location design under facility disruption risks. *Transportation Research Part B: Methodological, 50*, 90-103.

Wilson, M. C. (2007). The impact of transportation disruptions on supply chain performance. *Transportation Research Part E: Logistics and Transportation Review, 43*(4), 295-320.

Zeng, B., An, Y., Zhang, Y., & Kim, H. (2010, July). A reliable hub-spoke model in transportation systems. *In Proceedings of the 4th international symposium on transportation network reliability, Minneapolis, Minnesota, USA (pp. 22-23).*

12

Cost models of additive manufacturing: A literature review

G. Costabile[a], M. Fera[b*], F. Fruggiero[c], A. Lambiase[a] and D. Pham[d]

[a]University of Salerno - Department of Industrial Engineering - Via Giovanni Paolo II, Fisciano (SA) – Italy
[b]Second University of Naples - Department of Industrial and Information Engineering - Via Roma 29, Aversa (CE) – Italy
[c]University of Basilicata - School of Engineering - Via Nazario Sauro, 85, 85100 (PZ) – Italy
[d]Department of Mechanical Engineering - University of Birmingham – Edgbaston, Birmingham B15 2TT, UK

CHRONICLE	ABSTRACT
	From the past decades, increasing attention has been paid to the quality level of technological and mechanical properties achieved by the Additive Manufacturing (AM); these two elements have achieved a good performance, and it is possible to compare this with the results achieved by traditional technology. Therefore, the AM maturity is high enough to let industries adopt this technology in a more general production framework as the mechanical manufacturing industrial one is. Since the technological and mechanical properties are also beneficial for the materials produced with AM, the primary objective of this paper is to focus more on managerial facets, such as the cost control of a production environment, where these new technologies are present. This paper aims to analyse the existing literature about the cost models developed specifically for AM from an operations management point of view and discusses the strengths and weaknesses of all models.
Keywords: *Additive manufacturing* *Additive manufacturing cost model*	

1. Introduction

Nowadays, globalization, high competition and a shift towards buyers' market are some of the main challenges faced by the manufacturing industry. In this modern manufacturing environment, effective and flexible manufacturing processes are the foundation of successful in everyday businesses. Buyers look for innovative, customized and high-quality products but they do not want to pay high prices as the same time. Additionally, the economic lifespan of these products decreases with the necessity of having shorter time-to-market and shorter development cycles. Furthermore, the individualization of customer demands increases with an increase of different variants. One possibility to encounter these developments may be delivered by the production technology of Additive Manufacturing (AM) (Lindemann et al., 2015). According to ASTM (2012). "AM can be defined as a collection of technologies able to join materials to make objects from 3D model data, usually layer upon layer, as opposed to subtractive manufacturing methodologies". The technology was created in 1986 when Charles Hull received a patent

* Corresponding author.
E-mail: Marcello.FERA@unina2.it (M. Fera)

(Hull, 1986) for the production of 3D objects using stereolitography. Rapid prototyping (RP) was the natural application of this new production technique for many years. The improvement of mechanical characteristics and quality, the advent of new technology (fused deposition modelling and laser sintering) and the new materials (from polymers to metals) enable the RP to realize objects with the same characteristics of finished products realized by traditional manufacturing systems. In the AM technology, Rapid Tooling (RT) and Rapid Manufacturing (RM) are used differently to suite the customer's needs on the characteristics of the product.

Nowadays, there are different technologies for various types of materials, quality and energy sources; however, all of these have some actions in common:

- Create a design CAD model;
- Convert the CAD model into STL format;
- Slice the STL file into thin cross-sectional layers;
- Construct the model layer by layer;
- Clean and finish the model (i.e. support removal and surface treatments).

Additive technology has various advantages and disadvantages. Some of these have been identified by Lindemann et al. (2012) and are listed below:

Advantages:

- More flexible development;
- Freedom of design and construction;
- Less assembly;
- No production tool necessary;
- Less spare parts in stock;
- Less complexity in business because less parts to manage;
- Less time-to-market for products;
- Faster deployment of changes.

Disadvantages:

- High machine and material costs;
- Quality of parts is in need of improvement;
- Rework is often necessary (support structures);
- Building time depends on the height of the part in the building chamber.

In addition to advantages and disadvantages, AM has a deep impact on manufacturing systems requiring different approaches in design and operations management. It is possible to realize high shape complexity without increasing the production costs (contrary to traditional technology). Freedom of design impacts the weight of the object that can be made lighter. Reduction of weight has impacts on lifecycle cost, material cost and energy consumption in the production phase.

Production lead time and supply chain are the other important aspects upset by AM.

Even if each aspect is crucial in the manufacturing systems, the impact on the costs is the most important aspect that a decision maker has to analyse before choosing a new technology. To understand AM advantages, it is necessary to analyse its impact on production management area. Nowadays, the high costs of the machines and materials make technology more expensive than traditional ones, and its use seems to be good only for a low volume of production (Ruffo et al., 2006). Furthermore, some researchers have based their studies on cost models of additive technology based on different costs structures of AM.

Every cost models of AM focuses on a specific technology aspect like large scale production, time and energy estimate, relevant activities involved and sensitivity analysis on the parameters of the costing model.

This study aims to explore the most relevant cost models defined about the AM. We will examine similarities and differences in all of them, showing their strengths and weaknesses.

2. Literature review method

Before starting with the literature review, it is important to develop the method being applied in the literature review. The number of papers on AM to examine is very big; therefore, it is possible to approach that problem by defining a method that can analyse and select the papers automatically.

The procedure is divided into the following several steps:

- Definition of the keywords;
- Collection of papers from the main international scientific papers' database;
- Analysis of the papers' characteristics among the first 100 papers sorted by relevance;
- Selection of the most interesting theme defined by the keyword;
- Eventual knowledge lack of literatures.

The following keywords were investigated: (i) additive manufacturing overview and additive manufacturing technology, (ii) additive manufacturing cost models and additive manufacturing business model, (iii) additive manufacturing mechanical properties and additive manufacturing material, (iv) additive manufacturing supply chain, (v) additive manufacturing sustainability and (vi) additive manufacturing lifecycle cost.

2.1. Review on general aspects of AM

Fig. **1** and Table 1 show the total and the trend of the papers accepted from 1997 to date and the trend of curves shows a growth for each keyword selected. Until today, researchers have focused on general aspects and technological characteristics of AM (see

Fig. **2**). Lifecycle cost, sustainability, supply chain and production costs have been less focused on.

Table 1

Papers from 1997 to date, per keyword (data source: ScienceDirect)

Keywords	Number of papers
AM overview, AM technology	34165
AM cost model, AM business model	3001
AM mech. properties, AM material	18825
AM supply chain	7285
AM sustainability	2466
AM lifecycle cost	1022
Total	66764

General aspects of AM (Bikas et al., 2015; Bogers et al., 2016; Brans, 2013; Cardaropoli et al., 2012a; Ferro et al., 2016; Gao et al., 2015; Go & Hart, 2016; Hedrick et al., 2015; Kianian et al., 2016; Lindemann et al., 2015; Mellor et al., 2014; Muita et al., 2015; Nickels, 2016; Rayna & Striukova, 2016; Scholz et al., 2016; Winkless, 2015; Wits et al., 2016; Wong & Hernandez, 2012); industry (Brettel et al., 2016; Gaub, 2015; Fera & Macchiaroli, 2010; Stock & Seliger, 2016); implications of its use (Bogers et al., 2016; Huang et al.; Newman et al., 2015; Rayna & Striukova, 2016); example of its application (Caiazzo et al., 2013; Cardaropoli et al., 2012b; Gupta et al., 2016; Huang et al.; Uhlmann et al., 2015; Wits et al., 2016); flexibility (Brettel et al., 2016; Cox et al., 2016) and technology selection (Newman et al., 2015) allow us to contextualize AM in an actual production system. There are many papers on mechanical characteristics, microstructures and properties (Brugo et al., 2016; Hinojos et al., 2016;

Huynh et al.; List et al., 2014; Ma et al., 2016; Naghieh et al., 2016; Ordás et al., 2015; Palanivel et al., 2016; Park & Rosen; Park et al., 2014; Quan et al., 2016; Shamsaei et al., 2015; Thompson et al., 2015; Wang et al., 2015, 2016; Yang et al., 2015) and on structural imperfections (Cardaropoli et al., 2012c; Cheng & Chou, 2015; Dietrich & Cudney, 2011; Nouri et al., 2016; Witherell et al., 2016) but, in our research, we found few papers on the themes of sustainability (Bechmann, 2014; Belkadi et al., 2015; Burkhart & Aurich, 2015; Chen et al., 2015; Ford & Despeisse; Gebler et al., 2014; Giret et al., 2015; Gupta et al., 2016; Huang et al.; Le Bourhis et al., 2014; Nyamekye et al., 2015; Sreenivasan et al., 2010); lifecycle cost (Cozmei & Caloian, 2012; Gebler et al., 2014; Fera & Machiaroli, 2009; Nyamekye et al., 2015; Petek Gursel et al., 2014; Watson & Taminger; Würtz et al., 2015); supply chain (Barz et al., 2016; Bogers et al., 2016; Emelogu et al.; Gress & Kalafsky, 2015; Jia et al., 2016; Khajavi et al., 2014; Mellor et al., 2014; Nyamekye et al., 2015; Pinkerton, 2016; Scott & Harrison, 2015; Silva & Rezende, 2013; Thomas, 2015) and cost models (Achillas et al., 2015; Alexander et al., 1998; Baumers et al., 2016; Klahn et al., 2015; Lim et al., 2012; Mellor et al., 2014; Piili et al., 2015; Sahebrao Ingole et al., 2009; Schröder et al., 2015; Stucker, 2012; Weller et al., 2015). All these aspects are topics of interest for our research group and, among these aspects, after solving technological problems (materials, tolerances and mechanical characteristics), production cost is the most important matter to be analysed. For this reason, we decide to extend the study to older cost models before AM existed. Older models, in fact, were created to calculate the cost of RP. In the following paragraph, we will analyse the most important work on the technology focusing on the approach and point of view of each author and giving a critical observation on their works.

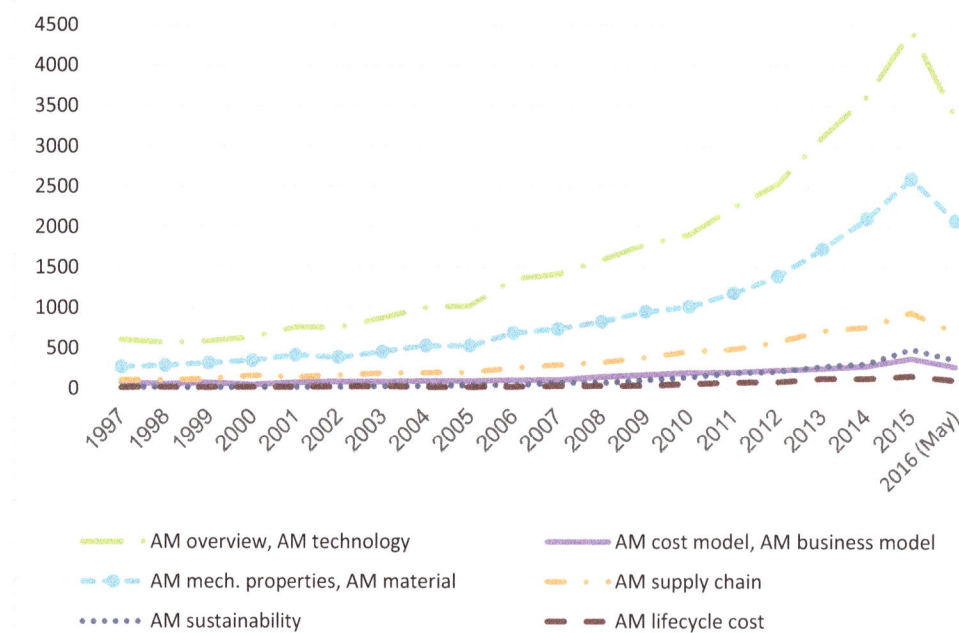

Fig. 1. Number of papers per year and per keyword

Hopkinson and Dicknes (2003) (HD) are among the first who realized an analysis of the AM costs. Initially, the technology was mainly used for RP and RT. However, the authors provided for a development of technology that would allow the realization of finished products in large scale. They reported a cost analysis to compare the traditional manufacturing method of injection moulding (IM) with layer manufacturing processes (stereolithography, fused deposition modelling and laser sintering) in terms of the unit cost for parts made in various quantities (see the example in Fig. 3). The results showed that, for some geometries, up to relatively high production volumes (in an order of a thousand pieces), it is more advantageous to use the layer manufacturing methods. According to HD, AM offers clear advantages compared to traditional technology, such as the lack of moulds and the facility of creating very complex geometries, where in some cases are not achievable just with injection moulding.

The absence of tooling also reduces a significant amount in the product development process at an early stage.

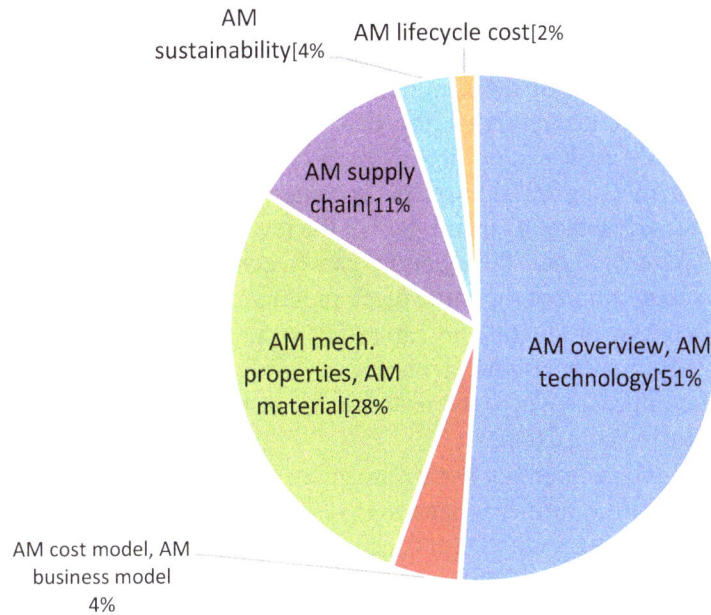

Fig. 2. Mix of papers on Additive Manufacturing

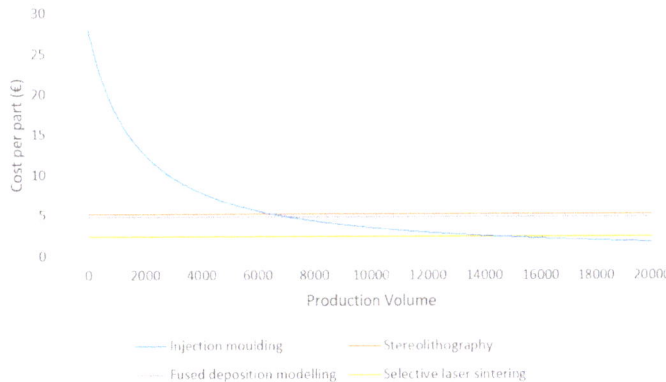

Fig. 3. Cost comparison for the lever by different processes

The costs of the parts were broken down into machine costs, labour costs and material costs.

$$Total\ cost\ per\ part = Machine\ cost\ per\ part + Labour\ cost\ per\ part + Material\ cost\ per\ part$$

These costs were calculated assuming that a machine produces one part constantly for one year. The material used for production is polyamide.

Observations

The proposed model provides an approximation of the costs in different additive technologies. The work was realized when the technology had not matured; later, different aspects of Hopkinson and Dickens' research were further developed and improved by other researchers. The list of the observations is given below:

- The model does not consider the recycling of non-sintered powder (Ruffo et al., 2006);
- Production volume is not originated by market demand, but it is obtained by multiplying production rate [parts/h] with machine uptime. This assumption is misleading because the cost model should also calculate production costs for different productive contexts where the market demand is a variable and the machine does not work under the same conditions.
- The hypothesis of production of the same component for a whole year cancels one of the main advantages of the additive technologies: the simultaneous production of different parts (Baumers, 2012). Furthermore, the break-even point of the production costs for the lever of Laser Sintering (LS) compared to the Injection Moulding (IM) is achieved for a volume of about 14000 pieces. This suggests an economic advantage of using the LS for lower volumes; however, with the production rate stated at 17.66 parts per hour, this quantity can be achieved in about 37 days. That seems to be in conflict with the assumption of the same piece production for a full year.
- HD calculate unit cost of production in saturation conditions of the machine chamber. For this reason, the technology does not show economies of scale. Nevertheless, it is necessary to analyse production costs even in cases where the chamber is not full. The curve of production costs should have a deflection (Ruffo et al., 2006).
- Power consumption is considered but not included because of its low impact in total costs.
- The model does not consider further processing (i.e. surface finishing, often necessary for addictive technologies). To compare SLS with IM it is necessary realize objects with the same characteristics.
- HD set the machine uptime to 90% of the total time (365 days × 24 hours), like IM, for high-volume manufacturing systems. In this way, they set the operation time as 7884 hours (328.5 days/year). This assumption leads to an overestimation of the machine working time. The approach of working over 46 weeks per year, twenty-four hours per day, seems to be extremely high.
- The costs for injection moulding were obtained by quotes for unit cost of tooling plus unit costs for each moulding produced. It is unclear how the authors calculate the unit cost per part. The authors do not include typical industrial cost factors like energy cost, machine depreciation and labour costs.
- The authors neglected the impact of RM on lead time and time-to-market. Usually, for IM, the time-to-market is high; in fact, many weeks are necessary to build a moulding tool for a new product, while for the RM, this time is drastically lower, due to the possibility to start (if the raw material is available) directly with the production of the part and not needing the tools for moulding; even so, it is important to note that the cycle time for RM is, normally, longer than the IM one. These aspects have not been considered in IM–RM comparison of this paper.

Hopkinson and Dickens were the first researchers who analysed the RP costs. The hypothesis of a large scale production shifts the focus away from prototyping to manufacturing usage of additive technologies. Our observations show a rather rough economic model. Probably it is due to the incomplete understanding of technology potentiality and because the technology has offered performance very lowers than today.

Some of the observations made on the HD model were analysed and resolved by Ruffo et al. (2006). The model considers a high impact of the overhead costs of the technology analysed. Ruffo et al. (2006) analysed the production of the same object (lever) used by Hopkinson and Dickens, that is obtained by laser sintering. The HD model assumed an allocation of indirect costs on the basis of annual production volume: this caused a constant unit cost when measured with the quantity produced. According to Ruffo et al. (2006), this assumption is incorrect. Injection moulding, in fact, amortizes the cost of the mould in the initial phase of production, showing a deflection of the curve cost (like in Fig. 3). In the same way LS amortizes the cost of the machine in the initial production phase. Ruffo et al. (2006) asserted that cost curve for both technologies must have the same trend. The model offers a breakdown cost structure in various activities (activity based costing). This approach comprises a definition of the activities involved, the calculation of the costs of each activity and in the summing of each cost. Activity costs are then split into direct and indirect costs and are attributed to a single part with a full costing system. The only direct cost is relative to the material used. The costs of labour, machine, production overhead and administration overhead are indirectly allocated. Labour and machine maintenance are considered to be indirect as they are paid annually with contracts. Total cost of a single build is the sum of direct and indirect costs. The direct costs depend on the amount of material used and indirect costs depend on the time of construction:

$$Cost\ of\ a\ build = Direct\ costs + Indirect\ costs$$

where

Direct costs: material used;
Indirect costs: labour, machine and overheads.

The main differences compared to the previous model are listed below:

- Labour was considered by HD as a direct cost adding its cost directly to the part produced. New model adds the machine operator salary indirectly to the product. Its allocation is proportional to the working time of the machine.
- Material recycle is not considered in the HD model.
- Machine utilization set in HD was 90% versus the more realistic 57% of the new model.
- Recycle of material is possible but with limitations due to the thermal treatment of the powder.

This approach, like HD, provides the definition of cost curve for different production volumes. The positioning of the object in the machine chamber is like a 3D matrix in which every element represents an object to build. Starting by the first object we can add the next one until the line is full. Then, to add more objects, we have to start a new line horizontally (on the same layer). When all lines are full, to add more objects, we have to start a new layer (on the top of the previous full layer). When all the available layers are full, if we want to produce more parts, we have to start a new bed (new chamber) repeating the previous phases. Unlike the Hopkinson and Dickens study, which shows a constant cost for the LS parts, the cost curve (Fig. 4) has a deflection for low production volumes and a change in the costs curve tendency whenever one of three following situation arises:

1. It is necessary to use a new row (line) in the x direction for the addition of a part.

2. It is necessary to add a new vertical layer for the addition of a part.

3. It is necessary to start a new bed for the addition of a part.

This trend is justified by extra time and materials necessary to produce more parts. The saw tooth trend is caused by the impact of a fixed time element for every build (warm-up and cool-down) and every layer (powder deposition time). Increasing the number of parts in every layer and every build, the effect of fixed time consumption (and consequently of costs) will be lower. Instead, the increase of the layer number (i.e. adding only one object in a new layer) produces a negative effect on the costs. We have the same negative situation when we start a new bed containing only one part.

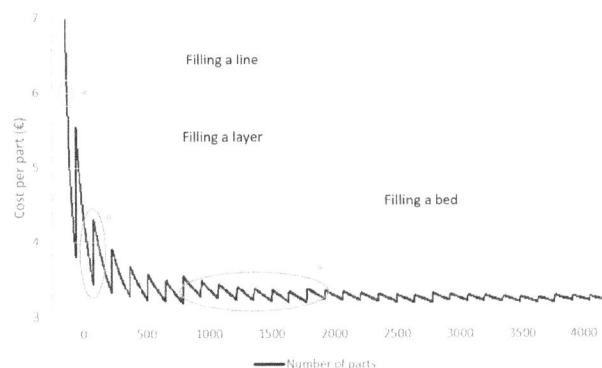

Fig. 4. Production curve for the lever (Laser Sintering)

The cost of the single piece, according to the new model, is significantly higher than the cost calculated by Hopkinson and Dickens (2006). The comparison evidences a cost underestimation of the old model. For high production volumes, the curve tends to stabilize because of the indirect cost splitting on a higher number of parts.

Observations

The possibility to recycle the powder and a more accurate analysis of the overhead costs make the Ruffo et al. (2006) model more accurate than HD model. Moreover, the trend of the saw tooth curve of the unit cost means that the model is very sensitive to the number of parts produced focusing on the build saturation issue. The observations are listed below:

- Energy consumption is considered for the first time by Ruffo et al. (2006), even if its cost has been inserted between overhead costs. This approach is unusual by having the possibility to refer the energy consumption to the single part produced so to its cost.
- Moreover, in this model, no post-processing was taken in account (like the previous model).
- Lead time and time-to-market have not been considered by comparing RM and IM (like the previous model);
- Since Ruffo et al. (2006) consider the same IM unit cost of HD, it is worth noting that this issue was not solved by them;
- Ruffo et al. (2006) set the machine utilization to 57% of the total time (by contact with industrial partners). In this way, they defined working time in 100 hours/week, for 50 weeks per year. Even if this approach seems more reliable than the previous HD assumption, they do not consider possible failure and maintenance time. HD uses the 'uptime' terminology, while Ruffo et al. (2006) use 'utilization'. In each case, the operating time of the machine is obtained by multiplying the total time (365 days × 24 hours) and the machine utilization (or uptime).

The studies by Baumers (2012) and Baumers et al. (2012) are the first examined ones, in detail, the economic and energetic aspects and also the time necessary to realize the AM construction. The highlights of his work are listed below:

- An activity based cost (ABC) estimator of the type devised by Ruffo et al. (2006) is employed, but energy costs are grouped as direct;
- An estimate of total build time;
- An accurate analysis of energy consumption.

According to Baumers, high indirect costs of AM (time dependent allocated) and the presence of fixed element of time consumption (for each layer and for each build) make the analysis of the build unused capacity problem very important. In this way, it is possible to reduce the effect of indirect costs. Hopkinson and Dickens assumed no excess of capacity because the chamber of the machine is always full. However, Ruffo et al. (2006) also based their estimates on average unit cost on much smaller production quantities and the assumption is that any excess capacity remains unused. From the perspective of economic theory, average cost functions are seen as cost/quantity combinations that are technically efficient. This means that the minimum cost is obtained from the input used. In this contest, Ruffo and Hague (2007) note that "in reality manufacturers set every build with the highest packing ratio possible", which means there is an incentive to completely fill build volumes with products. Another important observation of Baumers et al. (2012) is that break-even costing models may not be able to capture the capabilities of geometrically less restrictive manufacturing processes to create a complex product. Furthermore, in direct competition with conventional mass production, AM faces the disadvantage of not being able to offer the scale economies available to conventional manufacturing processes. For their research, Baumers et al. (2012) employed an activity-based cost estimator of the type devised by Ruffo et al. (2006). The cost estimate for the build is constructed by combining data on the total indirect and direct costs incurred. Indirect costs, expressed as a cost rate measured per machine hour, contains costs arising from administrative and production overheads, production labour and machine costs (including depreciation). Unlike Ruffo et al. (2006), energy cost is grouped as a direct cost.

The total cost for each build can be expressed as follows

$$C_{build} = \left(\dot{C}_{Indirect} \times T_{Build}\right) + \left(w \times P_{Raw\ material}\right) + \left(E_{Build} \times P_{Energy}\right)$$

where

$\dot{C}_{Indirect}$:   indirect cost per machine hour;
T_{Build}:   total build time;
w: total weight of the part in the build (including support structure);
$P_{Raw\ material}$: price per kg of raw material;
E_{Build}:   total energy consumption per build;
P_{Energy}:   mean price of electricity.

The time and energy estimator and the grouping of energy in direct costs make the cost model more accurate than previous ones. Baumers et al. (2012), however, did not consider other activities that are indirectly connected to the phase of building but are still relevant from the economic point of view (post-processing and material removal). In many cases, it is necessary as a further phase, for example, to enhance mechanical property or to improve surface quality. As we will see later in this study, these activities will be analysed by other researchers.The estimate of the building time is obtained by combining fixed time consumption per build (warm-up and cool-down), layer dependent time consumption (time necessary to add powder) and laser deposition time for the sintering of the powder:

$$T_{Build} = T_{Job} + \left(T_{Layer} \times n\right) + \sum_{z=1}^{z}\sum_{y=1}^{y}\sum_{x=1}^{x} T_{Vovel\ xyx}$$

where

T_{Job}:   fixed time consumption per build;
n: number of layer;
T_{Layer}:   fixed time consumption per layer;
$T_{Voxel\ xyz}$:   time needed to process each voxel;

In the analysis of energy consumption, Baumers et al. (2012) divided total energy between consumption for each job, single layer energy consumption, geometry dependent energy consumption and a constant base line level of energy consumption throughout the build:

$$E_{Build} = E_{Job} + \left(\dot{E}_{Time} \times T_{Build}\right) + \left(E_{Layer} \times l\right) + \sum_{z=1}^{z}\sum_{y=1}^{y}\sum_{x=1}^{x} E_{Vovel\ xyx}$$

where

E_{Job}:   fixed energy consumption per build;
$\dot{E}_{Time}$:   fixed energy consumption rate;
E_{Layer}:   fixed energy consumption per layer;
l: number of layer;
$E_{Voxel\ xyz}$:   energy needed to process each voxel;

Observations

The observation on previous works (HD and Ruffo et al.) and the quality of the build time and energy estimator are further important steps. These observations concur to better understand AM performance in terms of costs and potentiality.

Even if Baumers et al. (2012) group energy costs as direct (more accurate costing), their cost model is the same as the type devised by Ruffo et al. (2006); therefore, the observations derived from the previous cost model (see Ruffo et al.) are still valid. Lindemann et al. (2012) asserted that Hopkinson and Dickens and Ruffo et al. (2006) chose a similar approach for the calculation of costs in their models and each of them set a specific emphasis on a certain topic. In sum, we could say that every existing costing model has advantages and disadvantages; however, no single model satisfies all criteria. Thus, there is a need to combine the strengths of existing models without including their weaknesses and develop a new costing model that is suitable for the calculation of today's AM. Before developing a cost model, all relevant cost processes of the AM production process, have been investigated and modelled with Event-driven Process Chains. As a calculation method, a "Time Driven Activity Based Costing" approach has been adopted. "Time Driven" means that the allocation depends on the duration of the activities. For the estimation of relevant cost processes, Lindemann et al. (2012) define four main processes:

- Building job preparation;
- Building job production;
- Sample parts and support manually removing;
- Post processing to enhance material properties.

The main processes were selected to be able to represent different cost centres. This facilitates the calculation and makes it easier to adopt the model to different production environments.

The framework of machine cost per build defined by Lindemann et al. (2012) is structurally similar to the previous framework defined by Ruffo et al. (2006). The differences are due to:

- Electrical energy cost and gas cost per hour grouped as direct costs;
- Fixed costs per every build (labour costs and gas costs).

The material cost is defined in the same way defined by Ruffo.

For completeness we show the formulation of the costs structure stated by the authors. The cost of a single build is obtained by summing the activities costs (A_i) involved:

$$Cost \ of \ a \ build = \sum_i A_i$$

Observations

Lindemann et al. (2012) are among the first researchers who included the post-processing activity in the costing model. This activity includes, for example, quality control, surface treatment and support removal. Some of the items included in the post processing activity are unavoidable in all additive processes. For this reason, the idea of considering the cost of the post processing phases helps to better understand the economic aspects related to the technology.

According to Rickenbacher et al. (2013), AM processes are interesting candidates for the replacement of conventional production processes like cutting or casting. The integration of AM processes into a production environment requires a cost-model that allows the estimation of the real costs of a single part, although it might be produced in the same build job together with other parts of different geometries. The highlights of the cost model proposed by Rickenbacher et al. (2013) are listed below:

- Cost calculation of single part in a build (assuming a contemporary production of different parts),
- Deep analysis of the steps involved in the process,
- Cost model including all pre- and post- processing steps linked to AM processes,
- Algorithm to calculate the time fraction for each part in the build job,
- Build time estimator derived by a linear regression on 24 different build jobs.

The cost model is based on the generic cost model of Alexander et al. (1998). The cost of the single part (P_i) is obtained summing the costs of the seven process steps which is defined below:

$$C_{tot}(P_i) = C_{Prep}(P_i) + C_{Buildjob}(P_i) + C_{Setup}(P_i) + C_{Build}(P_i) + C_{Removal}(P_i) + C_{Substrate}(P_i) \\ + C_{Postp}(P_i)$$

where

$C_{tot}(P_i)$:   total manufacturing costs,
P_i: part with ith geometry,
$C_{Prep}(P_i)$:   cost for preparing geometry data (orientation, support structures, etc.),
$C_{Buildjob}(P_i)$: cost for build job assembly,
$C_{Setup}(P_i)$:   machine set up costs,
$C_{Build}(P_i)$:   cost for building up the part,
$C_{Removal}(P_i)$: cost for removing the part from the SLM machine,
$C_{Substrate}(P_i)$:   cost to separate parts from substrate plate,
$C_{Postp}(P_i)$:   cost for post-processing.

Rickenbacher et al. (2013) also developed an algorithm to calculate the time fraction for each part in the build job. The algorithm allows calculating the layer-dependent fraction of total build time per part. In Fig. 5, we can see that in dark grey three parts are affected, two in the grey slice and so on. The allocation criterion of time could be proportional to the corresponding cross-section of the part or proportional to the amount of layers of the slices defined.

Fig. 5. Simultaneous build-up of multiple parts with different heights

The algorithm suggests optimizing the use of building space by simultaneously building up as much geometries with similar part height as possible.

To estimate build time, Rickenbacher et al. (2013) used a linear regression model derived from 24 different build jobs. He defines the following equation that allows to calculate the regression coefficients:

$$\sum_i T_{Build}(P_i) = a_0 + a_1 * N_L + a_2 \times V_{tot} + a_3 \times S_{Supp_{tot}} + a_4 \times \sum_i N_i + a_5 * S_{tot}$$

where
T_{Build}:   building time;
P_i: part with ith geometry;
$a_0, \dots, a_5$:   regression coefficients;
N_L: number of layers;
V_{tot}:   total volume of building job;
$S_{Supp_{tot}}$: total surface area of the support structures,
N_i: quantity of parts with ith geometry;
S_{tot}:   total surface area of the build job.

These regression coefficients ($a_0, \dots, a_5$) will be used to calculate the total build time of each part using the following equation:

$$T_{Build}(P_i) = \frac{a_0}{\sum_i N_i} + T_L P_i + a_2 * V(P_i) + a_3 \times S_{Supp}(P_i) + a_4 + a_5 \times S(P_i)$$

where

T_{Build}: building time;
P_i: part with ith geometry;
$a_0, \dots, a_5$: regression coefficients;
T_L: layer-dependent fraction;
V: volume of the part;
S_{Supp}: surface area of the support structures,
N_i: quantity of parts with ith geometry;
S: surface area of the part.

Observations

- Even if the cost model includes a detailed analysis of the pre- and post-processing related to the AM process, a possible material removal step has not been included.
- Like all previous cost models, Rickenbacher et al. (2013) assert that it is possible to have a lower unit cost of the parts by building more parts in the same build job.
- Machine's cost per hour [€/h] included in idle phases of the machine (*Setup* and *Removal* steps).
- Reliable computation of the effect of material change and additional work of using inert gas have been included in the machine set up costs by using the factors $F_{Inertgas}$ and $F_{Mat.Change}$.
- Proposed algorithm to allocate the time fraction of the total build time to each part realized is simple and effective.
- The authors consider labour cost as an hourly rate multiplied for the duration of each step where operator is necessary. The operator hourly cost for each step of production (90 €/h) is assumed equal for any kind of worker is involved. In our opinion, this should be a strong limitation of the generality of the model proposed because it is normal to have different labour hourly cost for different skills.
- The cost model does not take in account energy consumption and its cost. We disagree with this approach because this item is not negligible in metal AM processes.
- The authors do not explain which cost items is included in the machine's cost per hour.
- In our opinion, it is tough to assign a realistic weight to the way in which the authors predict the building time through a formula of estimation that is calculated with different parameters very different among themselves. The parameters are referred also to surface area of support structure and surface area of building job; these are related to a shape complexity element that is out of the game when the matter is focused on the AM and that is well known as a technology complexity free. Moreover, in the equation used for total build time, calculation does not derive explicit possible warm-up time and cool-down time. Due to the deep impact of these fixed elements of time consumption in some additive technologies, we think that it is correct to analyse these parameters. Moreover, it is very important to note that the authors do not validate their time estimation approach in anyway.

This cost model represents a good step forward in the effectiveness of cost estimation of AM. Even if we have doubts on quality of the time estimator, the deep analysis of pre- and post-processing and the algorithm defined to calculate the time fraction for each part in the build job are important tools for AM technology. The work of Schröder et al. (2015) states the AM business models in the literature review, the development of a business model that evaluates the process costs of AM technologies and a sensitivity analysis on the output of theur Excel tool. In their review, they analyse the main business models developed for the different AM processes. The outcome of their study is the summary of the requirement of the new cost model necessary to represent costs completely and realistically. The requirements of the new cost model are listed as follows:

- Integration of recycling and waste of material;
- Integration of support structures of products including a general appraisal for different AM processes;
- Calculation of the printing time;
- Maximum possible number of products that can be printed simultaneously in the workspace;
- Level of complexity of the product;
- Duration of the post-processing;
- Integration of modern quality management methods for the protection and monitoring of product and process quality.

For the development of their business model, Schröder et al. (2015) use an activity based costing. The relevant activities are defined using interviews submitted to a group of experts (small and medium companies having experience on AM technologies) and researchers on AM. The following seven main process steps were identified:

- Design & planning;
- Material processing;
- Machine preparation;
- Manufacturing;
- Post-processing;
- Administration and sales;
- Quality.

An Excel cost calculation tool has been developed to provide both costs and a sensitivity analysis as output. For completeness, we show the formulation of the costs structure stated by the authors. The cost of a single build is obtained by summing the activities costs (A_i) involved:

$$Cost\ of\ a\ build = \sum_i A_i$$

The sensitivity analysis is made to identify special economic effects of AM. The main findings are

- The most cost-influencing factors are the investment costs of the machine and its load factor;
- The post-processing of products with high quantities and small bodies has a big potential for process optimization;
- Economies of scale only exist for small products; products with big bodies are nearly independent from the ordered quantity.

Observations

Schröder et al. (2015) increased the number of the relevant activities included in their cost model. Design and planning were not included in each of the cost models analysed. Probably one of the most interesting is the design activity. AM, in fact, compared with subtractive technologies, requires extra design phases. This phase is not present for the case of building exactly the same part typically obtained by subtractive technologies; however, in most cases, AM is able to realize complex geometries that are not achievable using material removal technique. In all of these cases, costs of redesign have to be considered to have a more accurate cost of the finish product. This cost model gives certainly an accurate analysis of additive technologies. The relevant activities included, not directly related to the building phase, give an overview on additive processes that no models have been able to realize. However, we are unsure of the definition of administrations and sales activities. The cost model for AM should include only industrial costs.

Administrative and sales should be included in cost accounting with all other overhead costs. Furthermore, these cost items are invariant respect to the technology adopted; therefore, for this reason, if we want use the cost model to compare different technologies, it is not necessary to consider them.

3. Conclusions

Regardless of the technology in question (DMLS, EBM, LS, SLA and FDM), we can identify similar process phases that allow the definition of a single cost model valid for each of them. Some of involved factors are labour, machine, material, power source, warm-up time, build rate and energy consumption. By varying these factors, we obtain a different costs impact on the finished product.

Over time, every author adds something to the previous cost model allowing an increase of accuracy.

Older models, by comparing Laser Sintering with Injection Moulding, disprove affordability for large scale production (Hopkinson & Dicknes, 2003; Ruffo et al., 2006). Scarce understanding of the technology led to the development of crude cost models that do not consider effectively all involved variables (energy consumption and labour). Ruffo et al. (2006) and Hopkinson et al. (2003) used a microeconomic approach (break-even analysis). Both understood the potentiality of AM in chamber saturation condition, but they focused on large scale production (economic approach) instead of efficiency of the process (engineering approach). Subsequently different work, in this direction, was realized by Baumers et al. (Baumers et al., 2011a, 2011b). In their experiments, they investigated the impact of capacity utilization of the machine on time, energy consumption and total costs.

Over time, we have had an increase of the detail level of the business models defined. We pass from the initial crude cost models, through the accurate definition of time and energy consumption realized by Baumers et al. (2012) up to the Schröder et al. (2015) cost model that appears to be very accurate. Another important aspect to be considered is related to materials used in AM. Originally AM, or better Rapid prototyping, only allowed the production of polymeric object and it was compared with injection moulding. For large scale productions, IM shows a low and descending unit cost of the product. The absence of economies of scale in AM and the high costs involved in terms of machine and materials make the technology suitable only for low production volumes.

Related to materials, we have to acknowledge energy consumption. Hopkinson and Dickens assumed energy consumption cost as negligible for its low effect on total cost. At the time, additive processes were suitable only to realize polycarbonate and polypropylene objects. Subsequently, Ruffo et al. (2006) considered energy costs even if they were inserted between overhead costs. Higher energy consumption necessary to realize metallic objects made essential to take into account this cost item. In this context, Baumers et al. (2012) analysed the theme and inserted energy between direct costs. Older cost models do not consider any post processing (HD and Ruffo et al. and Baumers et al.). However, Lindemann et al. (2012), Rickenbacher et al. (2013) and Schröder et al. (2015) consider post-processing activities like surface treatments and quality controls. In some cases, AM can be a substitute of subtractive manufacturing, whereas in other cases, after the building process using AM, some mechanical characteristics of the parts need to be enhanced (i.e. rugosity and tolerances) (Atzeni & Salmi, 2012). It should be interesting to hypothesize a hybrid production model that includes the post-processing cost of AM, not only in terms of quality controls and support removal but in terms of material removal (Campbell et al., 2012; Manogharan et al., 2015). Existing cost models consider the activities directly connected to building process of AM; however, due to the fact that AM allows the production of end use product, it is important to include in the cost model all the activities involved to the calculation of the full cost of a finish part like, for example, redesign costs (Eleonora Atzeni et al., 2010; Hague et al., 2003) and material removal cost (Manogharan et al., 2016). For this reason, the definition of a new cost model appears necessary. As stated in the initial part of this paper, it aimed to define a critical analysis about the cost models for AM from an operations management point of view. Therefore, further observation regards the approach used by all the authors for defining their cost models. Each of their cost models measures additive systems as separate by their production systems of which they are part. No one takes in account aspects like, for example, demand of a product, production mix, lead time and delivery time. For example, the hypothesis of large scale production is made by considering the absolute measure of the production volume of a single machine with no attention to the general market size. Moreover, the hypothesis of cost reduction by increasing the capacity utilization of the chamber could be in contrast with the delivery time; for example, it is not possible to infinitely wait for the saturation achievement if a delivery has to be performed. To consider only additive process were correct if we wanted to measure an individual phase; however, in an integration point of view of AM in a conventional production system, we think that it should be correct to take in account also other aspects like some the previously cited operations, that can be before and after the work phase of AM. From an enterprise point of view, these models appear as scarcely related to the real needs of an industrial company that are generally underestimated from the theoretical approaches. Moreover, it is worth noting that all the observations on the existing cost models, and the synthesis of their strengths and weaknesses, can lay the foundations to define and build a new cost model that will help solve the open issues analysed here.

References

Achillas, C., Aidonis, D., Iakovou, E., Thymianidis, M., & Tzetzis, D. (2015). A methodological framework for the inclusion of modern additive manufacturing into the production portfolio of a focused factory. *Journal of Manufacturing Systems*, *37*, Part 1, 328–339. http://doi.org/10.1016/j.jmsy.2014.07.014

Alexander, P., Allen, S., & Dutta, D. (1998). Part orientation and build cost determination in layered manufacturing. *Computer-Aided Design*, *30*(5), 343–356. http://doi.org/10.1016/S0010-4485(97)00083-3

ASTM, 2012. (n.d.). ASTM F2792 - 12e1 Standard Terminology for Additive Manufacturing Technologies.

Atzeni, E., & Salmi, A. (2012). Economics of additive manufacturing for end-usable metal parts. *The International Journal of Advanced Manufacturing Technology*, *62*(9–12), 1147–1155.

Barz, A., Buer, T., & Haasis, H.-D. (2016). A Study on the Effects of Additive Manufacturing on the Structure of Supply Networks. *IFAC-PapersOnLine*, *49*(2), 72–77. http://doi.org/10.1016/j.ifacol.2016.03.013

Baumers, M. (2012). *Economic aspects of additive manufacturing: benefits, costs and energy consumption*. \copyright Martin Baumers. Retrieved from https://dspace.lboro.ac.uk/dspace-jspui/handle/2134/10768?sequence=3

Baumers, M., Dickens, P., Tuck, C., & Hague, R. (2016). The cost of additive manufacturing: machine productivity, economies of scale and technology-push. *Technological Forecasting and Social Change*, *102*, 193–201.

Baumers, M., Tuck, C., Bourell, D. L., Sreenivasan, R., & Hague, R. (2011). Sustainability of additive manufacturing: measuring the energy consumption of the laser sintering process. *Proceedings of the Institution of Mechanical Engineers, Part B: Journal of Engineering Manufacture*, *225*(12), 2228–2239.

Baumers, M., Tuck, C., Wildman, R., Ashcroft, I., & Hague, R. (2011). Energy inputs to additive manufacturing: does capacity utilization matter? *EOS*, *1000*(270), 30–40.

Baumers, M., Tuck, C., Wildman, R., Ashcroft, I., Rosamond, E., & Hague, R. (2012). Combined build-time, energy consumption and cost estimation for direct metal laser sintering. *From Proceedings of Twenty Third Annual International Solid Freeform Fabrication Symposium—An Additive Manufacturing Conference*, *13*. Retrieved from http://sffsymposium.engr.utexas.edu/Manuscripts/2012/2012-71-Baumers.pdf

Bechmann, F. (2014). Changing the future of additive manufacturing. *Metal Powder Report*, *69*(3), 37–40. http://doi.org/10.1016/S0026-0657(14)70135-3

Belkadi, F., Bernard, A., & Laroche, F. (2015). Knowledge Based and PLM Facilities for Sustainability Perspective in Manufacturing: A Global Approach. *Procedia CIRP*, *29*, 203–208. http://doi.org/10.1016/j.procir.2015.01.065

Bikas, H., Stavropoulos, P., & Chryssolouris, G. (2015). Additive manufacturing methods and modelling approaches: a critical review. *The International Journal of Advanced Manufacturing Technology*, 1–17.

Bogers, M., Hadar, R., & Bilberg, A. (2016). Additive manufacturing for consumer-centric business models: Implications for supply chains in consumer goods manufacturing. *Technological Forecasting and Social Change*, *102*, 225–239. http://doi.org/10.1016/j.techfore.2015.07.024

Brans, K. (2013). 3D Printing, a Maturing Technology. *IFAC Proceedings Volumes*, *46*(7), 468–472. http://doi.org/10.3182/20130522-3-BR-4036.00112

Brettel, M., Klein, M., & Friederichsen, N. (2016). The Relevance of Manufacturing Flexibility in the Context of Industrie 4.0. *Procedia CIRP*, *41*, 105–110. http://doi.org/10.1016/j.procir.2015.12.047

Brugo, T., Palazzetti, R., Ciric-Kostic, S., Yan, X. T., Minak, G., & Zucchelli, A. (2016). Fracture mechanics of laser sintered cracked polyamide for a new method to induce cracks by additive manufacturing. *Polymer Testing*, *50*, 301–308. http://doi.org/10.1016/j.polymertesting.2016.01.024

Burkhart, M., & Aurich, J. C. (2015). Framework to Predict the Environmental Impact of Additive Manufacturing in the Life Cycle of a Commercial Vehicle. *Procedia CIRP*, *29*, 408–413.

Caiazzo, F., Cardaropoli, F., Alfieri, V., Sergi, V., & Cuccaro, L. (2013). Experimental analysis of selective laser melting process for Ti-6Al-4V turbine blade manufacturing (Vol. 8677). Presented at the Proceedings of SPIE - The International Society for Optical Engineering. http://doi.org/10.1117/12.2010577

Campbell, I., Bourell, D., & Gibson, I. (2012). Additive manufacturing: rapid prototyping comes of age. *Rapid Prototyping Journal*, *18*(4), 255–258.

Cardaropoli, F., Alfieri, V., Caiazzo, F., & Sergi, V. (2012a). Dimensional analysis for the definition of the influence of process parameters in selective laser melting of Ti-6Al-4V alloy. *Proceedings of the Institution of Mechanical Engineers, Part B: Journal of Engineering Manufacture*, *226*(7), 1136–1142. http://doi.org/10.1177/0954405412441885

Cardaropoli, F., Alfieri, V., Caiazzo, F., & Sergi, V. (2012b). Manufacturing of porous biomaterials for dental implant applications through Selective Laser Melting. *Advanced Materials Research*, *535–537*, 1222–1229. http://doi.org/10.4028/www.scientific.net/AMR.535-537.1222

Cardaropoli, F., Caiazzo, F., & Sergi, V. (2012). Evolution of direct selective laser sintering of metals. *Advanced Materials Research*, *383–390*, 6252–6257. http://doi.org/10.4028/www.scientific.net/AMR.383-390.6252

Chen, D., Heyer, S., Ibbotson, S., Salonitis, K., Steingrímsson, J. G., & Thiede, S. (2015). Direct digital manufacturing: definition, evolution, and sustainability implications. *Journal of Cleaner Production*, *107*, 615–625. http://doi.org/10.1016/j.jclepro.2015.05.009

Cheng, B., & Chou, K. (2015). Geometric consideration of support structures in part overhang fabrications by electron beam additive manufacturing. *Computer-Aided Design*, *69*, 102–111. http://doi.org/10.1016/j.cad.2015.06.007

Cox, S. C., Jamshidi, P., Eisenstein, N. M., Webber, M. A., Hassanin, H., Attallah, M. M., … Grover, L. M. (2016). Adding functionality with additive manufacturing: Fabrication of titanium-based antibiotic eluting implants. *Materials Science and Engineering: C*, *64*, 407–415. http://doi.org/10.1016/j.msec.2016.04.006

Cozmei, C., & Caloian, F. (2012). Additive Manufacturing Flickering at the Beginning of Existence. *Procedia Economics and Finance*, *3*, 457–462. http://doi.org/10.1016/S2212-5671(12)00180-3

Dietrich, D. M., & Cudney, E. (2011). Impact of integrative design on additive manufacturing quality. *International Journal of Rapid Manufacturing*, *2*(3), 121–131. http://doi.org/10.1504/IJRAPIDM.2011.043454

Eleonora Atzeni, Luca Iuliano, Paolo Minetola, & Alessandro Salmi. (2010). Redesign and cost estimation of rapid manufactured plastic parts. *Rapid Prototyping Journal*, *16*(5), 308–317. http://doi.org/10.1108/13552541011065704

Emelogu, A., Marufuzzaman, M., Thompson, S. M., Shamsaei, N., & Bian, L. (n.d.). Additive Manufacturing of Biomedical Implants: A Feasibility Assessment via Supply-Chain Cost Analysis. *Additive Manufacturing*. http://doi.org/10.1016/j.addma.2016.04.006

Fera, M., & Macchiaroli, R. (2010). Use of analytic hierarchy process and fire dynamics simulator to assess the fire protection systems in a tunnel on fire.*International Journal of Risk Assessment and Management*, *14*(6), 504-529. http://doi.org/10.1504/IJRAM.2010.037087

Fera, M., & Macchiaroli, R. (2009). Proposal of a quali-quantitative assessment model for health and safety in small and medium enterprises. *WIT Transactions on the Built Environment, 108.* http://doi.org/10.2495/SAFE090121

Ferro, C., Grassi, R., Seclì, C., & Maggiore, P. (2016). Additive Manufacturing Offers New Opportunities in UAV Research. *Procedia CIRP*, *41*, 1004–1010. http://doi.org/10.1016/j.procir.2015.12.104

Ford, S., & Despeisse, M. (n.d.). Additive manufacturing and sustainability: an exploratory study of the advantages and challenges. *Journal of Cleaner Production.* http://doi.org/10.1016/j.jclepro.2016.04.150

Gao, W., Zhang, Y., Ramanujan, D., Ramani, K., Chen, Y., Williams, C. B., … Zavattieri, P. D. (2015). The status, challenges, and future of additive manufacturing in engineering. *Computer-Aided Design*, *69*, 65–89. http://doi.org/10.1016/j.cad.2015.04.001

Gaub, H. (2015). Customization of mass-produced parts by combining injection molding and additive manufacturing with Industry 4.0 technologies. *Reinforced Plastics.*

Gebler, M., Schoot Uiterkamp, A. J. M., & Visser, C. (2014). A global sustainability perspective on 3D printing technologies. *Energy Policy*, *74*, 158–167. http://doi.org/10.1016/j.enpol.2014.08.033

Giret, A., Trentesaux, D., & Prabhu, V. (2015). Sustainability in manufacturing operations scheduling: A state of the art review. *Journal of Manufacturing Systems*, *37, Part 1*, 126–140. http://doi.org/10.1016/j.jmsy.2015.08.002

Go, J., & Hart, A. J. (2016). A framework for teaching the fundamentals of additive manufacturing and enabling rapid innovation. *Additive Manufacturing*, *10*, 76–87. http://doi.org/10.1016/j.addma.2016.03.001

Gress, D. R., & Kalafsky, R. V. (2015). Geographies of production in 3D: Theoretical and research implications stemming from additive manufacturing. *Geoforum*, *60*, 43–52. http://doi.org/10.1016/j.geoforum.2015.01.003

Gupta, K., Laubscher, R. F., Davim, J. P., & Jain, N. K. (2016). Recent developments in sustainable manufacturing of gears: a review. *Journal of Cleaner Production*, *112, Part 4*, 3320–3330. http://doi.org/10.1016/j.jclepro.2015.09.133

Hague, R., Campbell, I., & Dickens, P. (2003). Implications on design of rapid manufacturing. *Proceedings of the Institution of Mechanical Engineers, Part C: Journal of Mechanical Engineering Science, 217*(1), 25–30.

Hedrick, R. W., Urbanic, R. J., & Burford, C. G. (2015). Development Considerations for an Additive Manufacturing CAM System. *IFAC-PapersOnLine*, *48*(3), 2327–2332. http://doi.org/10.1016/j.ifacol.2015.06.435

Hinojos, A., Mireles, J., Reichardt, A., Frigola, P., Hosemann, P., Murr, L. E., & Wicker, R. B. (2016). Joining of Inconel 718 and 316 Stainless Steel using electron beam melting additive manufacturing technology. *Materials & Design*, *94*, 17–27. http://doi.org/10.1016/j.matdes.2016.01.041

Hopkinson, N., & Dicknes, P. (2003). Analysis of rapid manufacturing—using layer manufacturing processes for production. *Proceedings of the Institution of Mechanical Engineers, Part C: Journal of Mechanical Engineering Science, 217*(1), 31–39.

Huang, R., Riddle, M., Graziano, D., Warren, J., Das, S., Nimbalkar, S., … Masanet, E. (n.d.). Energy and emissions saving potential of additive manufacturing: the case of lightweight aircraft components. *Journal of Cleaner Production.* http://doi.org/10.1016/j.jclepro.2015.04.109

Hull, C. W. (1986). *Apparatus for production of three-dimensional objects by stereolithography.* Google Patents. Retrieved from http://www.google.com/patents/us4575330

Huynh, L., Rotella, J., & Sangid, M. D. (n.d.). Fatigue behavior of IN718 microtrusses produced via additive manufacturing. *Materials & Design.* http://doi.org/10.1016/j.matdes.2016.05.032

Jia, F., Wang, X., Mustafee, N., & Hao, L. (2016). Investigating the feasibility of supply chain-centric business models in 3D chocolate printing: A simulation study. *Technological Forecasting and Social Change*, *102*, 202–213. http://doi.org/10.1016/j.techfore.2015.07.026

Khajavi, S. H., Partanen, J., & Holmström, J. (2014). Additive manufacturing in the spare parts supply chain. *Computers in Industry*, *65*(1), 50–63. http://doi.org/10.1016/j.compind.2013.07.008

Kianian, B., Tavassoli, S., Larsson, T. C., & Diegel, O. (2016). The Adoption of Additive Manufacturing Technology in Sweden. *Procedia CIRP*, *40*, 7–12. http://doi.org/10.1016/j.procir.2016.01.036

Klahn, C., Leutenecker, B., & Meboldt, M. (2015). Design Strategies for the Process of Additive Manufacturing. *Procedia CIRP*, *36*, 230–235. http://doi.org/10.1016/j.procir.2015.01.082

L. Rickenbacher, A. Spierings, & K. Wegener. (2013). An integrated cost-model for selective laser melting (SLM). *Rapid Prototyping Journal*, *19*(3), 208–214. http://doi.org/10.1108/13552541311312201

Le Bourhis, F., Dipartimento per gli affari giuridici e legislativi, Dembinski, L., Hascoet, J.-Y., & Mognol, P. (2014). Predictive Model for Environmental Assessment in Additive Manufacturing Process. *Procedia CIRP*, *15*, 26–31. http://doi.org/10.1016/j.procir.2014.06.031

Lim, S., Buswell, R. A., Le, T. T., Austin, S. A., Gibb, A. G., & Thorpe, T. (2012). Developments in construction-scale additive manufacturing processes. *Automation in Construction*, *21*, 262–268.

Lindemann, C., Jahnke, U., Moi, M., & Koch, R. (2012). Analyzing product lifecycle costs for a better understanding of cost drivers in additive manufacturing. In *23th Annual International Solid Freeform Fabrication Symposium–An Additive Manufacturing Conference. Austin Texas USA 6th-8th August*.

Lindemann, C., Reiher, T., Jahnke, U., & Koch, R. (2015). Towards a sustainable and economic selection of part candidates for additive manufacturing. *Rapid Prototyping Journal*, *21*(2), 216–227.

List, F. A., Dehoff, R. R., Lowe, L. E., & Sames, W. J. (2014). Properties of Inconel 625 mesh structures grown by electron beam additive manufacturing. *Materials Science and Engineering: A*, *615*, 191–197. http://doi.org/10.1016/j.msea.2014.07.051

Ma, Y., Cuiuri, D., Li, H., Pan, Z., & Shen, C. (2016). The effect of postproduction heat treatment on γ-TiAl alloys produced by the GTAW-based additive manufacturing process. *Materials Science and Engineering: A*, *657*, 86–95. http://doi.org/10.1016/j.msea.2016.01.060

Manogharan, G., Wysk, R. A., & Harrysson, O. L. A. (2016). Additive manufacturing-integrated hybrid manufacturing and subtractive processes: Economic model and analysis. *International Journal of Computer Integrated Manufacturing*, *29*(5), 473–488. http://doi.org/10.1080/0951192X.2015.1067920

Manogharan, G., Wysk, R., Harrysson, O., & Aman, R. (2015). AIMS – A Metal Additive-hybrid Manufacturing System: System Architecture and Attributes. *Procedia Manufacturing*, *1*, 273–286. http://doi.org/10.1016/j.promfg.2015.09.021

Mellor, S., Hao, L., & Zhang, D. (2014). Additive manufacturing: A framework for implementation. *International Journal of Production Economics*, *149*, 194–201.

Muita, K., Westerlund, M., & Rajala, R. (2015). The Evolution of Rapid Production: How to Adopt Novel Manufacturing Technology. *IFAC-PapersOnLine*, *48*(3), 32–37. http://doi.org/10.1016/j.ifacol.2015.06.054

Naghieh, S., Karamooz Ravari, M. R., Badrossamay, M., Foroozmehr, E., & Kadkhodaei, M. (2016). Numerical investigation of the mechanical properties of the additive manufactured bone scaffolds fabricated by FDM: The effect of layer penetration and post-heating. *Journal of the Mechanical Behavior of Biomedical Materials*, *59*, 241–250. http://doi.org/10.1016/j.jmbbm.2016.01.031

Newman, S. T., Zhu, Z., Dhokia, V., & Shokrani, A. (2015). Process planning for additive and subtractive manufacturing technologies. *CIRP Annals-Manufacturing Technology*, *64*(1), 467–470.

Nickels, L. (2016). Additive manufacturing: A user's guide. *Metal Powder Report*, *71*(2), 100–105. http://doi.org/10.1016/j.mprp.2016.02.045

Nouri, H., Guessasma, S., & Belhabib, S. (2016). Structural imperfections in additive manufacturing perceived from the X-ray micro-tomography perspective. *Journal of Materials Processing Technology*, *234*, 113–124. http://doi.org/10.1016/j.jmatprotec.2016.03.019

Nyamekye, P., Leino, M., Piili, H., & Salminen, A. (2015). Overview of Sustainability Studies of CNC Machining and LAM of Stainless Steel. *Physics Procedia*, *78*, 367–376. http://doi.org/10.1016/j.phpro.2015.11.051

Ordás, N., Ardila, L. C., Iturriza, I., Garcianda, F., Álvarez, P., & García-Rosales, C. (2015). Fabrication of TBMs cooling structures demonstrators using additive manufacturing (AM) technology and HIP. *Fusion Engineering and Design*, *96–97*, 142–148. http://doi.org/10.1016/j.fusengdes.2015.05.059

Palanivel, S., Dutt, A. K., Faierson, E. J., & Mishra, R. S. (2016). Spatially dependent properties in a laser additive manufactured Ti–6Al–4V component. *Materials Science and Engineering: A*, *654*, 39–52. http://doi.org/10.1016/j.msea.2015.12.021

Park, S., & Rosen, D. W. (n.d.). Quantifying effects of material extrusion additive manufacturing process on mechanical properties of lattice structures using as-fabricated voxel modeling. *Additive Manufacturing*. http://doi.org/10.1016/j.addma.2016.05.006

Park, S.-I., Rosen, D. W., Choi, S., & Duty, C. E. (2014). Effective mechanical properties of lattice material fabricated by material extrusion additive manufacturing. *Additive Manufacturing*, *1–4*, 12–23. http://doi.org/10.1016/j.addma.2014.07.002

Petek Gursel, A., Masanet, E., Horvath, A., & Stadel, A. (2014). Life-cycle inventory analysis of concrete production: A critical review. *Cement and Concrete Composites*, *51*, 38–48. http://doi.org/10.1016/j.cemconcomp.2014.03.005

Piili, H., Happonen, A., Väistö, T., Venkataramanan, V., Partanen, J., & Salminen, A. (2015). Cost Estimation of Laser Additive Manufacturing of Stainless Steel. *Physics Procedia*, *78*, 388–396. http://doi.org/10.1016/j.phpro.2015.11.053

Pinkerton, A. J. (2016). [INVITED] Lasers in additive manufacturing. *Optics & Laser Technology*, *78*, Part A, 25–32. http://doi.org/10.1016/j.optlastec.2015.09.025

Quan, Z., Larimore, Z., Wu, A., Yu, J., Qin, X., Mirotznik, M., … Chou, T.-W. (2016). Microstructural design and additive manufacturing and characterization of 3D orthogonal short carbon fiber/acrylonitrile-butadiene-styrene preform and composite. *Composites Science and Technology*, *126*, 139–148. http://doi.org/10.1016/j.compscitech.2016.02.021

Rayna, T., & Striukova, L. (2016). From rapid prototyping to home fabrication: How 3D printing is changing business model innovation. *Technological Forecasting and Social Change*, *102*, 214–224. http://doi.org/10.1016/j.techfore.2015.07.023

Ruffo, M., & Hague, R. (2007). Cost estimation for rapid manufacturing'simultaneous production of mixed components using laser sintering. *Proceedings of the Institution of Mechanical Engineers, Part B: Journal of Engineering Manufacture*, *221*(11), 1585–1591.

Ruffo, M., Tuck, C., & Hague, R. (2006). Cost estimation for rapid manufacturing-laser sintering production for low to medium volumes. *Proceedings of the Institution of Mechanical Engineers, Part B: Journal of Engineering Manufacture*, *220*(9), 1417–1427.

Sahebrao Ingole, D., Madhusudan Kuthe, A., Thakare, S. B., & Talankar, A. S. (2009). Rapid prototyping-a technology transfer approach for development of rapid tooling. *Rapid Prototyping Journal*, *15*(4), 280–290.

Scholz, S., Mueller, T., Plasch, M., Limbeck, H., Adamietz, R., Iseringhausen, T., … Woegerer, C. (2016). A modular flexible scalable and reconfigurable system for manufacturing of Microsystems based on additive manufacturing and e-printing. *Robotics and Computer-Integrated Manufacturing*, *40*, 14–23. http://doi.org/10.1016/j.rcim.2015.12.006

Schröder, M., Falk, B., & Schmitt, R. (2015). Evaluation of Cost Structures of Additive Manufacturing Processes Using a New Business Model. *Procedia CIRP*, *30*, 311–316.

Scott, A., & Harrison, T. P. (2015). Additive Manufacturing in an End-to-End Supply Chain Setting. *3D Printing and Additive Manufacturing*, *2*(2), 65–77.

Shamsaei, N., Yadollahi, A., Bian, L., & Thompson, S. M. (2015). An overview of Direct Laser Deposition for additive manufacturing; Part II: Mechanical behavior, process parameter optimization and control. *Additive Manufacturing*, *8*, 12–35. http://doi.org/10.1016/j.addma.2015.07.002

Silva, J. V. L., & Rezende, R. A. (2013). Additive Manufacturing and its future impact in logistics. *IFAC Proceedings Volumes*, *46*(24), 277–282. http://doi.org/10.3182/20130911-3-BR-3021.00126

Sreenivasan, R., Goel, A., & Bourell, D. L. (2010). Sustainability issues in laser-based additive manufacturing. *Physics Procedia*, *5*, 81–90.

Stock, T., & Seliger, G. (2016). Opportunities of Sustainable Manufacturing in Industry 4.0. *Procedia CIRP*, *40*, 536–541. http://doi.org/10.1016/j.procir.2016.01.129

Stucker, B. (2012). Additive manufacturing technologies: technology introduction and business implications. In *Frontiers of Engineering: Reports on Leading-Edge Engineering From the 2011 Symposium, National Academies Press, Washington, DC, Sept* (pp. 19–21).

Thomas, D. (2015). Costs, benefits, and adoption of additive manufacturing: a supply chain perspective. *The International Journal of Advanced Manufacturing Technology*, 1–20.

Thompson, S. M., Bian, L., Shamsaei, N., & Yadollahi, A. (2015). An overview of Direct Laser Deposition for additive manufacturing; Part I: Transport phenomena, modeling and diagnostics. *Additive Manufacturing*, *8*, 36–62. http://doi.org/10.1016/j.addma.2015.07.001

Uhlmann, E., Kersting, R., Klein, T. B., Cruz, M. F., & Borille, A. V. (2015). Additive Manufacturing of Titanium Alloy for Aircraft Components. *Procedia CIRP*, *35*, 55–60. http://doi.org/10.1016/j.procir.2015.08.061

Wang, X., Gong, X., & Chou, K. (2015). Scanning Speed Effect on Mechanical Properties of Ti-6Al-4V Alloy Processed by Electron Beam Additive Manufacturing. *Procedia Manufacturing*, *1*, 287–295. http://doi.org/10.1016/j.promfg.2015.09.026

Wang, Z., Palmer, T. A., & Beese, A. M. (2016). Effect of processing parameters on microstructure and tensile properties of austenitic stainless steel 304L made by directed energy deposition additive manufacturing. *Acta Materialia*, *110*, 226–235. http://doi.org/10.1016/j.actamat.2016.03.019

Watson, J. K., & Taminger, K. M. B. (n.d.). A decision-support model for selecting additive manufacturing versus subtractive manufacturing based on energy consumption. *Journal of Cleaner Production*. http://doi.org/10.1016/j.jclepro.2015.12.009

Weller, C., Kleer, R., & Piller, F. T. (2015). Economic implications of 3D printing: Market structure models in light of additive manufacturing revisited. *International Journal of Production Economics*, *164*, 43–56.

Winkless, L. (2015). Is additive manufacturing truly the future? *Metal Powder Report*, *70*(5), 229–232. http://doi.org/10.1016/j.mprp.2015.05.003

Witherell, P., Herron, J., & Ameta, G. (2016). Towards Annotations and Product Definitions for Additive Manufacturing. *Procedia CIRP*, *43*, 339–344. http://doi.org/10.1016/j.procir.2016.01.198

Wits, W. W., García, J. R. R., & Becker, J. M. J. (2016). How Additive Manufacturing Enables more Sustainable End-user Maintenance, Repair and Overhaul (MRO) Strategies. *Procedia CIRP*, *40*, 693–698. http://doi.org/10.1016/j.procir.2016.01.156

Wong, K. V., & Hernandez, A. (2012). A review of additive manufacturing. *ISRN Mechanical Engineering*, *2012*. Retrieved from http://downloads.hindawi.com/journals/isrn.mechanical.engineering/2012/208760.pdf

Würtz, G., Lasi, H., & Morar, D. (2015). Additive Manufacturing – Enabling Technology for Lifecycle Oriented Value-increase or Value-decrease. *Procedia CIRP*, *33*, 394–399. http://doi.org/10.1016/j.procir.2015.06.092

Yang, L., Harrysson, O., West, H., & Cormier, D. (2015). Mechanical properties of 3D re-entrant honeycomb auxetic structures realized via additive manufacturing. *International Journal of Solids and Structures*, *69–70*, 475–490. http://doi.org/10.1016/j.ijsolstr.2015.05.005

13

The role of uncertainty in supply chains under dynamic modeling

M. Fera[a]*, F. Fruggiero[b], A. Lambiase[c], R. Macchiaroli[a] and S. Miranda[c]

[a]*Second University of Naples, Department of Industrial and Information Engineering, Via Roma 29, 81031 Aversa (CE), Italy*
[b]*University of Basilicata, School of Engineering, Via Ateneo Lucano 10, 85100 Potenza, Italy*
[c]*University of Salerno, Department of Industrial Engineering, Via Giovanni Paolo II 132, 84084 Fisciano (SA), Italy*

CHRONICLE	ABSTRACT
Keywords: *Supply chain* *Order penetration point* *Uncertainty*	The uncertainty in the supply chains (SCs) for manufacturing and services firms is going to be, over the coming decades, more important for the companies that are called to compete in a new globalized economy. Risky situations for manufacturing are considered in trying to individuate the optimal positioning of the order penetration point (OPP). It aims at defining the best level of information of the client's order going back through the several supply chain (SC) phases, i.e. engineering, procurement, production and distribution. This work aims at defining a system dynamics model to assess competitiveness coming from the positioning of the order in different SC locations. A Taguchi analysis has been implemented to create a decision map for identifying possible strategic decisions under different scenarios and with alternatives for order location in the SC levels. Centralized and decentralized strategies for SC integration are discussed. In the model proposed, the location of OPP is influenced by the demand variation, production time, stock-outs and stock amount. Results of this research are as follows: (i) customer-oriented strategies are preferable under high volatility of demand, (ii) production-focused strategies are suggested when the probability of stock-outs is high, (iii) no specific location is preferable if a centralized control architecture is implemented, (iv) centralization requires cooperation among partners to achieve the SC optimum point, (v) the producer must not prefer the OPP location at the Retailer level when the general strategy is focused on a decentralized approach.

1. Introduction

Supply chain management (SCM) is generally divided into two main categories of issues: strategic and tactical. These two elements are present in many research approaches, but for the problem of locating the optimal order penetration point (OPP), the tactical and strategic decisions for supply chains (SCs) are merged. OPP, as defined by Olhager in 2003, is the one that 'sets the stage of the value chain of production, where a particular product is connected to a customer's request' (Olhager, 2003).

By the nature of the decision positioning of OPP, it is affected by uncertainty in each SC. Uncertainty management becomes the key driver to achieving optimal performance for a robust supply chain (SC)

* Corresponding author.
E-mail: Marcello.FERA@unina2.it (M. Fera)

(Petrovic, 2001). The uncertainty for the SC is linked to the stochastic behaviour of the volume of demand, customer requirements and changes in delivery times (Chauhan et al., 2009; Shukla et al., 2013). To meet the CS uncertainty, many authors have suggested using an approach based on cooperation between the different agents of the system. The economic externalities (due to the mismatch between information actors in the SC) can be reduced through a collaborative approach (Kwon et al., 2011; Hülsmann et al., 2008). The collaborative approach (that guarantees good results many times) can be used whenever a mathematical formulation and a simulation model are implemented in the design of an SC (Sambasivan et al., 2013; Efendigil & Önüt, 2012).

When the word 'collaborative' is used, it needs to be meant as the management of the relations between actors of the SCs, paying attention to both strategic and tactical issues. It represents the functional integration of many interdependent activities that are capable of being translated in terms of money flows connected to goods, services and generally to the SC processes. To use this way to understand and design SCs, many theoretical models are available. It is worth noting that in SCM, both physical and information management structures are influenced by the coordination level between the SCs actors. Through the application of these models, it is possible to achieve a level of optimization that alters the storage and flow of information; moreover, these models are able to incorporate the impact of information technology tools, leading to the possibility of designing a distinct set of node connections to simulate and translate the collaborative issues, thus defining new SC substructures.

The main effects of SC performance that a collaborative approach can provide are as follows:(i) a general risk reduction and the achievement of competitive advantages, (ii) inventory level reduction, (iii) total cost reduction, (iv) lower customer rotation and (v) reduction in delivery lead time and revenue enhancements. To achieve these results, the main variables to be analysed are the price, quantity, shipment conditions (ex-works, free on board, etc.) and time, which are the main levers to obtain a coordinated SC (Tsay, 1999). In the literature there are present several coordination mechanisms for the price: (i) quantity discounts (Fugate et al., 2006), (ii) revenues shared (Kanda & Deshmukh, 2008), (iii) refund and part-return policy (Padmanabhan & Png, 1997; Sahin & Robinson, 2002) and (iv) sharing tariffs of the two actors involved (Fugate et al., 2006). For all the other variables not related to the price, coordination mechanisms are mainly constructed on flexibility capacity (Eppen & Iyer, 1997), allocation rules, i.e. rules for allocating the capacity among the different Retailers (Cachon & Lariviere, 2001), exclusive dealing (Besanko & Perry, 1993), policy management of inventory for vendor management (Sahin & Robinson, 2002; Waller et al., 1999), revenues growing for products with high variance of demand and outsourcing cost (Cheung & Lee, 2002), planning, forecasting and replenishment managed in a collaborative way (Esper & Williams, 2003; Kanda & Deshmukh, 2008), quick response (Sahin & Robinson, 2002), efficient customer response (Lohtia & Subramaniam, 2004), postponement (Pagh & Cooper, 1998) and the order penetration as introduced earlier (Yu et al., 2001). It is important to note that different OP points have effects on the design of SCs and the tactical and strategic perspective in its management.

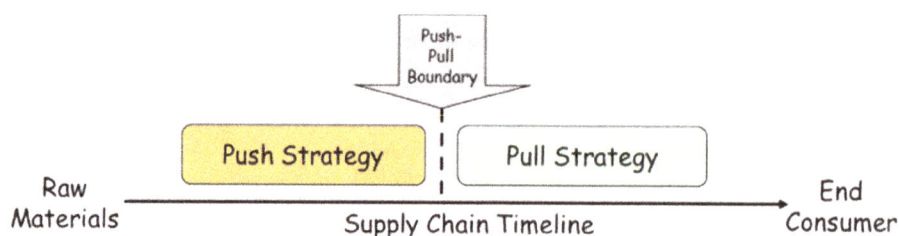

Fig. 1. Order Penetration Point as tactical and strategic link among supply

The OPP positioning in the value chain is one of the main strategic problems (with several tactical consequences) for a manufacturer or any other actor in the SC. For the reasons mentioned in the previous paragraphs, i.e. OPP is a part of the collaborative tactics), the OPP decision is strongly related to strategic collaboration with all the others actors in the SC. It influences the level of trust— related to the belief, decisions and actions in SCM (McEvily et al., 2003)—among SC echelons. The trust plays an essential role in the coordination process (Sako, 1994); indeed one of the most important research issues in this field is related to the idea of connecting all the actors in this "game" through a unique approach, translating the problem in a mathematical way.

This paper aims to build a system dynamics model to investigate the role of OPP in a collaborative strategy in order to assure performances reliable for the SCs. The authors want to suggest strategic and tactical choices in terms of OPP for the maximization of overall revenues and incomes that come from SCM.

2. Investigating SCs trends in literature references

The aim of this section is to identify the main paths—which are historically discussed and recognized— for SC uncertainty management. Authors present a bibliographic analysis to let the reader define the context (type of supply), mechanism (vertical integration among partners in centralized and decentralized forms) and tools (information exchange) of the simulation scenarios that will be implemented in the next sections of this paper.

For this literature analysis, the Web of Science (WoS as per Thompson Reuters) is used. Researches on WoS were done for many publication years (1970–2015) and for several topics, all related to the terms 'Uncertainty' and 'SC'. A selection among all the papers available was done, in such papers with more than 20 citations are included; this has led to the identification of 31 research papers, and each paper has a mean of 16.53 citations per year. A literature network was created to ease readers in their understanding, based on the static and dynamic interaction of knowledge. A main path analysis was performed (Colicchia & Strozzi, 2012), and a Pajek tool was used (http://pajek.imfm.si). The design of a SC is recognized in its strategic role historically (Lehtonen, 1998). The SC performances are monitored by the business strategy decision-makers, and generally, they are modified by the amount of trust and information sharing among the SC's actors (Bowersox & Calantone, 1998).

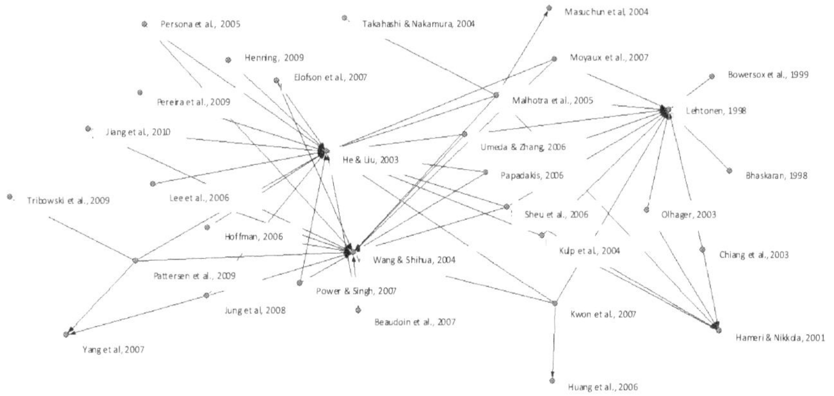

Fig. 2. Reporting on static and dynamic interaction in the management of SCs: Citation Network of relevant (>20 citations) trends for the WoS database in SC management

Interdependence between the actors' decisions; intensity of the relationships between actors; trust and information sharing between the actors; inventory system; information technology capabilities and the coordination structures (meant as SC architecture) are identified as strategic decisions for SCM, because they influence the enterprise in the long run. Collaboration in the SC is influenced by its structure, and because collaboration is difficult to implement, many times the echelons of the SC suffer from a lack of coordination mechanisms with several operative consequences even if it is demonstrated that coordination enhances the performances of both the supplier and the Retailer (Sheu et al., 2006). It is important to underline that in the past, the main elements to achieve collaboration were put up as evidence. Indeed, based on a survey conducted in the industrial districts of northern Taiwan, it is possible to understand that the position in the network and the firm sizes are key factors—in terms of business success—for coordination and consequently, for the management of the SC (Wang et al., 2004).

Before starting to speak about the uncertainty, it is important to present the main causes of instability and uncertainty in SCM. Generally, it is possible to speak of horizontal integration (i.e. when a resource is shared between several actors of SCs) and of vertical integration (when it is possible to connect with and align the raw material of an actor with its final product). These two main categories of strategic design issues lead to instability of the SC processes due to their own statistic behaviours. This paper analyses only the role of vertical integration in SC performances. Therefore, with regard to the instability of the SC due to the vertical tactics, it is worth noting that stable performances, of the SC, can be achieved through controlling production schedules, extended to all the SC actors through sharing and management of information (Henning, 2009). If the control of the production schedule is not performed, instability arises and leads to high average inventory levels (Bhaskaran, 1998).

Another possible approach to enhancing the performances of SCs is the adoption of direct market strategies that can improve the manufacturing profitability for the SC. Moreover, this strategy increases the capability of negotiation and cooperation to allow SC performances to mature. As result of the application of this strategy, it is important to note that it is accompanied by wholesale price reduction with great margins for the Retailer. This strategy is accompanied by a retailer's stocks equilibrium that is generally produced by establishing an adequate balance between the push and pull approaches to satisfy demand (Chiang et al., 2003). To apply these strategies, several tactics could be used. The balancing between the planning for Manufacturing To Order (MTO) and for Manufacturing To Stock (MTS) is one of these; in SCs, they contribute to create different production variabilities and variances in the amount of safety stock required to satisfy a fixed service level (Ma et al., 2004). Moreover, there is the demand respond tactic (i.e., pull), which can lead to lower risk exposure after a sales or production disruption (Papadakis, 2006). Therefore, in general, it is possible to say that the direct market strategy improves reliability in marketplace service, but it requires integration among partners.

Even if this paper considers vertical integration for SCs, let us introduce the main issues of the horizontal integration strategy. Horizontal integration is the most important contributor to the cost-containment of the SC management process (Won et al., 2007). To implement it, the integration of the data among trading partners is required to enable effective management of the SC (Power& Singh, 2007). The partnership enables knowledge creation, operational efficiency and information exchange; collaboration and cooperation are fertile opportunities for advancing in SC optimization (Malhotra et al., 2005). Moreover, the partnership in SCM may reduce the risks of the bullwhip effect. To understand better which are the levers amenable for implementing a horizontal strategy, it is important to remember the decision-making model for supplier selection published by Cheng et al. (2006); this fuzzy model identifies the price, quality and delivery performance as the relevant variables for strategic selection of a partner between SC actors (Cheng et al., 2006).

Coordinating the different points of view between the Retailer (buyer driven) and the manufacturer (supplier driven) may help uncertainty management (Yang et al., 2007). Obviously, this difference between the SC actors is possible only if a cluster creation is made (i.e., between collectors, connectors, collaborators and so on). Therefore, the possibility of creating clusters can easily lead to a SC change. In fact, based on empirical data involving 281 Australian organizations, the structural changes and close collaboration with trading partners are required to realize a cost-effective perspective on SCs. The creation of an echelon as a collective customer in SCs can be a key factor in improving the overall performances in terms of service level and cost for centralized SCs (Elofson & Robinson, 2007).

In general, it is possible to affirm that the coordination that leads to a centralized ordering system and vendor re-order point system in pull operational systems has a positive impact on supply performance

(Umeda & Zhang, 2006). Coordination provides the possibility of managing and reducing the stochasticity of the process, which influences storage capacity between the echelons in the SC. In fact, for example, the coefficients of variation for the lead times were recognized as related to WIP (it increases when WIP increases) (Pettersen & Segerstedt, 2009). It is worth noting that the inventory level and the customer demand are recognised as the main contributors to the stochastic nature of SCs(Beaudoin et al., 2007).

To confirm the importance of customer demand in SCM, it is possible to cite a paper that demonstrated that the demand forecast can positively influence performances of small SCs (lower than four echelons) (Yan et al., 2003). Moreover, it is demonstrated that the continuous updating of the demand information can benefit the SC cost (that is demonstrated to be convex and differentiable with respect to order quantity). Moreover, another possibility of reducing the stochastic behaviour of the SCs is to manage and adjust the bidding behaviour; this management can help to respond effectively to changes in the SC and in the demand in the market (He et al., 2003).

On continual investigation of the role of OPP in SCs, the importance of demand volume and its volatility (generally expressed as Coefficient of Variation) for the production and the delivery lead-time becomes evident (Olhager, 2003). For enterprises, this is the main issue tied to the inventory problem (Hameri & Nikkola, 2001). It is worth noting that the investment required to achieve the minimum inventory level (that is the main effect in the post-OPP operations) and the maximum manufacturing efficiency (that is the main effect in the pre-OPP operations) affects customer service and also has effects on the operations costs. To facilitate improvement in the previously mentioned elements (i.e. the inventory level and the manufacturing efficiency), it is important to share information regarding the inventory levels and customers' needs, which are generally are associated with higher measures of manufacturer performance (Kulp et al., 2004). The capacity to control both the previous variables and the ones related to them is typical of the hybrid control mechanism that generally is required (Takahashi & Nakamura, 2004). Starting from the external conditions of the market, it is possible to outline situations in which a pure push strategy outperforms a pull strategy in terms of customer service level and throughput. Even if, in general, the pull strategy reduces inventory levels in the SC (Masuchum et al., 2004). So, it is possible to affirm that the OPP location is relevant for the success of SCM.

In the general context of SC optimisation, it important to always remember the role of the bullwhip effect (Moyaux et al., 2007) that is influenced by several variables, such as the inventory management, lot-sizing, market supply, operations uncertainty and information sharing process.

The performances of the SCs obtained through integration are differentially associated with manufacturer performance. For example, vendor inventory management (in terms of collaborative planning or replenishment among all SC actors) is positively related to the increase of the margins for the manufacturer. Another example of relations between variables is the one between the higher prices of the Wholesaler and the lower stock-outs level of the Retailer and manufacturer. Moreover, collaboration between the SC actors for the creation of a new product or service is positively related to the performances of the entire SC (Hoffman et al., 2006). In this context of relations, great importance has to be laid on IT strategies, which can make the SC more robust and resilient to the uncertainties of the market and foster and help order management (Pereira, 2009). These relations are also observed on the online market where the navigability, price savings and security are equally important to create an SC success (Lee et al., 2006); in the online SC's case, it is important to consider that to face the uncertainty and obsolescence in the products sold, the consignment stock strategy is generally suggested for implementation (Persona et al., 2005).

In general, in the past few years, great emphasis has been laid on applying the IT revolution to the SC world (e.g., collaborative planning, collaborative forecasting and replenishment (CPFR) and so on); this new tool for collaboration in SCs alters the interaction between the enterprise and the supplier/customers. IT is the main facilitator of information sharing between all SC actors (and in this paper, it is assumed to be present). A minimum amount of information sharing between the manufacturer and the third-party logistics provider (3PL) is considered to be the most important factor to implement decentralized planning fully (Jung et al., 2008).

After IT solutions, another technology that can help in the integration of information all along the SC is Radio Frequency Identification (i.e., RFiD); this technology is demonstrated to help in the standardization of the supply chain and in SC Event Management (SCEM) (Tribowski et al., 2009). Another hypothesis at the base of our model is that this technology is adopted in the proposed centralised coordination mechanisms. This choice is due to the will of the authors to reinforce the collaboration concepts between all the actors of the SC that can lead an increase in each player's profit in the SC (Jiang et al., 2010).

2.1. Managing uncertainty in the SC

The problem of uncertainty in SCs produces risk, and it is related to SCs' performances. Uncertainty in SCs is the final manifestation of factors influencing supply scenarios (Olhager, 2003). The main issues, on which the researchers focus, are: (i) SC inventory management, (ii) vendor selection, (iii) transport planning, (iv) production/distribution planning and (v) procurement-production-distribution planning. All these issues are reported to be planning strategy problems for SCs. Uncertainty in planning is the cause of errors, and generally, it influences the variability in SC performances, such as the variation in production and delivery lead times and the variation in stock levels (choosing a right mix between the push and pull strategies). Consequentially, it is possible to affirm that uncertainty management is correlated with the problem of effects measuring the strategic decisions in SC management and of its accuracy.

So, this literature review aims at individuating the contributions by the international scientific literature for the planning methods used to understand and manage uncertainty in SCs. Moreover, this study introduces the concept of measuring the effect of strategic decisions.

Numerous methods for uncertainty management in SCs use fuzzy logic (FL) as a tool. Many contributions use FL, introducing measures of performances for the SCs that are detectable and theoretically based. FL is used for modelling uncertainty and to understand and manage its effects on SCs oriented to the flow-process management and driven by the market. Peidro et al. (Peidro et al., 2009) used a fuzzy linear programming model to simulate the behaviours of a SC. Other authors used FL to face the inventory management issue (Petrovic et al., 1998; Petrovic et al., 1999; Giannoccaro et al., 2003; Carlsson & Fullér, 2002). FL also contributed to modelling the vendor selection strategies (Kumar et al., 2004; Amid et al., 2006). Similar investigations use FL for modelling uncertainty in transportation (e.g., for the time and quantity) between several echelons in SCs (Chanas et al., 1993; Julien, 1994; Liu & Kao, 2004; Liang, 2006). Moreover, other fields of application of FL in SCs are the planning of the correct production-distribution system and the investigation of the effects of different strategies for procurement-production-distribution (Sakawa et al., 2001; Liang, 2007; Selim et al., 2008; Aliev et al., 2007; Chen & Chang, 2006; Torabi & Hassini, 2008).

Apart from the fuzzy logic modelling of the SCs parameters, to manage and measure the dependencies among supply partners, another approach to facing this theme is offered by the game theory (GT) approach, which is consistently used. Due to its nature, GT is always contextualized to a specific field, and it is possible to affirm that many contributions from literature report having applied GT in SCs. GT is used to reproduce the behaviours of hypothetical players (the SCs subjects) that try to modify their decisions, basing them on the decisions of the other players if the game is collaborative or standing alone if the game is not collaborative. GT's role in balancing several decision variables and effects for the SC performances is widely reported by many literature sources (Xiao et al., 2010; Ni & Li, 2012; Yin & Nishi, 2012; Lenga & Parler, 2012; Zhang & Huang, 2010; Esmaeili et al., 2009).

The system dynamics simulation technique is reported to be the main technique to represent, simulate and estimate the behaviours of the several subjects involved in an SC; even if in 2012, Tako and Robinson (Tako & Robinson, 2012) wrote a literature review regarding the techniques used in the SC context. Making a comparison between discrete event simulation (DES) and system dynamics (SD), they revealed that both DES and SD are good for the purpose of representing the different elements characterizing an SC. Actually, Lee and Chung (2012) used SD to understand and find possible representation and simulation scenarios of the inventory levels for the several players of a SC. Also, SD was used to represent the information sharing process in a SC (Feng, 2012). Moreover, it is possible to affirm that SD and FL may be able to report the role of strategy for demand forecasting for SCs (Campuzanoa et al., 2010). In general, as evident from the previous sentences, a comparison between DES and SD is possible; here, the authors want to mark the different application fields of these two methods. SD is recognized as the best technique to discuss regarding the added value of strategic and tactical decisions in SCs (Ashayeri & Lemmes, 2006), while the DES technique, using typical simulation software, is recognized as the leader in SC simulation of case-based scenarios (Lee et al., 2002; Persson & Araldi, 2009). Recently, a combination of DES and agent-based modelling was found to be the solution that allows the overcoming of the problem of simulating the dependencies between several partners in SCs (Long & Zhang, 2014).

In general, it is important to say that SCM is not just related to determining a consistent approach with the specific characteristics of the SC in analysis, but it also requires the identification of the correct performance measures. It is worth noting that SCs are quite difficult to measure, and this is for several reasons, such as the identification of the key points to be measured (financial and nonfinancial facets), social aspects, environmental aspects, technical parameters, customer satisfaction and product availability, etc. Between all these aspects, it is important to note that in the past few years, the environmental impact has grown in importance with the development of green and sustainable concepts for SCs (Sarkis et al., 2011). To make possible the involvement of all these aspects for SC management, the tools to be used are Balanced Scorecard (BS) and economic evaluation methods, which consider the costs and revenues from an SC system (Li et al., 2005; Pettersson & Segerstedt, 2013; Bhagwat & Sharma, 2007; Tracht et al., 2013). In our survey, we decided to investigate in-depth the financial methods, such as the net present value (NPV), to assess the goodness of a specific SC. In particular, NPV is used as an objective function to be maximized for SC optimization as also done by several literature sources (Chen, 2012; Bogataj et al., 2011; Naim, 2006). The NPV application enhances the strategic role of any SC decisions, reinforcing the concept in which any decision in an SC has a strategic influence in terms of investment. Indeed, the strategic facets are measured using a financial parameter, such as the annuity stream or NPV (Grubbström, 1986).

This paper discusses strategic decisions for the management of uncertainty in SCs. This paper uses one of the most used tools (as it is evident from the literature mentioned above) for modelling an SC and its behaviour rules, i.e. SD, and for interpreting the effects of the model, it uses NPV.

3. Modelling SCs scenarios under system dynamics

To understand better the relation between all the actors of an SC, it is fundamental to introduce the following rule: 'the operations of each actor of the SC interact with the immediately preceding member in the process of order placing'. Moreover, if there is a closed interaction between the actors, the SC is defined as being decentralized, and a system for the information sharing has to be built. In contrast, when a scheme of operations between several actors of a SC does not act in a closed loop, the transmission of real-time demand data from the Retailer up to the first supplier is required (so a centralised system for sharing information is needed). It is important to say that in this, the presence of a suitable mechanism is assumed, for example, an electronic data interchange (EDI) system for enabling the sharing of demand information from the Retailer to the single actor considered. For instance, if we speak about information sharing between Retailers and Distributors, we mean that this Distributor has access to the real-time demand experienced at the Retailers' level. Therefore, in most cases, the SCs are characterised by the absence of closed-loop relations, so they use a 'centralized' system for the information in conformity with the previous definition.

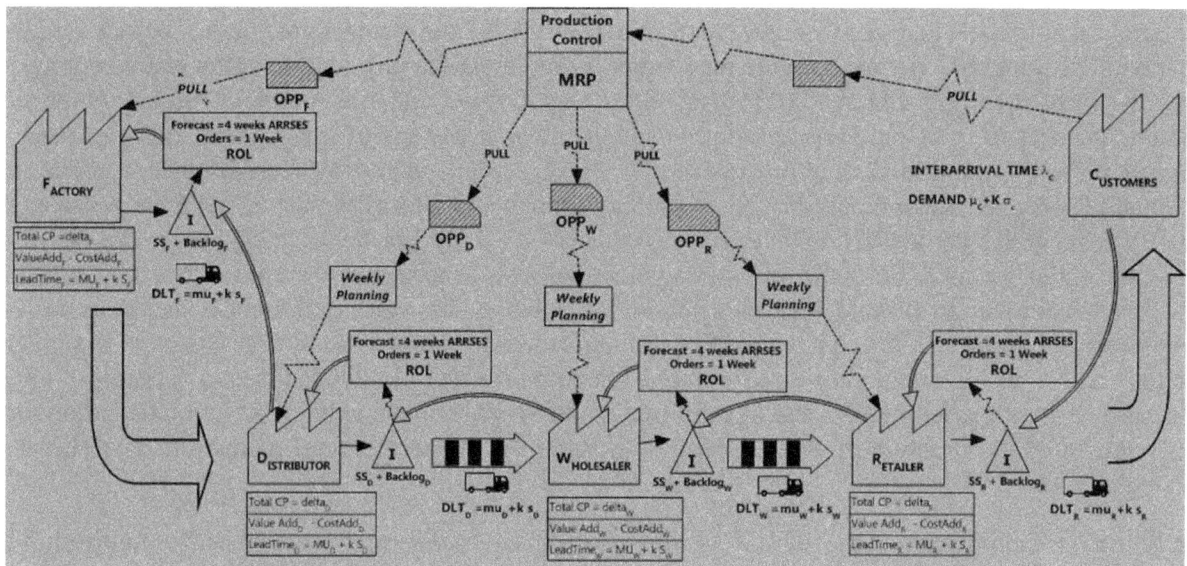

Fig. 3. Mixing strategies for MTS and MTO penetration in SCs

If the SC analysed is not characterised by the last situation (i.e. there is a closed-loop scheme for the relation), each member of the SC uses forecasts for its own demand, including safety stock, and after that, it proceeds with order placement according to a demand forecast for its immediate downstream echelon. Each echelon deals with the maximization of the individual annuity stream measured in terms of NPV.

For this purpose, since the information sharing level is a fundamental factor in the simulation experiments presented in this paper, it is important to introduce the maximum and minimum levels of information sharing; this variable can range from 'no-information' to 'complete' information sharing.

Another fundamental issue to be discussed is related to the stochastic behaviour of the main SC variables. The traditional inventory models in SCs often use normal distributions and only focus on the forecast accuracy, not paying attention to the stock units' locations, and the main SC variables, i.e. demand and lead time. In general, SC models consider the impact of several stocking scenarios, i.e. upstream and downstream for the actor considered. This kind of analysis generally leads to the creation of a stock that is capable of absorbing the variations of the variables before cited.

In the simulation model, a multi-echelon production system is considered. An SC is composed of four echelons. At the material flow level, each level consists of one inventory. Each echelon forecasts downstream demand (towards the end-consumer/s). At echelon n, the input at time t is the order rate (δ_t^n), which is determined by feeding forward the forecast of sales (γ_t^n) and feeding back the error in inventory and the work in progress (wip_t^n) to raise the actual value of the stream, which is calculated through the cash flow (NPV_t^n) of stocks. Direct (management and maintenance costs and financial charges) and indirect costs (shortage and obsolescence costs), generally indicated as C(t) (Fig.6) related to the inventory, which are able to have a financial facet through the real payment (net value of inv_t^n & wip_t^n) (Grubbström & Thorstenson, 1986; Grubbström & Kingsman, 2004) compared with the revenues actualized (*i-rate* in Fig.6).

Because of Lead Time effects (LT_t^n), errors in wip_t^n occur if there is an accumulation of orders ($\sum_{t=1}^{T} \delta_t^n$) that have been placed on the echelon n and these are not yet completed at the time the client requires. The transportation costs are different according to the demand and location of the supplier. The orders arrive into the system at price p, which has been negotiated previously between the echelons. The error in wip_t^n will generate, as said before, an accumulation of stocks inv_t^n that influences the required service level (LS^n). It is important to say that all the problems involved in SC management will be dumped on the stock levels; in fact, forecast models try to reduce these effects on stocks, reducing the errors in lot-sizing, and generally, the forecast is upgraded in a closed-loop approach, adjusting the forecasting solution to the actual data available. Passing now to the other main actor in the single echelon, it is worth noting that the production department is dependent on the upstream levels, i.e. by the n-1 levels of echelons positioned behind it and by the flow of goods that is then based on the forecast and service level assumptions, which are dependent on the revenues and costs of the business. The replenishment strategy influences the amount of stocks in store at the echelon n level, and this, consequently, changes the purchasing cost for the customer from the supplier, limiting the future customers' requests if these facets affect the price, which may increase and/or some stock-outs may occur.

Investments in SCs are generally put on a horizon of several years, so according to the time of investment, we get an annuity stream in SCM.

In the system, it is assumed that demand arrives according to a Poisson process with rate λ. Independent item demand rates are assumed to vary daily (t) and with sporadic surges in demand; to create a general distribution law, the Lyapunov Central Limit conditions are applied (Billingsley, 1986). Upon arrival, demand is normally distributed with mean μ and standard deviation σ that could be managed according to a k value based on Service Level (i.e., SL_i with $n=1,...,4$) considerations (Rice, 1995).

Provisions for handling the uncertainty of domain are incorporated in decisions based on forecasting. A forecast horizon less than a product's total Lead Time (i.e., LT) is assumed. LT_n changes between echelons. It is assumed in normal probability distribution. It is mainly divided into two blocks: 1) Production Lead Time (i.e., PLT_n with $n= 1,...,4$) and Delivery Lead Time from/to echelon (i.e., DLT_n with $n=1, ..., 4$) (Bertrand J.W.M. and Wortmann J.C., 1981). $PTL_n + DLT_n$ forces echelon n to make production decisions based on the forecasted consumption. Mean (i.e., mu_n) and standard deviation (i.e., s_n) at echelon n are in the value stream map of figure 3. Demand at all points of supply is assumed with no trend and Adaptive Response Rate Single Exponential Smoothing is applied (i.e., ARRSES). The Tracking Signal measurement is obtained and looped to keep the forecast unbiased as changes in the pattern data occur. Continuous inventory review is applied, which means inventory is monitored continuously and orders can be placed at any time according to the replenishment strategy. This implies variability of quantity and order time intervals.

Each echelon has to deal with the purchasing cost paid to the supplier after receipt of material; holding cost in proportion to the stored quantity per time; backorder the cost paid to the customer in proportion to the quantity per time; sales revenues received from customers in proportion to the delivered quantity and stock-out costs. Decisions at echelon n depend on the production decisions made by the supplier at echelon *n-1,* the stocking decision assumed by the manufacturer at echelon n and on the demand made by the client at echelon *n+1.* The intent is to adjust production and stocks to fit actual customer demand as it materializes (Fisher et al., 1994). Stocks at echelon n are defined on the basis of SL_n indicating the deterministic costs of purchasing, maintenance and set-up and of the probabilistic costs of shortage and obsolescence. Infinite stocks capacity is assumed. Stocks manifest as value in the annuity stream (AS), then the NPV approach is used (Grubbström & Thorstenson, 1986). Stock and Flow diagrams are used (i.e., System Dynamics) to model simulation. A Casual Loop Diagram (i.e., CLD) at generic echelon i is reported in Fig. 7. Here, the characterization of the replenishment strategy is applied according to the position of the order in the supply. A control panel has been organized in order to tune the system's parameters and control DoE plans. The Taguchi plan is elaborated and schematically reported in interaction plots and synoptic tables (see Fig. 9 and Fig. 10). Whenever control mechanisms are not applied, it is possible to confirm the presence of the bullwhip effect (Fig. 4). If demand contains high uncertainty, the pull approach remains the best strategy for Retailers in the decentralized approach (Fig.

5). Applying the VMI strategy, considerable advantages in terms of the annuity stream could be gained from the Wholesaler.

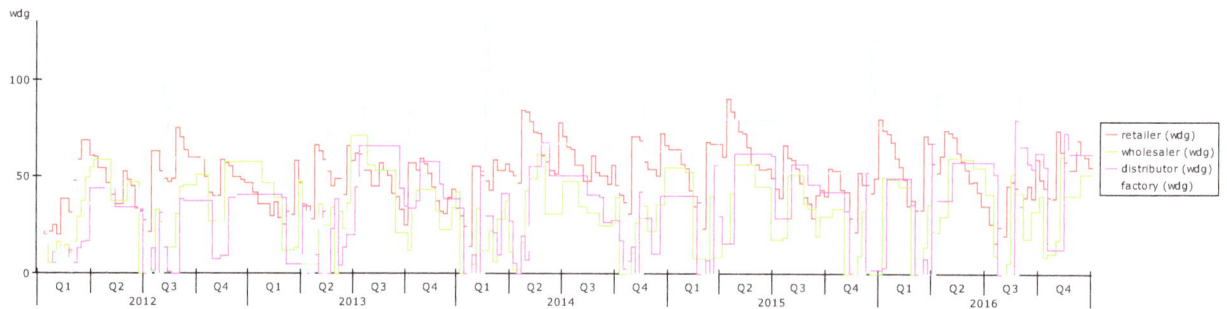

Fig. 4. A Bullwhip Scenario under decentralized Management control (Q.ty report)

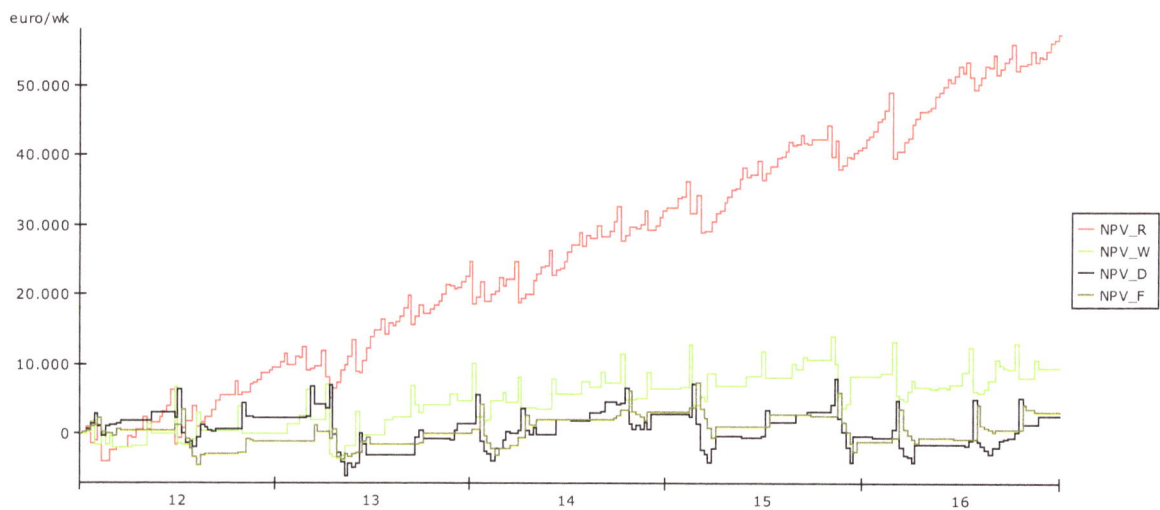

Fig. 5. Reporting about annuity stream under the condition of Bullwhip

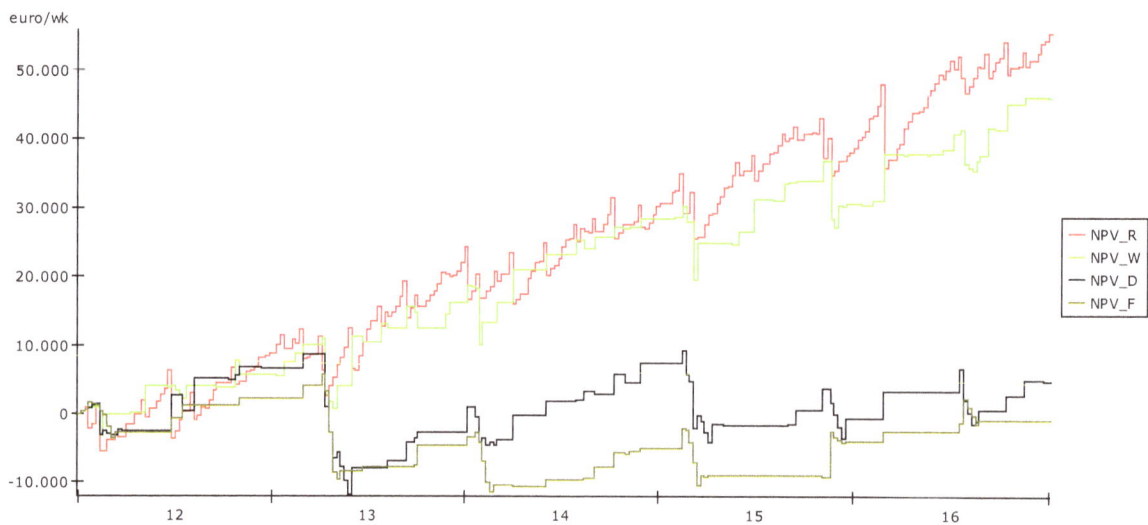

Fig. 6. Reporting about Annuity Stream under coordination based on VMI strategy

Here, customer demand is changed in order to set conditions, as per EU recommendations (European Commission, 2003), for small- and medium-sized enterprises. According to the total balanced sheet, demand by the customer for 10^2 unit/week realized an annual turnover of streams in the supply of ≤ 2 M€ (i.e., micro supply), demand for 10^3 unit/week produces 2 M€ $\leq$ turnover $\leq$ 10 M€ (i.e., small supply), and finally, demand 10^4 unit/week realizes 2 M€ $\leq$ turnover $\leq$ 50 M€ (i.e., medium supply).

An alternative location of OPP has been evaluated, including a measure of alteration in performance, i.e. Δ_{NPV_n} with n alternatively F–factory, D-Distributor, W-Wholesaler or R-Retailer, based on a fixed acceptance, i.e., Δ_{acpt}, rate.

$$\Delta_{NPV_{F,D,W,R}} = \frac{\max_{F,D,W,R}(NPV_{F,D,W,R}) - NPV_{F,D,W,R}}{\Delta_{acpt}} \leq 1 \qquad (1)$$

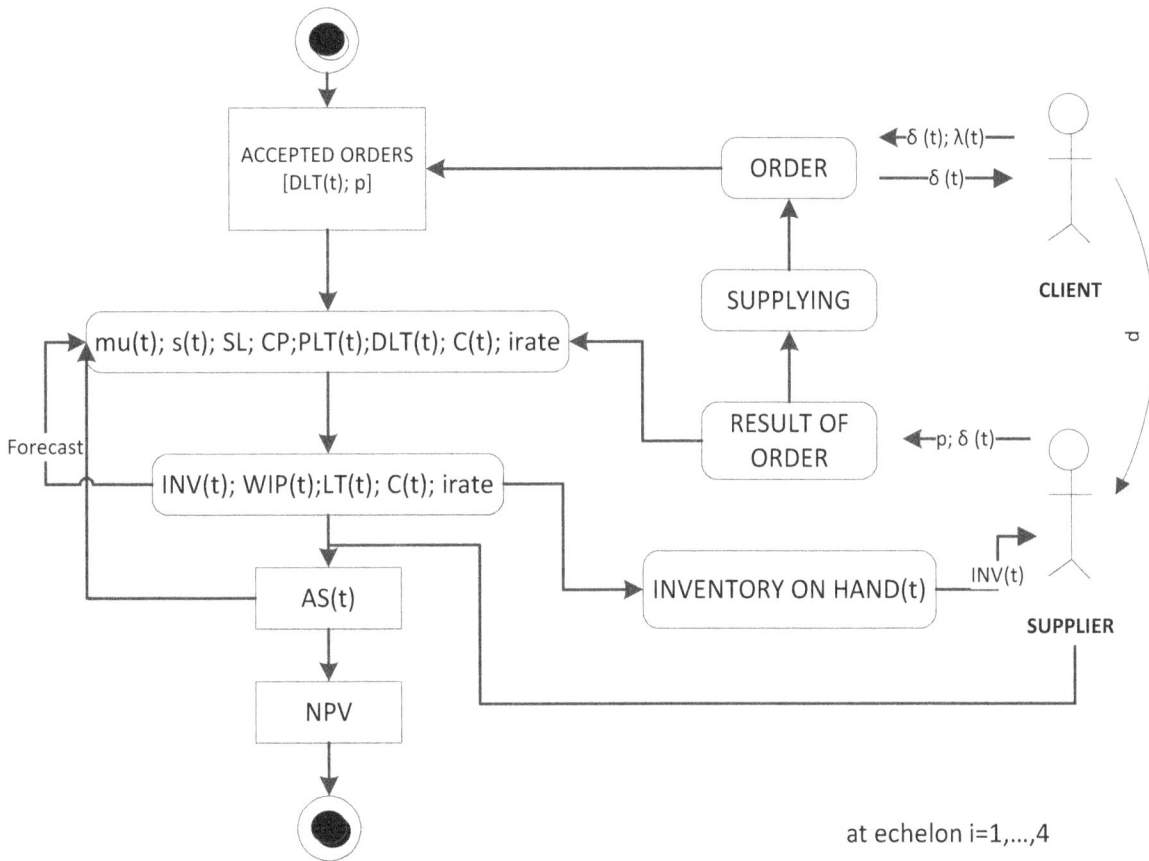

Fig. 7. Control architecture for SCs management: CENTRALISED CONTROL [OBJ = $\text{MAX}_t(\sum_{i=1}^4 \text{NPV}_i(t))$] & DECENTRALISED CONTROL [OBJ$_i$ = $\text{MAX}_t(\text{NPV}_i(t))$]] - where t is the time of strategy investment; I is the generic echelon with i=1,…,4.

The values in the table are reported according to eq.1. Δ_{acpt} means acceptable values of variation, which is fixed at 10% as per the literature (Borgonovo, 2004). IND is set to whatever $\Delta_{NPV} \leq 0.1$. Time simulation and the number of replications was chosen as per Law and Kelton's (2000) (Law, 2000) approach with a Bonferroni correction due to multiple performance measures (Quinzi, 2004). The DoE limit in the analysis is set as per literature (Montgomery, 2012). NPV after 4 years from the strategy investment is considered.

In this analysis, authors accept the consideration of Harris (1913) who defined inventory holding costs and its related strategy in supply as an opportunity cost related to the customer/system satisfaction. Besides the capital inventory cost, there may be other out-of-pocket expenses, such as transportation, management, storage, spoilage, shrinkage and insurance to be accounted for the supply. These are included in the payment cost. Moreover, the reward from sales is included for revenue evaluation. Thus, using the consideration of Grubbström (Grubbström & Kingsman, 2004), according to the time of investment, these become the cash flow of revenues and payment, and they are then transformed into an annuity stream. This yields the NPV of configuration at issue at the level *i* into supply. NPV is intended to be the equivalent cash flow generating the same NPV of a zeroth/first order approximation of the annuity stream (i.e., Annual Cost: AC).

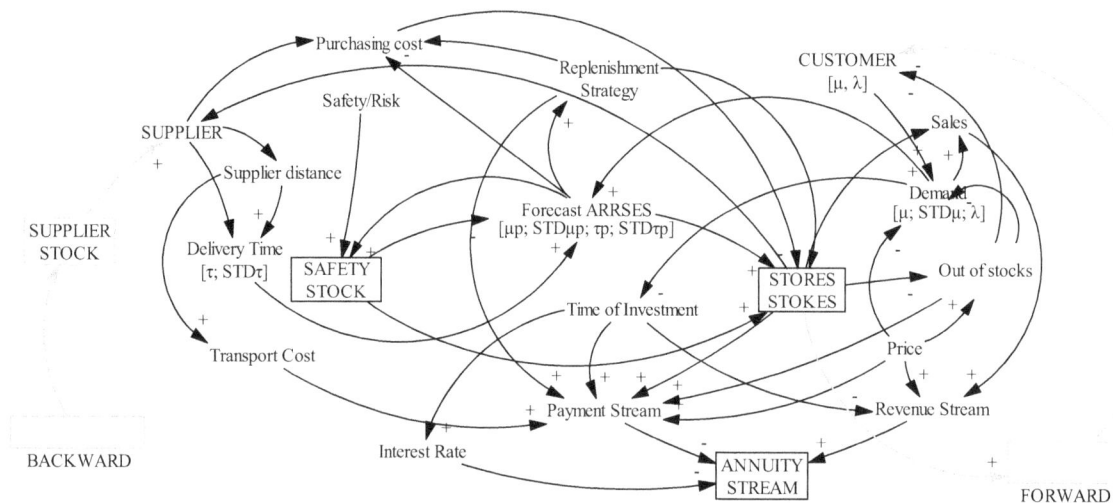

Fig. 8. CLD of generic echelon i of SC - Backward & Forward relation in grey characters

4. The role of OPP in echelon strategy whenever the decentralized approach comes out

First, let us consider the Retailer. If the Retailer only incurs inventory-related costs (that is, the costs related to order variance are constant and zero, as a consequence of which there is no stock-out probability, and under the condition of non-obsolescence of the product, the minimum is the warehousing cost) the position of OPP does not depend on demand (in terms of quantity). Retailers could get optimal performance with OPP varying randomly backward into supply (Fig. 9). This could beget considerable advantages under the condition of coalition between echelons of supply. Increasing product demand volatility driven in second quartile probability, Retailers could reduce performance in NPV by transmitting information on the customer's order into the supply. Under this condition, the backward-moving OPP does not remarkably influence Retailer costs and strategies for optimisation. This means that if echelons in supply are able to control customer preferences in unseasonal and high-frequency demand, Retailers do not need to worry about OPP location. Pull strategies are preferable for Retailers under the condition of high-volume demand. Here, if volatility in demand moves casually around the trend, the Wholesaler is in the preferable position of order penetration. Otherwise, whatever the increase in seasonality, OPP moves backwards up to the Factory under the condition of unstable delivery time. If an inadequate forecast generates a stock-out amount, Retailers prefer a joint purchase order quantity strategy under the condition of limited orders (demand nearby the first quartile in probability) and average standard deviation (after second quartile) in demand. As volatility in demand increases, the position of OPP moves backward to the Wholesaler, while accepting the Distributor and Factory positions with limited (under 5% variation) reduction in the annuity stream. Flexibility in the position of the order is guaranteed here under limited variation (under optimum of about 7%) of NPV. Thus, the push strategy is preferable for supply when the demand volatility is above the second percentile. The Wholesaler gets leadership in supply (it maintains the key role regarding information) when demand volatility is above the third percentile. Increasing the stock-out cost and standard deviation in demand (above the 50th percentile) in a situation in which Warehouse costs influence the Retailer's cash flow, the Factory gets the optimum position of OPP. Variation in NPV for the Retailer is remarkable if supply is organized with an OPP that moves forward/backward when the demand is over-planned. Push strategies are preferable for the Retailer whenever the volatility in demand is above the second quartile; a push/pull strategy for supply is suggested when STD is above the 75th percentile. OPP in the Wholesaler is preferable for the growth of cash flow for the Retailer. OPP moves backwards until the Distributor and the Factory if market origins stock-out of upper medium entity and STD of upper 75th percentile and product originate from medium to upper Warehouse costs. These factories make preference in pull strategies.

If supply performances are investigated in order to preserve the Wholesaler in the chain, under the condition of limited volatility of demand and when limited products (either of them under the first quartile in probability) are moving in supply, the Distributor has to maintain the central information role in order penetration (Fig. 9). Flexibility in the OPP strategy is maintained under conditions of limited volume of product in supply. If demand manifests a trend and seasonality that could considerably increase volume of demand from standard planned conditions, with different configurations of costs (i.e., Warehouse and stock ones) and with the upper first quartile of instability of demand (i.e., STD), supply considerably gains from optimality, setting OPP. It moves backwards from the Wholesaler to the Factory as STD increases. The Retailer is in the preferable position of OPP when demand maintains stability over time. While increasing the influence of stock-outs in terms of cost per unit per time, the Warehouse prefers the pull approach. Moving OPP backwards into supply remains the optimal strategy in the Wholesaler's interest under the condition of volatility in demand. The Wholesaler has to transmit information about ordering backwards to the Distributor and then to the Factory whenever the lead-time variability increases and demand becomes chaotic. The pull strategy is required whenever it is not possible to plan demand.

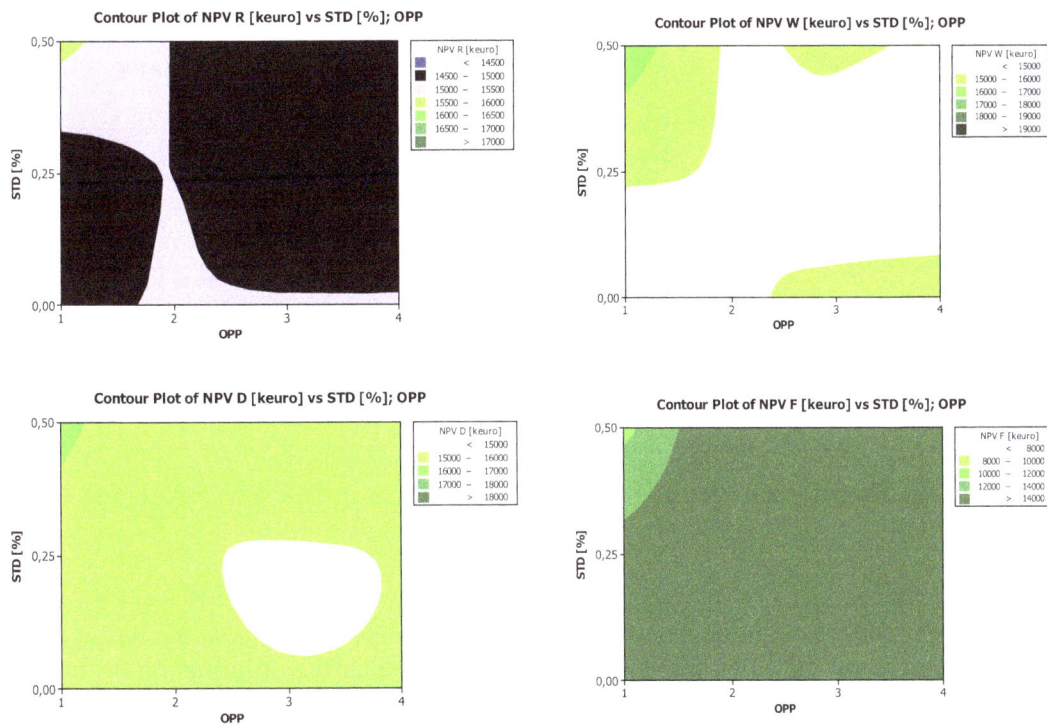

Fig. 9. Annuity Stream contour plot with different OPP location in: 1=RETAILER; 2=WHOLESALER; 3=DISTRIBUTOR; 4=FACTORY under different incidence of holding and stock-out (demand in small range) amount

For the Distributor in decentralised SC under limited demand for the product and stable planning conditions, a stock-out does not influence its profit (Fig. 9). The push strategy is preferable for the Distributor when the volume of the product increases in an upward trend and seasonality effects occur. In markets with a considerable quantity of the product (above 10,000 weekly, a medium SC is set), and in situations of unstable forecasting conditions, OPP has to move backwards to supply until the Factory echelon if the Distributor's profits are going to be optimised. If stability of demand occurs, the push strategy could be alternatively applied to supply if the Distributor wants to optimize its performances in terms of NPV. The Distributor can exchange information, and it need not be worried about them, with echelons in supply in any condition under the 75th percentile of demand volume. Here, there is a reduction of NPV greater than generally maintained (under 9%), and it is limited to a 5% reduction of NPV when cutbacks in stock-out cost per unit of the product in time occur (under the 50th percentile) and forecasting conditions are arranged (predictability and volatility of demand and lead time is around the 50th percentile). Speculative strategies between actors in supply are possible in cases in which the company bets on a single scenario.

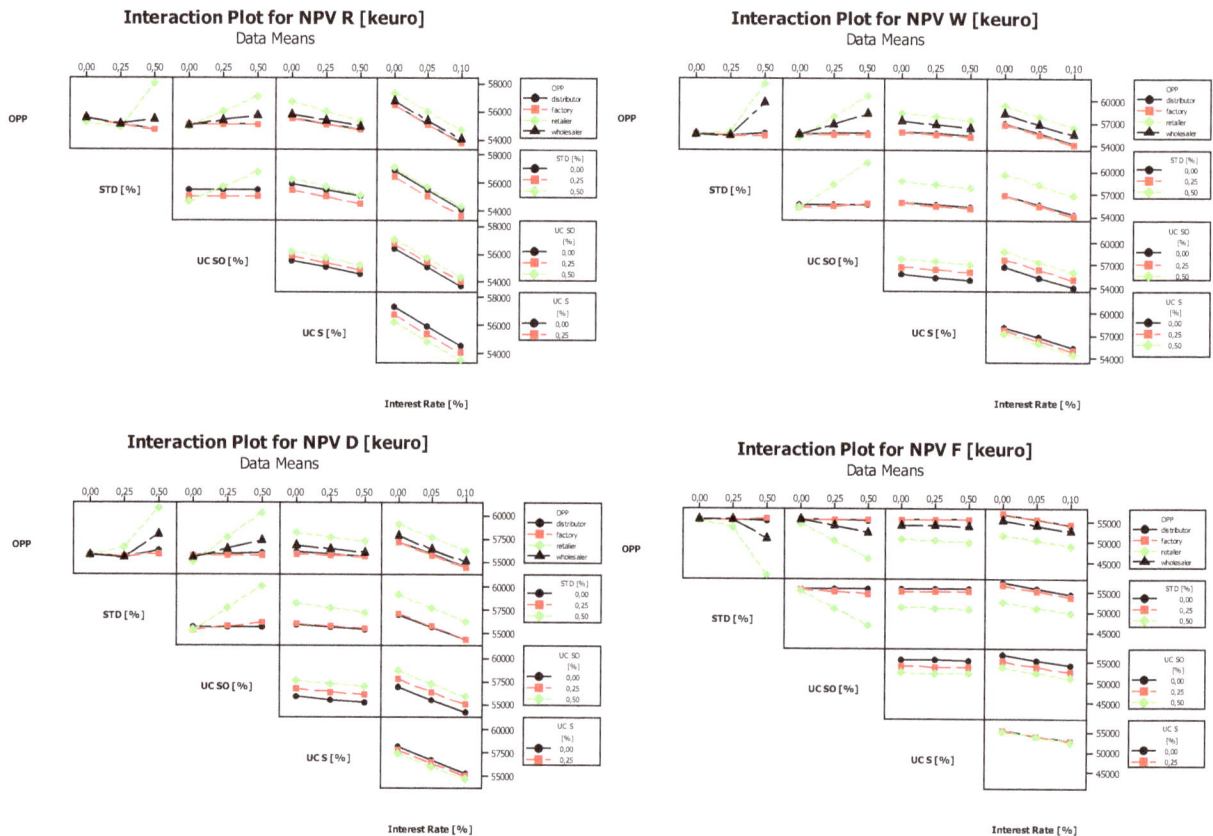

Fig. 2. Interaction Plot of main factors for NPV performances in DECENTRALIZED CONTROL ARCHITECTURE

Increasing probability to incur in stock-out, and whatever this manifests cost that are almost equal to product cost, the Factory remains the key element (i.e., the leader) in supply, and it has to maintain the knowledge of order, so the Distributor needs to worry about the information it transmits and communicates—while improving a pure pull strategy in supply. As discussed above, under any condition of unpredictability of demand and lead time, the position of order penetration has to be maintained at the Retailer's end if the Factory wants to reduce its payment streams and aims at improving its annuity stream. This means that the Factory has to use the push approach if the forecast in demand could fail (Fig.9). Here, the Factory is going to influence—a non-cooperation game is set up— the echelon's performance as well as overall rewards. This notwithstanding, the Factory would be able to optimize its net profits. This strategy guarantees the Factory security from bullwhipping. From limited to low variation in demand under predictable delivery times (under the 50th percentile in probability), the OPP location does not exert remarkable influence over the Factory streams (under 10% of the optimal variation of NPV is manifested). This does not make any difference in performance if demand is stable in time. Moreover, considering the incidence of warehousing in terms of cost as the cost of the product is growing, moving forward into supply and is reported in comparison with stock-outs. Mainly fixed in terms of the cost involved in purchase, maintaining safety stocks is unable to remarkably condition performance of the Factory in terms of the annuity stream. The pull strategy is generally not preferable for the Factory in any condition of demand and instability of market. OPP could be moved backwards or forward in supply under limited (under 7%) variation of the annuity stream.

5. The role of OPP in echelon strategy whenever centralized approach is coming out

Under a centralized control mechanism, the inventory decisions are made by a single decision-maker whose objective is to minimize system-wide cost (Fig.10).

Holding cost (UC_S in Fig. 10) influences the NPV of supply if stock-out costs are of a limited amount (≤0.25 €/unit*time)). If the product matches the limited amount of obsolescence and shortage cost does not occur, in a stable and trusty market, the demand rate does not influence the position of OPP that could be maintained at any echelon of supply if marginal changes in optimization performances are accepted.

In stable markets (with no correlation between events, this permits reference to the first quartile of cumulative standard deviation of demand), OPP does not influence global revenues if products have limited obsolescence (under the first quartile in cost per unit per time). The stability of the market is not relevant if the product requires high costs in maintaining and stock-out probability is limited. Here, OPP requires to be settled at the lowest level of supply (i.e., Retailers). Fixed warehousing costs with limited holding cost (0.25 ≤ Wh_c ≤0.5) in a small or medium market allows OPP fluctuations even though an optimal collocation of the pull/push hinge is admitted. In this case, the position of OPP moves forward from the Factory to the Retailer in a stable (first quartile in demand variation) market with limited unit stock-out cost. In cooperative supplies handling a huge amount of products (cumulative demand rate near 10^4 unit/week—medium enterprises), there exists optimum OPP. Volatility in demand influences Warehouse costs while preferring batch orders although seasonality in demand occurs. Here, pull supplies, under conditions of unstable market and relevant shortage cost per unit per time (i.e., OPP in the Factory), is required for the optimisation of overall performances. A pure push strategy in supply is required whenever stock-out cost is of a limited amount of unit per time (first quartile). OPP moves backwards as uncertainty in market increases. Here, the Warehouse cost influences the OPP location: the Retailer is preferable over the Factory location.

The performance of cost in supply is mainly influenced by stock and stock-out costs. The latter requires the careful definition of operations and a replenishment strategy while influencing inventory strategy. OPP movement from upstream to downstream is required as global contingency influences the stability of the market. There are two major driving forces shifting the OPP backwards: (i) reducing uncertainty in delivery time and demand and (ii) reducing the cost of stocks by optimising inventory strategies. In these cases, when the lead time is controllable (variance under one unit point) or alternatively fixed, work in process and the risk of obsolescence and reliance on forecast reduces. A reduction in manufacturing efficiency as well as in forecasting performances (optimal values in smoothing have to be set) improves probability (i.e., obsolescence and stock-out) to obtain optimal performance while shifting OPP forward.

6. Conclusion

In this paper, a dynamic system has been assumed in which operational strategies are mutually influenced between echelons. A casual loop approach is set. A full plan of experiment has been discussed. It was set with the inclusion of variations in lead times and loss of sales according to increases in holding and stock-out costs. The system has been investigated by characterizing small and medium SC using real case study analysis. Taguchi plans were implemented and interaction plots reported after sensitiveness analysis. The characterization of scenarios is summarized in tables under different arrangements. In our four-echelon SC model, the best results come from the cooperative approach in which players move backwards or forward information about customer orders according to the optimization of the annuity stream. When players act solely, the supply is decentralised and there is an optimal collocation of OPP that could considerably influence the performance of other players. This is in contrast with the Adam Smith thought because of the mutually influencing representation of scenarios. Altruistic behaviour is sometimes suggested but mainly the Retailer dominates scenarios. Nash equilibrium could be arranged in this situation.

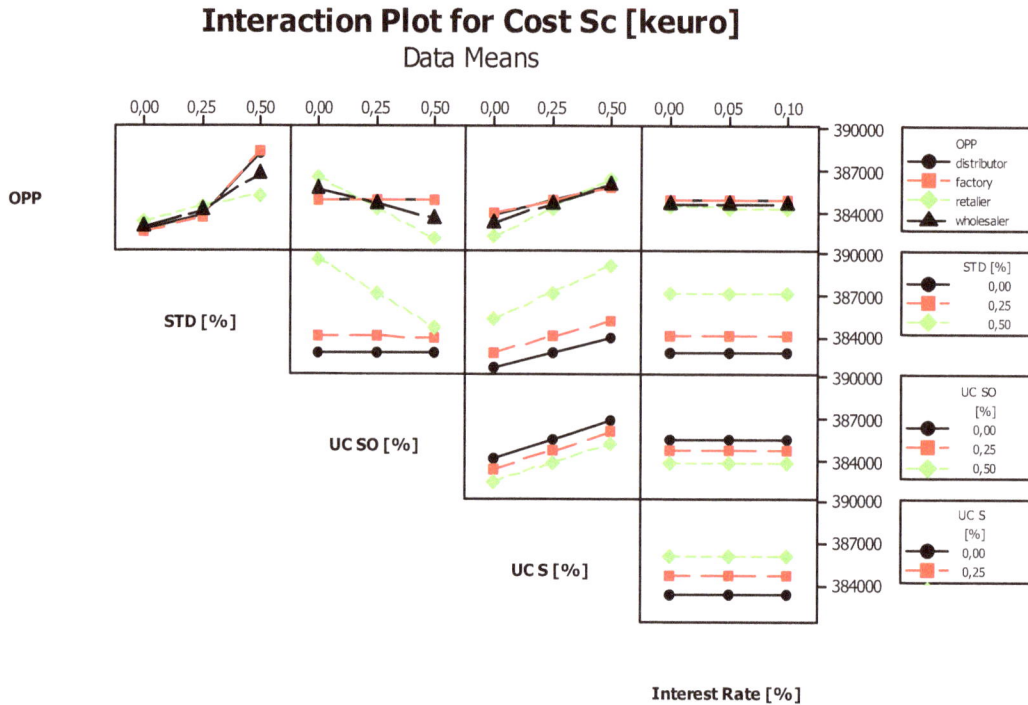

Fig. 11. *Sensitiveness analysis under different simulation scenarios in four echelons of SCs with centralised control strategy: Annuity Stream (cost) and NPV reports are reported under different configuration scenarios.*

Legend cell: [Position of the OPP opt, alternative collocation OPP$\Delta_{NPV_{F,D,W,R}}$]- IND= no preference between echelons. $\Delta_{NPV_{F,D,W,R}}$ evaluated according eq.1 with fixed 10% of Δ_{acpt}.

Stokes in the SC is the buffer of goods and services that players offer to the customer. Also, stokes is the amount of annuity stream connected to stochastic amounts of order at stochastic intervals. The inventory pattern is the payment/reward amount that the management decides according to a defined replenishment strategy.

Players in the SC should make re-engineering efforts to minimize volatility in customers' order and lead time while creating the most accurate forecast strategy. This generally supports growth in the annuity stream, and it optimizes the value of investment in time. An optimal optimisation strategy in a centralized SC requires players to first understand their own demand. Then, based on the characteristics of the product and its related market (i.e., price, demand rate and volatility and stock-out cost), the correct strategy in inventory management can be established. Confirming risk pooling concepts, a centralized control approach manifests lower inventory levels without affecting the service level.

The strategy in inventory prefers the push approach (OPP for the Retailer) and cooperation between players in supply is not required if uncertainty in deliveries and unpredictability occur as forecasted. Higher demand uncertainty leads to a SC based on realised demand: pull customer-driven supply. This is true whenever small and medium quantities are moved and the incidence of stock-out cost and Warehouse management costs is limited to a small percentage per unit per time in comparison with product cost. Normally, small and medium SCs suffer from scarce capital resources, and this can condition the inventory strategy. In this condition, moving OPP backwards from the Retailer to the Distributor and eventually to the Factory will help in dealing with great stock-out and Warehouse management costs. Alternative strategies, moving from push to pull, can be admitted in micro-centralized SCs with limited volatility in demand and lead times.

The Factory has to know the costumer's requirements and trend (i.e., OPP for the Factory) whenever a competitive (decentralized) SC is set under conditions of almost chaotic environments. Here, the Factory does not absolutely prefer a Retailer-driven SC. As the Factory treats raw materials and consequently, has the lowest price of the product in supply, it prefers be organized in the push approach (i.e., OPP for the Distributor or the Wholesaler) whenever small and medium quantities are moved under limited forecasting conditions.

The Distributor generally prefers a Retailer-driven SC under uncertainty in small- and medium-sized environments and whenever stock-out draws on the upper medium value of volume per time. Here, the Distributor has to share information about customers' orders in a timely and accurate manner while suffering a Factory-driven supply if optimization in its performances is to manifest.

OPP for the Retailer remains the preferable strategy for individual Wholesaler optimisation and whenever the Retailer forecasts to follow the current demand under different configurations of supply in terms of volatility in demand and uncertainty of the domain.

Reasonable performances of the SC could be achieved from altruistic Retailers when stability in demand occurs and stock-out cost does not remarkably influence the annuity stream. Re-engineering efforts at the Factory level are not required in decentralized supply under stable predictability conditions. Altruistic behaviour is suggested for the Distributor and the Wholesaler to gain individually optimal performance in terms of NPV. The Retailer is generally the leader in the SC, but this sometimes does not guarantee optimal global performances in a centralised approach.

References

Amid, A., Ghodsypour, S. H., & O'Brien, C. (2006). Fuzzy multiobjective linear model for supplier selection in a supply chain. *International Journal of Production Economics*, *104*(2), 394-407.

Ashayeri, J., & Lemmes, L. (2006). Economic value added of supply chain demand planning: A system dynamics simulation. *Robotics and Computer-Integrated Manufacturing*, *22*(5), 550-556.

Aliev, R. A., Fazlollahi, B., Guirimov, B. G., & Aliev, R. R. (2007). Fuzzy-genetic approach to aggregate production–distribution planning in supply chain management. *Information Sciences*, *177*(20), 4241-4255.

Bhagwat, R., & Sharma, M. K. (2007). Performance measurement of supply chain management: A balanced scorecard approach. *Computers & Industrial Engineering*, *53*(1), 43-62.

Beaudoin, D., LeBel, L., & Frayret, J. M. (2006). Tactical supply chain planning in the forest products industry through optimization and scenario-based analysis. *Canadian Journal of Forest Research*, *37*(1), 128-140.

Bertrand, J. W., & Wortmann, J. J. (1981). *Production control and information systems for component-manufacturing shops* (Doctoral dissertation, Elsevier Scientific Publishing Company).

Besanko, D., & Perry, M. K. (1993). Equilibrium incentives for exclusive dealing in a differentiated products oligopoly. *The RAND Journal of Economics*, *24*(4), 646-667.

Bhaskaran, S. (1998). Simulation analysis of a manufacturing supply chain.*Decision Sciences*, *29*(3), 633.

Billingsley, P. (1986). *Probability and Measure, 2nd ed.* New York: p. 371: Wiley.

Bogataj, M., Grubbström, R. W., & Bogataj, L. (2011). Efficient location of industrial activity cells in a global supply chain. *International journal of production Economics*, *133*(1), 243-250.

Borgonovo, E., & Peccati, L. (2004). Sensitivity analysis in investment project evaluation. *International Journal of Production Economics*, *90*(1), 17-25.

Bowersox, D. J., & Calantone, R. J. (1998). Global logistics. *Journal of International Marketing*, 83-93.

Cachon, G. P., & Lariviere, M. A. (2001). Contracting to assure supply: How to share demand forecasts in a supply chain. *Management science*, *47*(5), 629-646.

Carlsson, C., & Fullér, R. (2002). A fuzzy approach to taming the bullwhip effect. In *Advances in Computational Intelligence and Learning* (pp. 247-262). Springer Netherlands.

Campuzano, F., Mula, J., & Peidro, D. (2010). Fuzzy estimations and system dynamics for improving supply chains. *Fuzzy Sets and Systems*, *161*(11), 1530-1542.

Chen, C. T., Lin, C. T., & Huang, S. F. (2006). A fuzzy approach for supplier evaluation and selection in supply chain management. *International Journal of Production Economics*, *102*(2), 289-301.

Chen, P. Y. (2012). The investment strategies for a dynamic supply chain under stochastic demands. *International Journal of Production Economics*, *139*(1), 80-89.

Cheung, K. L., & Lee, H. L. (2002). The inventory benefit of shipment coordination and stock rebalancing in a supply chain. *Management Science,48*(2), 300-306.

Chiang, W. Y. K., Chhajed, D., & Hess, J. D. (2003). Direct marketing, indirect profits: A strategic analysis of dual-channel supply-chain design. *Management science*, *49*(1), 1-20.

Colicchia C. & Strozzi, F. (2012). Supply chain risk management: A new methodology for a systematic. *Supply Chain Management: An International Journal,17*(4), 403 - 418.

Chauhan, S. S., Dolgui, A., & Proth, J. M. (2009). A continuous model for supply planning of assembly systems with stochastic component procurement times. *International Journal of Production Economics, 120*(2), 411-417.

Chanas, S., Delgado, M., Verdegay, J. L., & Vila, M. A. (1993). Interval and fuzzy extensions of classical transportation problems. *Transportation Planning and Technology*, *17*(2), 203-218.

Chen, S. P., & Chang, P. C. (2006). A mathematical programming approach to supply chain models with fuzzy parameters. *Engineering Optimization*, *38*(6), 647-669.

Efendigil, T., & Önüt, S. (2012). An integration methodology based on fuzzy inference systems and neural approaches for multi-stage supply-chains.*Computers & Industrial Engineering*, *62*(2), 554-569.

Elofson, G., & Robinson, W. N. (2007). Collective customer collaboration impacts on supply-chain performance. *International Journal of Production Research*, *45*(11), 2567-2594.

Eppen, G. D., & Iyer, A. V. (1997). Backup agreements in fashion buying—the value of upstream flexibility. *Management Science*, *43*(11), 1469-1484.

Esper, T. L., & Williams, L. R. (2003). The value of collaborative transportation management (CTM): its relationship to CPFR and information technology.*Transportation Journal*, *42*(4), 55-65.

European Commission, (2003). *Commission Recommendation concerning the definition of micro, small and medium -sized enterprises,* s.l.: Official Journakl of the European Union.

Esmaeili, M., Aryanezhad, M. B., & Zeephongsekul, P. (2009). A game theory approach in seller–buyer supply chain. *European Journal of Operational Research, 195*(2), 442-448.

Esmaeili, M., Aryanezhad, M. B., & Zeephongsekul, P. (2009). A game theory approach in seller–buyer supply chain. *European Journal of Operational Research, 195*(2), 442-448.

Feng, Y. (2012). System Dynamics Modeling for Supply Chain Information Sharing. *Physics Procedia, 25*, 1463 – 1469 .

Fisher, M. L., Hammond, J. H., Obermeyer, W. R., & Raman, A. (1994). Making supply meet demand in an uncertain world. *Harvard Business Review, 72*, 83-83.

Fugate, B., Sahin, F., & Mentzer, J. T. (2006). Supply chain management coordination mechanisms. *Journal of Business Logistics, 27*(2), 129-161.

Giannoccaro, I., Pontrandolfo, P., & Scozzi, B. (2003). A fuzzy echelon approach for inventory management in supply chains. *European Journal of Operational Research, 149*(1), 185-196.

Grubbström, R. W., & Kingsman, B. G. (2004). Ordering and inventory policies for step changes in the unit item cost: a discounted cash flow approach.*Management science, 50*(2), 253-267.

Grubbström, R. W., & Thorstenson, A. (1986). Evaluation of capital costs in a multi-level inventory system by means of the annuity stream principle.*European Journal of Operational Research, 24*(1), 136-145.

Hameri, A. & Nikkola, J., 2001. Order penetration point in the paper supply chain. *Paperi Ja Puu-Paper and Timber, 83*(4), 299-302.

Harris, F. M. (1913). How Many Parts to Make at Once. *Factory, The Magazine of Management10:2, 135–136, 152. Reprinted in Operations Research 38:6 (1990)*, pp. 947-950.

He, M., Leung, H. F., & Jennings, N. R. (2003). A fuzzy-logic based bidding strategy for autonomous agents in continuous double auctions. *IEEE Transactions on Knowledge and data Engineering, 15*(6), 1345-1363.

Henning, G. P. (2009). Production scheduling in the process industries: current trends, emerging challenges and opportunities. *Computer Aided Chemical Engineering, 27*, 23-28.

Hoffman, J. M., Shah, N. D., Vermeulen, L. C., Schumock, G. T., Grim, P., Hunkler, R. J., & Hontz, K. M. (2006). Projecting future drug expenditures-2006. *American Journal of Health-system Pharmacy, 63*(2), 123-138.

Hülsmann, M., Grapp, J., & Li, Y. (2008). Strategic adaptivity in global supply chains—competitive advantage by autonomous cooperation. *International Journal of Production Economics, 114*(1), 14-26.

Iannone, R., Miranda, S., & Riemma, S. (2007). Supply chain distributed simulation: An efficient architecture for multi-model synchronization. *Simulation Modelling Practice and Theory, 15*(3), 221-236.

Iannone, R., Miranda, S., Riemma, S., Sarno, D. (2010), A model for vendor selection and dynamic evaluation, IFIP WG 5.7 *International Conference on Advances in Production Management Systems: New Challenges, New Approaches, APMS 2009, Bordeaux*. IFIP Advances in Information and Communication Technology, (338) AICT, 283-290.

Iannone, R., Martino, G., Miranda, S., & Riemma, S. (2015). Modeling fashion retail supply chain through causal loop diagram. *IFAC-PapersOnLine, 48*(3), 1290-1295.

Jiang, G., Hu, B., & Wang, Y. (2010). Agent-based simulation of competitive and collaborative mechanisms for mobile service chains. *Information Sciences,180*(2), 225-240.

Julien, B. (1994). An extension to possibilistic linear-programming. *Fuzzy Sets and Systems, 64*, 194-206.

Jung, H., Chen, F. F., & Jeong, B. (2008). Decentralized supply chain planning framework for third party logistics partnership. *Computers & Industrial Engineering, 55*(2), 348-364.

Kanda, A., & Deshmukh, S. G. (2008). Supply chain coordination: perspectives, empirical studies and research directions. *International Journal of Production Economics, 115*(2), 316-335.

Kumar, M., Vrat, P., & Shankar, R. (2004). A fuzzy goal programming approach for vendor selection problem in a supply chain. *Computers & Industrial Engineering, 46*(1), 69-85.

Kulp, S. C., Lee, H. L., & Ofek, E. (2004). Manufacturer benefits from information integration with retail customers. *Management Science, 50*(4), 431-444.

Kwon, O., Im, G. P., & Lee, K. C. (2011). An agent-based web service approach for supply chain collaboration. *Scientia Iranica, 18*(6), 1545-1552.

Lanzillotto, A., Martino, G., Gnoni, M. G., Iannone, R. (2015). Impact analysis of a cross-channel retailing system in the Fashion industry by a simulation approach. *14th International Conference on Modeling and Applied Simulation*, MAS 2015, Bergeggi; Italy; 21 September 2015 through 23 September 2015, 79-88.

Law, A. M., Kelton, W. D., & Kelton, W. D. (1991). *Simulation modeling and analysis* (Vol. 2). New York: McGraw-Hill.

Lee, Y. H., Cho, M. K., Kim, S. J., & Kim, Y. B. (2002). Supply chain simulation with discrete–continuous combined modeling. *Computers & Industrial Engineering, 43*(1), 375-392.

Lee, Y., & Kozar, K. A. (2006). Investigating the effect of website quality on e-business success: An analytic hierarchy process (AHP) approach. *Decision support systems, 42*(3), 1383-1401.

Lee, C. F., & Chung, C. P. (2012). An inventory model for deteriorating items in a supply chain with system dynamics analysis. *Procedia-Social and Behavioral Sciences, 40*, 41-51.

Lehtonen, J. M. (1998). Analysis of the Order Penetration Point Alternatives in the Nordic Paper Industry Supply Chains. In *Strategic Management of the Manufacturing Value Chain* (pp. 277-286). Springer US.

Leng, M., & Parlar, M. (2010). Game-theoretic analyses of decentralized assembly supply chains: Non-cooperative equilibria vs. coordination with cost-sharing contracts. *European Journal of Operational Research, 204*(1), 96-104.

Liang, T. (2006). Distribution planning decisions using interactive fuzzy multi-objective linear programming. *Fuzzy Sets and Systems, 157*, 1303–1316.

Liang, T. (2007). Applying fuzzy goal programming to production/transportation planning decisions in a supply chain. *International Journal of Systems Science, 38*, 293-304.

Liu, S. T., & Kao, C. (2004). Solving fuzzy transportation problems based on extension principle. *European Journal of operational research, 153*(3), 661-674.

Li, S., Rao, S. S., Ragu-Nathan, T. S., & Ragu-Nathan, B. (2005). Development and validation of a measurement instrument for studying supply chain management practices. *Journal of Operations Management, 23*(6), 618-641.

Lohtia, R., & Subramaniam, R. (2004). Efficient consumer response in Japan: Industry concerns, current status, benefits, and barriers to implementation.*Journal of Business Research, 57*(3), 306-311.

Long, Q., & Zhang, W. (2014). An integrated framework for agent based inventory–production–transportation modeling and distributed simulation of supply chains. *Information Sciences, 277*, 567-581.

Montgomery, D.C. (2012). *Design and Analysis of Experiments.* 8th edition ed. s.l.:Wiley & Sons Inc.

Ma, J., Nozick, L. K., Tew, J. D., Truss, L. T., & Costy, T. (2004). Modelling the effect of custom and stock orders on supply-chain performance. *Production Planning & Control, 15*(3), 282-291.

Malhotra, A., Gosain, S., & Sawy, O. A. E. (2005). Absorptive capacity configurations in supply chains: gearing for partner-enabled market knowledge creation. *MIS Quarterly*, 145-187.

Masuchun, W., Davis, S., & Patterson, J. W. (2004). Comparison of push and pull control strategies for supply network management in a make-to-stock environment. *International Journal of Production Research, 42*(20), 4401-4419.

McEvily, B., Perrone, V., & Zaheer, A. (2003). Trust as an organizing principle. *Organization Science, 14*(1), 91-103.

Miranda, S., Fera, M., Iannone, R., & Riemma, S. (2015). A multi-item constrained EOQ calculation algorithm with exit condition: a comparative analysis. *IFAC-PapersOnLine, 48*(3), 1314-1319.

Moyaux, T., Chaib-draa, B., & D'Amours, S. (2007). Information sharing as a coordination mechanism for reducing the bullwhip effect in a supply chain.*IEEE Transactions on Systems, Man, and Cybernetics, Part C (Applications and Reviews), 37*(3), 396-409.

Naim, M. (2006). The impact of the net present value on the assessment of the dynamic performance of e-commerce enabled supply chains. *International Journal of Production Economics, 104*, 382–393.

Ni, D., & Li, K. W. (2012). A game-theoretic analysis of social responsibility conduct in two-echelon supply chains. *International Journal of Production Economics, 138*(2), 303-313.

Olhager, J., 2003. Strategic positioning of the order penetration point. *International Journal of Production Economics, 85*, 319–329.

Peidro, D., Mula, J., Poler, R., & Verdegay, J. L. (2009). Fuzzy optimization for supply chain planning under supply, demand and process uncertainties. *Fuzzy Sets and Systems, 160*(18), 2640-2657.

Persson, F., & Araldi, M. (2009). The development of a dynamic supply chain analysis tool—Integration of SCOR and discrete event simulation. *International Journal of Production Economics, 121*(2), 574-583.

Petrovic, D., Roy, R., & Petrovic, R. (1998). Modelling and simulation of a supply chain in an uncertain environment. *European Journal of Operational Research, 109*(2), 299-309.

Petrovic, D., Roy, R., & Petrovic, R. (1999). Supply chain modelling using fuzzy sets. *International Journal of Production Economics, 59*(1), 443-453.

Pettersson, A. I., & Segerstedt, A. (2013). Measuring supply chain cost. *International Journal of Production Economics, 143*(2), 357-363.

Tracht, K., Niestegge, A., & Schuh, P. (2013). Demand planning based on performance measurement systems in closed loop supply chains. *Procedia CIRP, 12*, 324-329.

Padmanabhan, V., & Png, I. P. (1997). Manufacturer's return policies and retail competition. *Marketing Science, 16*(1), 81-94.

Pagh, J. D., & Cooper, M. C. (1998). Supply chain postponement and speculation strategies: how to choose the right strategy. *Journal of Business Logistics, 19*(2), 13.

Papadakis, I. S. (2006). Financial performance of supply chains after disruptions: an event study. *Supply Chain Management: An International Journal, 11*(1), 25-33.

Pereira, J. (2009). The new supply chain's frontier: Information management. *International Journal of Information Management,* Volume 29, p. 372–379.

Persona, A., Grassi, A., & Catena, M. (2005). Consignment stock of inventories in the presence of obsolescence. *International Journal of Production Research, 43*(23), 4969-4988.

Petrovic, D., 2001. Simulation of supply chain behaviour and performance in an uncertain environment. *International Journal of Production Economics,* Volume 71, pp. 429 - 438.

Pettersen J.A. & Segerstedt A., 2009. Restrictedwork-in-process:A study of differences between Kanban and CONWIP. *Int. J. Production Economics,* Volume 118, pp. 199-207.

Power, D., & Singh, P. (2007). The e-integration dilemma: The linkages between Internet technology application, trading partner relationships and structural change. *Journal of Operations Management, 25*(6), 1292-1310.

Quinzi, A., 2004. A Sequential Stopping Rule For Determining The Number Of Replications Necessary When Several Measures Of Effectiveness Are Of Interest. *Proceedings of Tenth U.S. Army Conference on Applied Statistics.*

Rice, J. (1995). *Mathematical Statistics and Data Analysis (Second ed.).* s.l.: Duxbury Press, ISBN 0-534-20934-3.

Sakawa, M., Nishizaki, I., & Uemura, Y. (2001). Fuzzy programming and profit and cost allocation for a production and transportation problem. *European Journal of Operational Research, 131*(1), 1-15.

Sambasivan, M., Siew-Phaik, L., Mohamed, Z. A., & Leong, Y. C. (2013). Factors influencing strategic alliance outcomes in a manufacturing supply chain: role of alliance motives, interdependence, asset specificity and relational capital. *International Journal of Production Economics, 141*(1), 339-351.

Shukla, N., Choudhary, A. K., Prakash, P. K. S., Fernandes, K. J., & Tiwari, M. K. (2013). Algorithm portfolios for logistics optimization considering stochastic demands and mobility allowance. *International Journal of Production Economics, 141*(1), 146-166.

Sahin, F., & Robinson, E. P. (2002). Flow coordination and information sharing in supply chains: review, implications, and directions for future research. *Decision sciences, 33*(4), 505-536.

Sako, M., 1994. *Supplier relationships and innovation. In: Dodgson, M., Rothwell, R.,.(Eds), the Handbook of Industrial Innovation.* Vermont, USA: Edward Elgar, Aldershot, Hants.

Sarkis, J., Zhu, Q., & Lai, K. H. (2011). An organizational theoretic review of green supply chain management literature. *International Journal of Production Economics, 130*(1), 1-15.

Selim, H., Araz, C., & Ozkarahan, I. (2008). Collaborative production–distribution planning in supply chain: a fuzzy goal programming approach.*Transportation Research Part E: Logistics and Transportation Review, 44*(3), 396-419.

Sheu, C., Rebecca Yen, H., & Chae, B. (2006). Determinants of supplier-retailer collaboration: evidence from an international study. *International Journal of Operations & Production Management, 26*(1), 24-49.

Takahashi K. & Nakamura H. (2004). Push, pull, or hybrid control in supply chain management. *International Journal Of Computer Integrated Manufacturing,* 17(2), pp. 126-140.

Tako, A. A., & Robinson, S. (2012). The application of discrete event simulation and system dynamics in the logistics and supply chain context. *Decision Support Systems, 52*(4), 802-815.

Tribowski, C., Goebel, C., & Günther, O. (2009, March). EPCIS-based supply chain event management– a quantitative comparison of candidate system architectures. In *Complex, Intelligent and Software Intensive Systems, 2009. CISIS'09. International Conference on* (pp. 494-499). IEEE.

Tsay, A. (1999). The quantity flexibility contract and supplier–customer incentive. *Management Science,* 45(10), p. 1339–1358.

Torabi, S. A., & Hassini, E. (2008). An interactive possibilistic programming approach for multiple objective supply chain master planning. *Fuzzy Sets and Systems, 159*(2), 193-214.

Umeda, S., & Zhang, F. (2006). Supply chain simulation: generic models and application examples. *Production planning & control, 17*(2), 155-166.

Waller, M., Johnson, M. E., & Davis, T. (1999). Vendor-managed inventory in the retail supply chain. *Journal of business logistics, 20*(1), 183.

Wang, Y., Chang, C. W., & Heng, M. S. (2004). *The levels of information technology adoption, business network, and a strategic position model for evaluating supply chain integration* (Doctoral dissertation, California State University, Long Beach, College of Businessn).

Won Lee, C., Kwon, I. W. G., & Severance, D. (2007). Relationship between supply chain performance and degree of linkage among supplier, internal integration, and customer. *Supply Chain Management: An International Journal, 12*(6), 444-452.

Xiao, T., Jin, J., Chen, G., Shi, J., & Xie, M. (2010). Ordering, wholesale pricing and lead-time decisions in a three-stage supply chain under demand uncertainty. *Computers & Industrial Engineering, 59*(4), 840-852.

Yang, T., Chen, M.-C. & Li, H.-C. (2007). Evaluating the supply chain performance of IT-based inter-enterprise collaboration. *Information & Management, 44*(6), 524-534.

Yan, H., Liu, K., & Hsu, A. (2003). Optimal ordering in dual-supplier system with demand forecast updates. *Production and Operations Management, 12*(1), 30-45.

Yin, S., & Nishi, T. (2012). Game theoretic approach for global manufacturing planning under risk and uncertainty. *Procedia CIRP, 3*, 251-256.

Yu, Z., Yan, H., & Edwin Cheng, T. C. (2001). Benefits of information sharing with supply chain partnerships. *Industrial management & Data systems,101*(3), 114-121.

Zhang, X., & Huang, G. Q. (2010). Game-theoretic approach to simultaneous configuration of platform products and supply chains with one manufacturing firm and multiple cooperative suppliers. *International Journal of Production Economics, 124*(1), 121-136.

Permissions

All chapters in this book were first published in IJIEC, by Growing Science; hereby published with permission under the Creative Commons Attribution License or equivalent. Every chapter published in this book has been scrutinized by our experts. Their significance has been extensively debated. The topics covered herein carry significant findings which will fuel the growth of the discipline. They may even be implemented as practical applications or may be referred to as a beginning point for another development.

The contributors of this book come from diverse backgrounds, making this book a truly international effort. This book will bring forth new frontiers with its revolutionizing research information and detailed analysis of the nascent developments around the world.

We would like to thank all the contributing authors for lending their expertise to make the book truly unique. They have played a crucial role in the development of this book. Without their invaluable contributions this book wouldn't have been possible. They have made vital efforts to compile up to date information on the varied aspects of this subject to make this book a valuable addition to the collection of many professionals and students.

This book was conceptualized with the vision of imparting up-to-date information and advanced data in this field. To ensure the same, a matchless editorial board was set up. Every individual on the board went through rigorous rounds of assessment to prove their worth. After which they invested a large part of their time researching and compiling the most relevant data for our readers.

The editorial board has been involved in producing this book since its inception. They have spent rigorous hours researching and exploring the diverse topics which have resulted in the successful publishing of this book. They have passed on their knowledge of decades through this book. To expedite this challenging task, the publisher supported the team at every step. A small team of assistant editors was also appointed to further simplify the editing procedure and attain best results for the readers.

Apart from the editorial board, the designing team has also invested a significant amount of their time in understanding the subject and creating the most relevant covers. They scrutinized every image to scout for the most suitable representation of the subject and create an appropriate cover for the book.

The publishing team has been an ardent support to the editorial, designing and production team. Their endless efforts to recruit the best for this project, has resulted in the accomplishment of this book. They are a veteran in the field of academics and their pool of knowledge is as vast as their experience in printing. Their expertise and guidance has proved useful at every step. Their uncompromising quality standards have made this book an exceptional effort. Their encouragement from time to time has been an inspiration for everyone.

The publisher and the editorial board hope that this book will prove to be a valuable piece of knowledge for researchers, students, practitioners and scholars across the globe.

List of Contributors

Susanta Kumar Gauri
SQC & OR Unit, Indian Statistical Institute, 203, B. T. Road, Kolkata-700108, India

Surajit Pal
SQC & OR Unit, Indian Statistical Institute, 110, Nelson Manickam Road, Chennai- 600029, India

Eliana M. Toro
Facultad de Ingeniería Industrial, Universidad Tecnológica de Pereira. Pereira, Colombia

John F. Franco
Universidade Estadual Paulista Júlio de Mesquita Filho, UNESP, Ilha Solteira, Brazil

Mauricio Granada Echeverri
Programa de Ingeniería Eléctrica, Facultad de Ingenierías, Universidad Tecnológica de Pereira., Pereira, Colombia

Frederico G. Guimarães
Department of Electrical Engineering, Universidade Federal de Minas Gerais, UFMG, Belo Horizonte, Brazil

Ramón A. Gallego Rendón
Programa de Ingeniería Eléctrica, Facultad de Ingenierías, Universidad Tecnológica de Pereira., Pereira, Colombia

N. Rincon-Garcia
Transportation Research Group, University of Southampton, Southampton, UK
Department of Industrial Engineering, School of Engineering, Pontificia Universidad Javeriana, Bogota, Colombia

B.J. Waterson and T.J. Cherrett
Transportation Research Group, University of Southampton, Southampton, UK

Dionicio Neira Rodado and Fabricio Andrés Niebles Atencio
Departamento de Ingeniería Industrial, Universidad de la Costa. Calle 58 # 55 - 66. Barranquilla, Colombia

John Willmer Escobar
Departamento de Ingeniería Civil e Industrial, Pontificia Universidad Javeriana. Calle 18 No 118-250, Cali (Colombia)

Rafael Guillermo García-Cáceres
Profesor Investigador, Universidad Antonio Nariño, Bogotá, Colombia. Colombia

Dhiraj P. Rai
Sardar Vallabhbhai National Institute of Technology, Surat – 395007, India

Priyabrata Sahoo and Asish Bandyopadhyay
Mechanical Engineering Department, Jadavpur University, Kolkata, India, 700032

Ashwani Pratap
Mechanical Engineering Department, Indian Institute of Technology Patna, India, 801103

Eleonora Bottani
Department of Industrial Engineering, University of Parma, viale G.P.Usberti 181/A, 43124 – Parma, Italy

Piera Centobelli and Roberto Cerchione
Department of Industrial Engineering, University of Naples Federico II, P.le Tecchio 80, 80125 – Naples, Italy

Lucia Del Gaudio and Teresa Murino
Department of Chemical, Materials and Industrial Production Engineering, P.le Tecchio 80, 80125 – Naples, Italy

Hamid Tebassi, Mohamed Athmane Yallese and Salim Belhadi
Mechanics and Structures Research Laboratory (LMS), May 8th 1945 University, Guelma 24000, Algeria

Francois Girardin
Laboratoire Vibrations Acoustique, INSA-Lyon, 25 bis avenue Jean Capelle, F-69621 Villeurbanne Cedex, France

Tarek Mabrouki
Université de Tunis El Manar, Ecole Nationale d'Ingénieurs de Tunis (ENIT), 1002, Tunis, Tunisie

Bala Murali Gunji and B. B. V. L. Deepak
Department of Industrial Design, National Institute of Technology- Rourkela,769008, India

M V A Raju Bahubalendruni and Bibhuti Bhusan Biswal
Department of mechanical engineering, GMR Institute of Technology –Andrapradesh, 532127, India

Hassan Rezazadeh and Amin Khiali-Miab
Department of Industrial Engineering, Faculty of Mechanical Engineering, University of Tabriz, Iran

S. K. Chaharsooghi and Farid Momayezi
Department of Industrial & Systems Engineering, Tarbiat Modares University, Tehran, Iran

Nader Ghaffarinasab
Department of Industrial Engineering, University of Tabriz, Tabriz, Iran

G. Costabile and A. Lambiase
University of Salerno - Department of Industrial Engineering - Via Giovanni Paolo II, Fisciano (SA) – Italy

M. Fera
Second University of Naples - Department of Industrial and Information Engineering - Via Roma 29, Aversa (CE) – Italy

F. Fruggiero
University of Basilicata - School of Engineering - Via Nazario Sauro, 85, 85100 (PZ) – Italy

D. Pham
Department of Mechanical Engineering - University of Birmingham – Edgbaston, Birmingham B15 2TT, UK

M. Fera and R. Macchiaroli
Second University of Naples, Department of Industrial and Information Engineering, Via Roma 29, 81031 Aversa (CE), Italy

F. Fruggiero
University of Basilicata, School of Engineering, Via Ateneo Lucano 10, 85100 Potenza, Italy

A. Lambiase and S. Miranda
University of Salerno, Department of Industrial Engineering, Via Giovanni Paolo II 132, 84084 Fisciano (SA), Italy

Index